Fish Species in Environmental Risk Assessment Strategies

Issues in Toxicology

Editor-in-chief:
Diana Anderson, *University of Bradford, UK*

Series editors:
Michael D. Waters, *Michael Waters Consulting, USA*
Timothy C. Marrs, *Edentox Associates, UK*
Alok Dhawan, *CSIR-Indian Institute of Toxicology Research, India*

Titles in the Series:
1: Hair in Toxicology: An Important Bio-Monitor
2: Male-mediated Developmental Toxicity
3: Cytochrome P450: Role in the Metabolism and Toxicity of Drugs and other Xenobiotics
4: Bile Acids: Toxicology and Bioactivity
5: The Comet Assay in Toxicology
6: Silver in Healthcare
7: *In Silico* Toxicology: Principles and Applications
8: Environmental Cardiology
9: Biomarkers and Human Biomonitoring, Volume 1: Ongoing Programs and Exposures
10: Biomarkers and Human Biomonitoring, Volume 2: Selected Biomarkers of Current Interest
11: Hormone-Disruptive Chemical Contaminants in Food
12: Mammalian Toxicology of Insecticides
13: The Cellular Response to the Genotoxic Insult: The Question of Threshold for Genotoxic Carcinogens
14: Toxicological Effects of Veterinary Medicinal Products in Humans: Volume 1
15: Toxicological Effects of Veterinary Medicinal Products in Humans: Volume 2
16: Aging and Vulnerability to Environmental Chemicals: Age-related Disorders and their Origins in Environmental Exposures
17: Chemical Toxicity Prediction: Category Formation and Read-Across
18: The Carcinogenicity of Metals: Human Risk Through Occupational and Environmental Exposure
19: Reducing, Refining and Replacing the Use of Animals in Toxicity Testing
20: Advances in Dermatological Sciences
21: Metabolic Profiling: Disease and Xenobiotics
22: Manganese in Health and Disease
23: Toxicology, Survival and Health Hazards of Combustion Products
24: Masked Mycotoxins in Food: Formation, Occurrence and Toxicological Relevance
25: Aerobiology: The Toxicology of Airborne Pathogens and Toxins
26: Chemical Warfare Toxicology, Volume 1: Fundamental Aspects

27: Chemical Warfare Toxicology, Volume 2: Management of Poisoning
28: Toxicogenomics in Predictive Carcinogenicity
29: Human Stem Cell Toxicology
30: The Comet Assay in Toxicology, 2nd edition
31: Computational Systems Pharmacology and Toxicology
32: Ecotoxicology and Genotoxicology: Non-traditional Terrestrial Models
33: Ecotoxicology and Genotoxicology: Non-traditional Aquatic Models
34: Oxidative Stress and Redox Signalling in Parkinson's Disease
35: Nanotoxicology: Experimental and Computational Perspectives
36: Regulatory Toxicology in the European Union
37: Formaldehyde: Exposure, Toxicity and Health Effects
38: Chemical Health Threats: Assessing and Alerting
39: The Micronucleus Assay in Toxicology
40: Nanoparticle–Protein Corona: Biophysics to Biology
41: Big Data in Predictive Toxicology
42: Challenges in Endocrine Disruptor Toxicology and Risk Assessment
43: Bisphenol A: A Multi-modal Endocrine Disruptor
44: Marsupial and Placental Mammal Species in Environmental Risk Assessment Strategies
45: Bird and Reptile Species in Environmental Risk Assessment Strategies
46: Amphibian Species in Environmental Risk Assessment Strategies
47: Perspectives and Challenges of Hair Analysis
48: Neurotransmitters and Toxicology
49: Male-mediated Developmental Toxicity: 2nd Edition
50: Fish Species in Environmental Risk Assessment Strategies

How to obtain future titles on publication:
A standing order plan is available for this series. A standing order will bring delivery of each new volume immediately on publication.

For further information please contact:
Book Sales Department, Royal Society of Chemistry, Thomas Graham House, Science Park, Milton Road, Cambridge, CB4 0WF, UK
Telephone: +44 (0)1223 420066, Fax: +44 (0)1223 420247
Email: booksales@rsc.org
Visit our website at books.rsc.org

Fish Species in Environmental Risk Assessment Strategies

Edited by

Guillermo Eli Liwszyc
University of Helsinki, Finland
Email: guillermo.liwszyc@kolumbus.fi

and

Marcelo L. Larramendy
National University of La Plata, Argentina
Email: marcelo.larramendy@gmail.com

ROYAL SOCIETY OF CHEMISTRY

Issues in Toxicology No. 50

Print ISBN: 978-1-83767-341-4
PDF ISBN: 978-1-83767-371-1
EPUB ISBN: 978-1-83767-372-8
Print ISSN: 1757-7179
Electronic ISSN: 1757-7187

A catalogue record for this book is available from the British Library

The Royal Society of Chemistry is a charity, registered in England and Wales, Number 207890, and a company incorporated in England by Royal Charter (Registered No. RC000524), registered office: Burlington House, Piccadilly, London W1J 0BA, UK, Telephone: +44 (0) 20 7437 8656.

For further information see our website at www.rsc.org

Printed in the United Kingdom by CPI Group (UK) Ltd, Croydon, CR0 4YY, UK

Preface

This volume is the last one of a series of four books to be published by the Royal Society of Chemistry. First in the list is *Marsupial and Placental Mammal Species in Environmental Risk Assessment Strategies*,[1] *Bird and Reptile Species in Environmental Risk Assessment Strategies*[2] followed by *Amphibian Species in Environmental Risk Assessment Strategies*.[3]

The chapters included in this book are just a mere enumeration of some practical examples. There are many more species that can be used as non-conventional experimental models and the list can certainly be expanded further by different research groups and academics all over the world. While animal laboratory models do have their merits, we certainly hope that many more scientists realise that it is more important to tackle subjects related to the status of the environment using autochthonous, or non-target species, which are truly exposed to locally employed xenobiotic agents. In many cases, some wild and domestic household species also share the same physical space with humans and thus enable us to identify which contaminants we are directly exposed to. This research plan would also increase the chances of independent scientists and research institutions in getting access to grants and attracting the attention of local authorities, who would be more interested in financing projects that really are within their sphere of political interest and pride, not to mention their country or province/district.

Without running the risk of being repetitive, we would like to recap on some of the important concepts as previously mentioned in 2017 in the two volumes already published by the RSC, *i.e.*, *Ecotoxicology and Genotoxicology – Non-traditional Terrestrial Models* and *Ecotoxicology and Genotoxicology – Non-traditional Aquatic Models*, by the same editor group.[4,5] We strongly recommend the perusal of both volumes, which are not overlapping subjects, in order to gain the full benefit of this series and have a more holistic and stimulating approach to the matter.

Issues in Toxicology No. 50
Fish Species in Environmental Risk Assessment Strategies
Edited by Guillermo Eli Liwszyc and Marcelo L. Larramendy

Published by the Royal Society of Chemistry, www.rsc.org

We share our planet with thousands of living species. Understanding how environmental stressors affect them will also help us in our quest for solutions to many of the problems humanity is facing at present and which will be more relevant in the near future. Thinking out of silos and cross-linking medical with applied science will lead to a faster pace of understanding of many of the factors relating to the emergence of a disease and thus help us in our search for finding a cure.

New emerging studies in several reputable publications (*e.g.*, Worldwide Trends 1975–2015) of common worldwide human diseases such as arterial hypertension, among others, seem to suggest that not only classical and lifestyle factors should be addressed in the search for adequate treatment.[6] From a medical perspective, it is not only a disease of affluence. One in eight deaths worldwide is due to this condition and its corresponding factors (heart, kidney disease and stroke). These studies seem to suggest that a closer look of other, up till now unrelated factors, is required. Early-life nutrition and exposure to air pollution, heavy metals and even noise are factors implicated, that may push blood pressure up later in life. Plastic and microplastics have been deliberately been excluded from this volume as there are many publications that refer to the matter with much more depth than we could possibly achieve, except for a single chapter, so as not to set aside its importance among the list of pollutants.

Many researchers have contributed to the publication of this book. We hope that it serves as a herald in order to bolster enthusiasm for the use of native or easily available local species in order to widen our knowledge on the subject. We would like to especially thank the authors of the chapters for their positive response, their time and contribution making the compilation of this series of four books possible. The task had been further complicated by the global pandemic of Covid-19, which we seem to have left behind, but not its long-term unforeseen consequences, only recently starting to emerge. Many, if not most of our institutions had to close their doors or cease all academic activity for long periods of time during the pandemic. Access to data banks, field trips, *etc.* were curtailed; nevertheless, against all odds, we were able to compile this piece of work which we hope would be of interest, use, but above all, for continuous nurture of the appetite for future lines of research. Last, but not least, our most sincere gratitude for the trust deposited upon us by the Royal Society of Chemistry in this project. We hope to have fulfilled the reader's expectations.

Guillermo Eli Liwszyc
University of Helsinki, Finland

Marcelo L. Larramendy
Universidad Nacional de La Plata, Argentina
Consejo Nacional de Investigaciones Científicas y Técnicas (CONICET), Argentina

References

1. M. L. Larramendy and G. E. Liwszyc, *Marsupial and Placental Mammal Species in Environmental Risk Assessment Strategies*, Royal Soc. Chem., Cambridge, United Kingdom, 2022.
2. G. E. Liwszyc and M. L. Larramendy, *Bird and Reptile Species in Environmental Risk Assessment Strategies*, Royal Soc. Chem., Cambridge, United Kingdom, 2023.
3. M. L. Larramendy and G. E. Liwszyc, *Amphibian Species in Environmental Risk Assessment Strategies*, Royal Soc. Chem., Cambridge, United Kingdom, 2023.
4. M. L. Larramendy, *Ecotoxicology and Genotoxicology: Non-traditional Terrestrial Models*, Royal Soc. Chem., Cambridge, United Kingdom, 2017.
5. M. L. Larramendy, *Ecotoxicology and Genotoxicology: Non-traditional Aquatic Models*, Royal Soc. Chem., Cambridge, United Kingdom, 2017.
6. NCD Risk Factor Collaboration, *Lancet*, 2016, **384**, 37–55.

Contents

Issues in Toxicology No. 50
Fish Species in Environmental Risk Assessment Strategies
Edited by Guillermo Eli Liwszyc and Marcelo L. Larramendy

Published by the Royal Society of Chemistry, www.rsc.org

CHAPTER 1

General Aspects – Current and Further Perspectives

MARCELO L. LARRAMENDY[a,b*] AND
GUILLERMO ELI LIWSZYC[c]

[a]Cátedra de Citología, Facultad de Ciencias Naturales y Museo, Universidad Nacional de La Plata, Calle 64 No. 3, B1904AMA La Plata, Argentina; [b]Consejo Nacional de Investigaciones Científicas y Técnicas (CONICET), Buenos Aires, Argentina; [c]Department of Pathology, University of Helsinki, Helsinki, Finland
*E-mail: marcelo.larramendy@gmail.com

By definition, it is accepted worldwide that environmental risk assessment is a scientific practice that identifies and evaluates the likelihood of an activity or substance to threaten the environment in general, and/or living organisms, natural habitats and ecosystems, in particular. In other words, it is a term used to describe the overall process or method allowing (a) the identification of hazards and risk factors that have the potential to cause harm (hazard identification) and (b) the analysis and evaluation of the risk associated with that particular hazard (risk analysis and risk evaluation). It also suggests all the appropriate ways to eliminate or minimize the hazard, or control the risk when the hazard cannot be eliminated (risk control).

The Food and Agriculture Organization of the United Nations (F.A.O.) predicted a few years ago that the global population will be over 9.1 billion by the middle of the 21st century. Accordingly, food production will have to increase by about 70% above current levels to maintain pace with demand. Disruptions in transport and production of basic commodities

Issues in Toxicology No. 50
Fish Species in Environmental Risk Assessment Strategies
Edited by Guillermo Eli Liwszyc and Marcelo L. Larramendy

Published by the Royal Society of Chemistry, www.rsc.org

due to unforeseen circumstances, such as the present war in Ukraine, have revealed once more the caveats of this model. One plausible method for achieving this target would be to increase yields and the amount of arable land. However, the conversion of natural forests and/or other wild habitats engenders a number of well-known negative impacts on climate change and global biodiversity. Furthermore, it is accepted worldwide that such an expansion of agriculture could be responsible for approximately 12% of global warming. Several anthropogenic activities introduce extensive amounts of chemicals and pollutants into the environment on a daily basis. In this sense, the use of fossil fuels, mining, manufacturing and construction industries, among others, are the main contributors to water, air and soil pollution. Increases in improper disposal of waste, chemicals, heavy usage of air-polluting equipment and machinery, asbestos handling, soil contamination and industrial noise need to be factored in too. Even our clothes, in many cases composed of artificial fibres that do not naturally degrade, when thrown away, or sent to some far-away country, as a sort of "help" from more economically privileged nations, find their way to landfills. In some cases, these are located close to the sea shores and eventually end up polluting the oceans. To this, we must add overfishing and overexploitation of marine resources, among others, which complicate matters further on.

Most of the xenobiotics in the different environmental compartments exert their effects through cytotoxicity and genotoxicity, interfering with metabolic pathways. Accordingly, current awareness of the real/potential hazards of pollutants in the environment has a high-ranking interest in the use of terrestrial and aquatic invertebrate and vertebrate species as indicators for monitoring pollutant-induced environmental deleterious effects. Notwithstanding, it is known that agrochemicals, among others, not only affect target organisms but concomitantly exert negative effects on non-target species as well.

In environmental risk assessment studies, there is an increasing interest in biomonitoring markers to provide practical indicators of biological exposure to pollutants. To achieve this goal, several endpoints for testing toxicity have been developed. Molecular techniques, plus cyto- and genotoxicity tests have been employed on aquatic and terrestrial organisms to assess the impact of pollution on contaminated areas (*in situ* assays) and for screening xenobiotics after direct or indirect exposure (*in vivo* assays).

This book covers many non-traditional fish species models employed to analyse the jeopardising effects of environmental pollutants, as well as examples of applications of models in integrated assessment solutions. We also hope it will provide answers or at least some clues to the inherent potential risk of the presence of these xenobiotics in our environment. To achieve this goal, a consideration of exposure, hazard, risk, mitigation, as well as short- and long-term monitoring is required. This information will give all interested parties, *e.g.*, regulators and safety adviser bodies, the tools needed for a better understanding of the matter in a broader context.

Many important ecosystems around the world are being constantly challenged due to growing human and industrial pressure exerted upon them. More than one million animals and plants are facing an unprecedented pattern of extinction. We are experiencing the largest biological disappearance wave since the extinction of the dinosaurs and humans could be the main cause of this event.

The use of various biomarkers in local, easily available species can be of use to evaluate the response of the biota to such pollutants. Several biological parameters mirror the interactions between toxic agents and biotic matrices. These are powerful tools that can be applied to environmental monitoring tests and studies. Their responses may reveal general deleterious effects to the organism in general, pinpointing alterations at the cellular, biochemical and molecular levels, as well as in higher levels of organization.

The health of our planet and its ecosystems has deteriorated severely. There is no question about this statement. Mankind has reached, for the first time in its history, an unprecedented level of food security and availability of consumer goods. Unfortunately, this comes with a price tag. The different geographical areas, species and substances studied show that no region, even if far away from polluting centres, is spared from this unfortunate phenomenon. We hope that the contributions of this book will stimulate further work along these lines, while at the same time raising several flags. As scientists, we have an obligation to present evidence of the damage that our planet, Earth, is being subject to. A transversal approach is needed, even in matters that do not seem to be related, but in reality are. An example of this is light pollution in populated areas, which disrupts the normal behaviour of birds and insects, altering their feeding habits and the pollination role they play in the ecosystem. This information, in turn, will help launch the debate at a civil society level, to change, in turn, many of its consumer patterns, press producers and manufacturers, while at the same time demanding more stringent action on behalf of politicians and law makers to protect this planet, our home, before it is too late.

Our global society needs to table down actions and set rules to evaluate and considerably reduce the real and potentially hazardous factors in the environment that can, as previously stated, result in health risks for all forms of life (including *Homo sapiens sapiens*). Despite major positive contributions in the field of health, due to the immense progress achieved in science, technology and industrialization, the interaction between environmental risk and health is an often intricate, not self-evident equation, that involves a variety of not only social, political and economic but also life style factors. This cannot be emphasized enough. Health depends on the good quality of environmental "basic ingredients", such as air, water, soil and food, among others. We believe that the ultimate challenge in this matter is to weigh-in short-term positive gains, while, at the same time, taking into account the long-term effects of substances used. Available information about the toxicity of heterogeneous xenobiotics, continuously released into human habitats, inadvertently, deliberately or by non-regulated industrial discharges on biological components of the environment, is inconclusive.

There is no clear-cut definition of the concept of environmental health. Various openings help us in the understanding of this concept. According to the World Health Organization (W.H.O.) from the United Nations, it is defined by "... all the physical, chemical and biological factors external to a person and all the related factors impacting upon behaviours. It encompasses the assessment and control of those environmental factors that can potentially affect health. It is targeted towards preventing disease and creating health-supportive environments ..." For the National Environmental Health Association, this concept refers to "the protection against environmental factors that may adversely impact human health or the ecological balances essential to long term human health and environmental quality, whether in the natural or man-made environment." A third definition by the National Institute of Environmental Health Science also involves the criterion that "the social environment encompasses life style factors like diet and exercise, socioeconomic status, and other societal influences that may affect health."

In general terms, our health and the health of many other species are negatively affected by broad categories of environmental hazards, namely, electromagnetic fields (produced by high-power lines, electrical wiring, appliances, mobile phones, computers, and TV sets, *etc.*), radiation (including nuclear fallout from weapons testing, fission materials from nuclear power plants and their respective accidents, leaking radioactive disposal sites, air travel and X-rays), toxic chemicals (*e.g.*, some organochlorines, phthalates, polybrominated flame retardants, perfluorinated substances, bisphenol-A) and several toxic metals, among others, which have been shown to have endocrine-disrupting properties, and soil mineral depletion as a complex environmental hazard. Plastic and microplastics have been deliberately been excluded from this volume as there are many publications that refer to the matter with much more depth that we could possibly achieve, except for a single chapter, so as not to set aside its importance among the list of pollutants.

By definition, health risk assessment in its quantitative and/or qualitative determinations includes variants such as the type of risk involved and the severity of response, within or without a probabilistic context. In this regard, risk-based methods of analysis play a strategic role in identifying and ranking adverse responses or the structure of the effects of exposure *vis-à-vis* environmental factors.

Many compounds can be hazardous if not used appropriately and may present a real risk to the environment by contaminating soil, water and air. Most of the pollutants in the different environmental compartments exert their effects through cytotoxic, genotoxic and metabolically toxic mechanisms. In pollution studies, there is an increasing interest in biomonitoring markers of biological exposure to pollutants. To achieve this goal, several end-points for the three above-mentioned factors have been applied to *in situ* and *in vivo* assays.

The use of species as indicators for monitoring pollutant-induced deleterious environmental effects will raise the current awareness of real and potential hazards. It is also known that most environmental pollutants not only affect target organisms but concomitantly exert negative effects on other non-target species as well.

Invertebrate and vertebrate animal models have been used for decades in acute and chronic toxicity tests for hazard identification. They can be very efficient screening systems that have a major role to play in toxicity research, because certain aspects of their biology, physiology and genetic characteristics make them suitable models in ecotoxicological and genotoxicological studies.

This volume aims to shed some light on the matter, whilst offering relevant tools for evaluating risk and providing a framework of practical discussions. These will, in turn, foster decisions and actions required to reduce the environmental health risk against environmental factors. This book presents some real-life examples, extending concepts (of hazardous factors) to living species that may stimulate new research lines and trends in the relevant fields.

Available information has been compiled from a diversity of sources, in an attempt to achieve a representative global and geographical balance, as far as possible, whilst at the same time aiming at high-quality studies. We believe that this book is unique in this sense.

The scientific community is starting to envisage that the time has come for a shift in emphasis away from basic towards applied research. While basic research is needed in order to shed more light on the fundamentals, applied research is required in order to find solutions to the many problems the world faces, *e.g.*, overpopulation, excessive use of the Earth's natural resources and pollution.

Many researchers from different parts of the world have contributed to the chapters of this book. We hope that this volume will meet the expectations and needs of all those interested in the environmental risk assessment field of study by the use of widely available species worldwide. Finally, we also hope that the examples included in the different chapters of this book will awaken the ability to search for new organisms in local and regional ecosystems to pursue further studies in ecotoxicology and genotoxicology.

CHAPTER 2

Biomarkers of Trace Element Toxicity in Fish: A New Paradigm in Environmental Health Risk Assessment

SURESH VIR SINGH RANA

Department of Toxicology, Choudhary Charan Singh University, Meerut 250 004, India
E-mail: sureshvs_rana@yahoo.com

2.1 Introduction

Broadly, biomarkers can be defined as physical parameters, anchored to a disease or a group of diseases. In medicine, these include blood lipids or serum transaminases, which can be directly related to cardiac or liver diseases, to name but a few. For environmental health purposes, biomarkers may include a wide range of exposure related indices, pollutant burden in tissues or body fluids, exposure-related morphological and biochemical changes in cells and tissues, protein and DNA ligands and gene expression changes, among others. They are further classified as biomarkers of exposure, effect and susceptibility.[1] Biomarkers of exposure estimate the actual internal dose that induces toxicity. Biomarkers of dose assess the interaction between toxicants and molecular targets/receptors. Biomarkers of effect predict health impairment such as necrosis or apoptosis. Biomarkers of susceptibility include intrinsic factors such as single nucleotide polymorphisms

Issues in Toxicology No. 50
Fish Species in Environmental Risk Assessment Strategies
Edited by Guillermo Eli Liwszyc and Marcelo L. Larramendy

Published by the Royal Society of Chemistry, www.rsc.org

(SNPs), or pre-existing diseases that influence the dose response. The employment of biomarkers in environment health risk assessment dates back to the early 1970s when blood lead concentration was found to reflect its neurotoxicity in children.[2] Later, in the 1980s, the concept of molecular epidemiology was proposed to link environment exposure to a toxin with disease.[3] In the 1990s, biomarkers for a wide range of pollutants, aromatic compounds, polycyclic aromatic hydrocarbons (PAH) and toxic metals were identified as well.[4–6] The biomedical landscape changed after the success of the first part of the ambitious human genome project, which was completed in 2003 and finally completed in 2022. It ushered a new era of omic technology. Today, scientists routinely compare genomic stretches among patients and healthy individuals searching for gene expression variations that could serve, *e.g.*, as biomarkers of several diseases, encompassing Alzheimer's, cancer or Parkinson's, to name but a few. The concept of multiple biomarkers that includes microRNA, mRNA, protein and gene expression profiles to predict a health risk has also been echoed.[1] Epidemiological data show that human exposure to heavy metals has been monitored in the past using biological monitoring tools. In addition to the estimation of their concentration in biological fluids, specific biomarkers of exposure to lead (δ aminolevulinic acid dehydratase), cadmium (β microglobulin) and arsenic (dimethylarsenic acid) have been identified in humans and rodents.[7] However, the need to minimize the use of laboratory animals in research has reinforced the development of alternate models for health risk assessment. Fish have successfully entered the fray as model organisms for conducting experimental toxicology studies. They offer a few advantages over other models. To name a few, fish have a short life span and high fecundity. They are affordable and easy to maintain under laboratory conditions.[8,9] Their eggs can be easily manipulated and fertilized externally and customized to desired experimental requirements. These attributes make fish a convenient model to test the toxicity of drugs and chemicals as well as their therapeutic efficacy. Additionally, the close genetic similarity between fish and humans is helpful in carrying out genetic manipulations. This chapter overviews the multiple fingerprints of heavy metal toxicity in fish that have ushered a new paradigm in environmental health risk assessment (see Figure 2.1).

2.2 Fish: An Experimental Model in Biomedicine

Evolutionary history shows that fish diverged from humans approximately 400 million years ago. However, with only a few differences at a molecular level, their contribution as an experimental model in research concerned with human genetics, anatomical and physiological responses, immunology, reproduction and pathogenesis of several human diseases, is justified. Popular models in aquatic toxicology include, among others, the goldfish *Carassius auratus* (Cypriniformes, Cyprinidae), the medaka *Oryzias latipes* (Beloniformes, Adrianichthyidae), the rainbow trout *Oncorhynchus mykiss* (Salmoniformes, Salmonidae), and small sized ornamental fishes *viz.* the

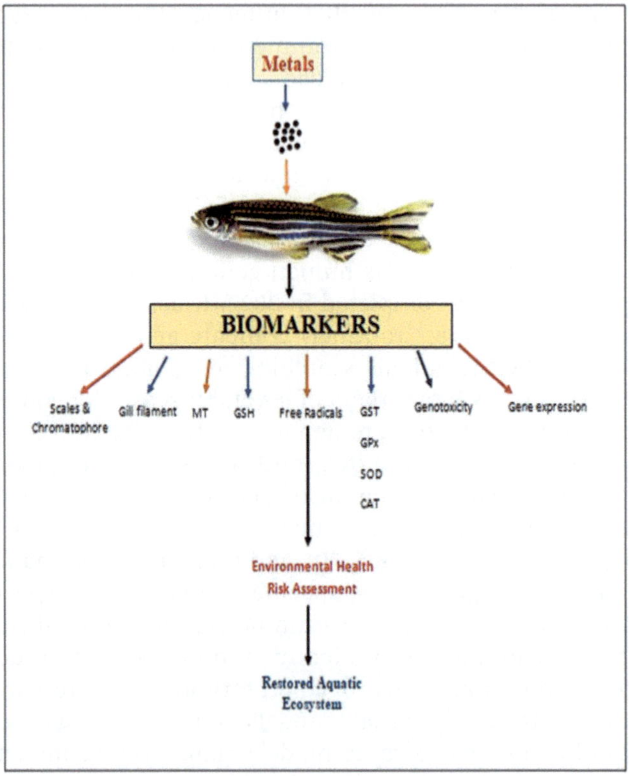

Figure 2.1 Scheme showing the possible biomarkers of toxic elements in fish.

swordtails *Xiphophorus* spp. (Cyprinodontiformes, Poeciliidae), the rivulus or mangrove rivulus *Rivulose* spp. (Cyrinodontiformes, Rivulidae), the mollies *Poecilia* spp. (Cyprinodontiformes, Poeciliidae) as well as other cyprinids and catfish (Siluriformes).[9] For *in vitro* studies, about 283 cell lines have been developed from fin fish around the globe.[10] Other platforms considered as alternatives to animal models include stem cells and other cell cultures,[11] 3D tissue models,[12] organs on chips, *in silico* prediction,[13] the round free-living nematode *Caenorhabditis elegans* (Rhabditida, Rabditidae) as well the worldwide fish model, the zebra fish, *Danio rerio* (Cypriniformes, Cyprinidae).[14,15]

The zebra fish, *D. rerio*, is today a typical model of selection in biomedical research. George Streisinger first introduced this species to study developmental genetics in the 1960s.[14] It shares genetic similarities with humans with approximately 70% of human disease genes having functional homology. It has been widely used as a candidate species for modelling human diseases *viz.* Duchenne muscular dystrophy (DMD), Parkinson's disease (PD), attention deficit hyperactivity disorder (ADHD) and obsessive-compulsive disorder. Recently, many drug discovery studies have been performed using "avatar models" of zebra fish. It has also been employed in xenotransplantation of human cancer cells for investigating drug efficacy, a step forward for a personalized drug treatment strategy.[16,17]

Other fish species that have been employed in biomedical/pharmacological/toxicological research include, *e.g.*, the spotted snakehead *Channa punctatus* (Anabantiformes, Channide), the African sharptooth catfish *Clarias gariepinus* (Siluriformes, Clariidae), the mummichog *Fundulus heteroclitus* (Cyprinodontifores, Funulidae), the shorthorn sculpin *Myoxocephalus scorpius* (Scorpaeniformes, Cottidae), the Mozambique tilapia *Oreochromis mossambicus* (Actinopterygii, Cichliformes), the Nile tilapia *O. niloticus* (Actinopterygii, Cichliformes) and two species of Salmoniformes (Actinopterygii, Actinopterygii), *e.g.*, the brown trout *Salmo trutta* and the salmon trout *S. irideus.*[10]

2.3 Heavy Metal Toxicity in Fish

Metals are naturally occurring elements in all ecosystems. They can exist in elemental or organic forms. They include transition metals, some metalloids, lanthanides and actinides. Volatile metals can be widely transported across different ecosystems.[18] Trace elements, *e.g.*, Zn, Fe, Cu, Mo, Mg, Mn and Se are classified as essential elements. The activity of many enzymes is dependent on these elements. Others, *viz.* Pb, Hg, Cr, Ni and As are extremely toxic, even in small doses. Since these are persistent and nonbiodegradable, they manifest bioaccumulation and biomagnification through the aquatic food chain. Fish from contaminated aquatic systems have been found to be good indicators of heavy metal pollution.[19] Factors that might affect their toxicity include, among others, metal solubility, pH and temperature of water, presence of other toxins and ecosystem complexity. Portals of their absorption include gills, body surface and digestive tract.[20] Consequently, bioaccumulation and/or metabolism takes place in tissues like gills, liver, kidney, brain and gonads.[21]

In general, metal toxicity is manifested through the following mechanisms, namely:

(i) Redox active elements generate free radicals by redox cycling.[22]
(ii) Redox inactive elements impair thiol containing antioxidants and enzymes.[23]
(iii) The Fenton reaction promotes the generation of free radicals.[24]
(iv) Redox sensitive transcription factors *viz.* activator protein (AP-1), p53, nuclear factor kappa B (NF κB) are activated. These factors regulate cell division, cell differentiation and apoptosis.[25]
(v) Several elements, *i.e.*, Pb, Hg, Cr, work as direct enzyme poisons.[26]
(vi) They induce gene expression changes.[27]

2.4 Scales and Chromatophores as Biomarkers of Metal Toxicity

Chromatophores are a particular category of cells in which the pigment granules can be dispersed or aggregated by movement within the cell. Mature chromatophores in fish are grouped as xanthophores (yellow), erythrophores (red), iridohores (reflective), leucophores (white) and melanophores (black/brown). Physiological colour change is used by the fish to camouflage. However, polluted water can also affect the chromatophores. A few studies on the effect of toxic metals on fish chromatophores have been made in the past. Singh and Dutta Munshi studied the effect of $HgCl_2$ (mercuric chloride) on melanophores of the Asian stinging catfish *Heteropneustes fossilis* (Siluriformes, Heteropneustidae)[28] and Chaudhary *et al.* demonstrated depigmentation in the rohu *Labeo rohita* (Cypriniformes, Cyprinidae).[29] Allen *et al.* postulated that fish chromatophores are reliable biomarkers of As exposure (see Figure 2.2).[30] Exposure of the goldfish *C. auratus* to a sublethal concentration of Hg induced changes in its pigmentation.[31] Exposure of *H. fossilis* to $CdCl_2$ (cadmium chloride) altered chromatophores' morphology from reticulate to punctuate-stellate type, as they lost their dendritic processes and aggregation of melanin in the centre.[32] These authors opined that fish chromatophores may serve as non-invasive and more accurate/precise biomarkers to indicate metal pollution and ecosystem health.

Ultrastructural changes in scales have also been considered as bioindicators of aquatic pollution. The herbivorous fresh water fish, the grass carp *Ctenopharyngodon idella* (Cypriniformes, Cyprinidae), showed significant changes in the ultrastructure of scales after exposure to agrochemicals including organophosphates and chlorpyriphos.[33] These changes included damaged radii, focused circuli and severely damaged lepidonts. The intensity of damage was reciprocated with dose and period of exposure. These authors did suggest that scales can be used as suitable bioindicators of

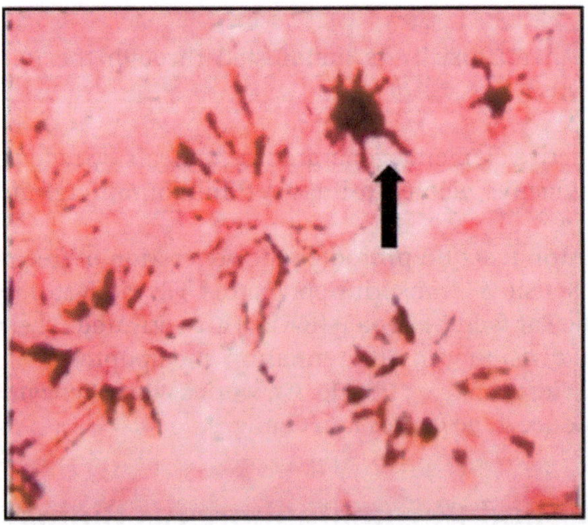

Figure 2.2 Punctuated chromatophores (black arrow) observed in the spotted snakehead *Channa punctata* (Anabantiformes, Channidae) after exposure to As_2O_3 arsenic trioxide for 90 days. Magnification 400×. Reproduced from ref. 30 with permission from Springer Nature, Copyright 2004.

aquatic pollution. A similar opinion was expressed by Hidayati *et al.* while making their studies on fish scales collected from the Porong River (Malaysia), known to be contaminated with Al and Fe.[34] The study suggested that fish scales can potentially be used as a rapid and valid biomarker of aquatic pollution, at least for Al and Fe.

2.5 Gills as a Biomarker of Metal Toxicity in Fish

Fish possess specialized organs for breathing – the gills. Most fish exchange oxygen and carbon dioxide through gills that are protected by the operculum. Water flows over the gills counter to blood flow ensuring a constant partial pressure gradient of oxygen and carbon dioxide over the duration of blood transit through the gills. Most of the xenobiotics enter the fish through gills. Their uptake is determined by many factors, the most important being the transfer capacity of the gills and the physical/chemical nature of the xenobiotic. The gills are composed of comb-like filaments, the gill lamellae that increase the surface area for gaseous exchange. In addition, fish gills are responsible for a number of critical functions, *i.e.*, osmoregulation, excretion of nitrogenous waste, pH regulation and hormone production. Though many excellent reviews describe the lesions induced by heavy metals in the gills of fish, the concept of gill health is not explicitly defined.[35] Gill health can be interpreted through a combination of behaviour, clinical signs, gross pathology, histopathology and ultrastructural changes.

Histopathological changes caused by exposure to toxic metals in the gills of fish have been treated as reliable markers of their toxicity. Recently, the putative relationship between histopathology and exposure to several trace elements *viz.* Al, As, Ba, Cd, Co, Cr, Cu, Fe, Mn, Mo, Ni, Pb, V and Zn has been studied in a demersal fish, the mojarra *Diapterus rhombeus* (Perciformes, Gerreidae) as well as in a benthic fish, the Brazilian batfish *Ogcocephalus vespertelio* (Lophiiformes, Ogcocephalidae) in Santos Bay (Brazil).[36] The demersal fish showed more potential than the benthic fish for risk assessment. These authors suggested incorporation of functional gill morphology in the management of an estuarine ecosystem. Gill morphology of the grass carp *C. idella* was also investigated after LC_{15} exposure to Cu, Cr, and Pb for 24, 48, 72 and 96 h.[37] Other reports include histopathological effects of Cr in the gills of the rohu *L. rohita*,[38] Ag and Cu nanoparticles on the gills of the Siberian sturgeon *Acipenser baerii* (Acipenseriformes, Acipenseridae)[39] and $HgCl_2$ (mercuric chloride) on the gills of *D. rerio.*[40] Recently, the effects of CdS (cadmium sulphide) nanoparticles on the ultrastructure of gills of a fresh water fish *C. punctatus* have also been studied (see Figures 2.3 and 2.4).[41]

The concept of an integrated biomarker response index (IBRv2) was recently proposed by El-Agri *et al.*[42] Their results indicated that effective biomarkers of heavy metal pollution on the Egyptian sole *Solea aegyptiaca* (Pleuronectiformes, Soleidae) were malondialdehyde and 8-hydroxy-2′-deoxyguanosine in the gills. Similarly, biochemical and morphological biomarkers in the gills of a Neotropical fresh water fish, the streaked prochilod, also known as sabalo, *Prochilodus lineatus* (Characiformes, Prochilodontidae)

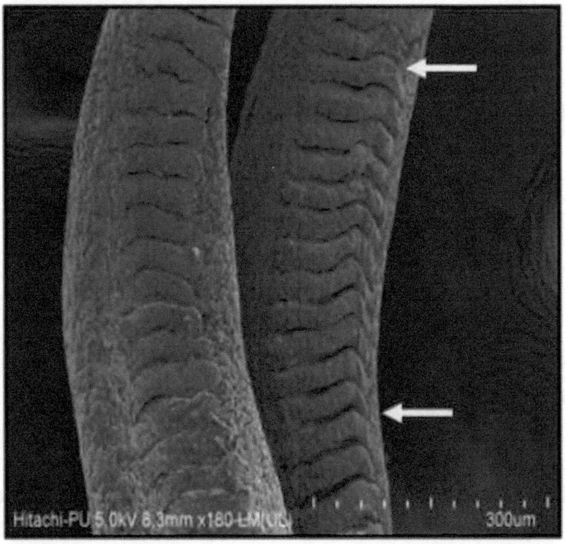

Figure 2.3 FESEM observations on the gill lamina show regular/smooth arrangement of lamellae (white arrows) in control Asian stinging catfish *Heteropneustes fossilis* (Siluriformes, Heteropneustidae) maintained under clean water conditions.

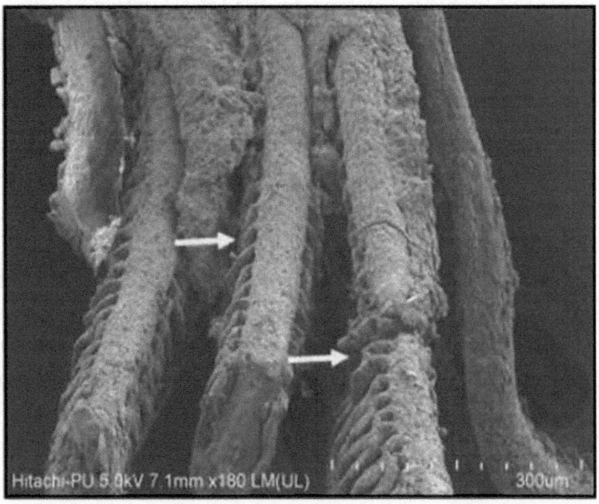

Figure 2.4 FESEM observations on the gill lamina of the Asian stinging catfish *Heteropneustes fossilis* (Siluriformes, Heteropneustidae) exposed to cadmium sulphide nanoparticles for 30 days showing irregular arrangement of lamellae (white arrows).

were simultaneously studied.[43] The authors concluded that a flavonoid metal-insecticide [Mg(hesp)$_2$(phen)], denominated as MgHP, at a high concentration affected the gills by changing glutathione S-transferase activity, GSH levels and histology of the gills. Multiple biomarker responses including serum biochemistry, oxidative stress, genotoxicity and histopathology in the spotted snakehead *C. punctatus*, exposed to waste water containing Fe, Mn, Zn, Co, Ni, Cu and Cr have also been studied.[44] Activity of ATPases in the gill, kidney and muscles of the Nile tilapia *O. niloticus* proved to be a sensitive biomarker after exposure to Cu and Zn.[45] All these studies favour the use of gills as part of a programme in determining ecosystem health.

2.6 Metallothionein

Metallothionein (MT) is a ubiquitous low molecular weight (2.0–7.0 kDa) protein. It was discovered in equine kidney cortex as a Cd binding protein responsible for the natural accumulation of this element in renal tissue.[46] It should be highlighted that 30% of the amino acid residues from MT are cysteines. Thus, the protein has a large number of sulfhydryl groups that can be coordinated by heavy metals (see Figure 2.5). MTs occur throughout the animal kingdom and are also found in higher plants, eukaryotic microorganisms and some prokaryotes.[47] There are four major forms of MT which differ slightly in their amino acid sequence and are coded by distinct genes. Among them, only one or two isoforms have been reported to occur in fishes.[48] Metallothioneins' apoprotein, thionein, can be induced after exposure to Cd, Cu, Hg and Zn. Since, there exists a well-established correlation between MT induction and intracellular concentration of heavy metals, it can be used as a potential biomarker of heavy metal contamination.[49] Liver and kidney have proven to be the best organs for MT analysis. MTs have been used as biomarkers of metal pollution around Hong Kong through their characterization in goldfish *C. auratus* and the Mozambique tilapia *O. mossambicus.*[50] Moreover, results on MT expression after As exposure in channel catfish

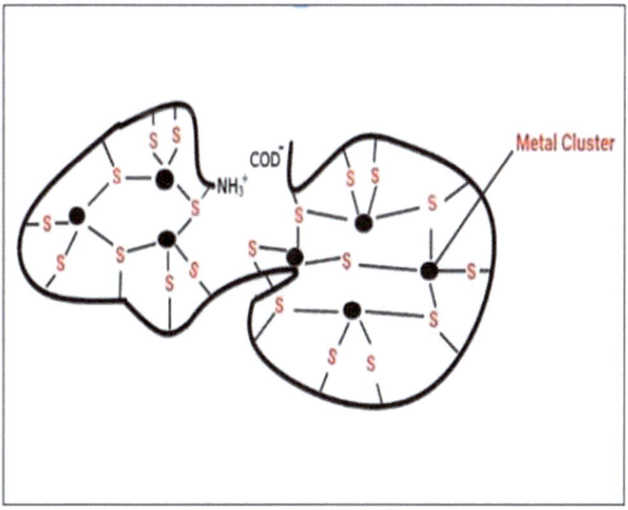

Figure 2.5 Schematic representation of the structure of metallothionein showing the possible sites of metal clusters.

Ictalurus punctatus (Siluriformes, Ictaluridae)[51] as well as Zn and Cu exposure in African catfish *C. gariepinus* have been used as biomarkers of aquatic pollution.[52] MT induction by silver nanoparticles has also been considered as a reliable marker in yellow perch *Perca flavescens* (Perciformes, Percidae).[53] Significant information has accumulated recently to show that MT(s) can be reliable molecular biomarkers of heavy metal toxicity in fish. A radioimmunoassay method for fish metallothioneins has also been developed.[54] However, several compounding factors related to the biology of fish *viz.* species, sex, age, morphometry, season and quality of water, *i.e.*, pH, temperature, salinity and presence of contaminants, may affect MT estimation.

2.7 Free Radicals

Free radicals are defined as an atom, molecule or ion having unpaired electrons. Metabolism of exogenous as well as endogenous substances leads to the formation of free radicals. They are implicated in the genesis of various pathologies and diseases in humans and animals. Toxic elements are known to cause toxicity though free radicals involving the peroxidation of membranous polyunsaturated fatty acids (PUFA). Several reviews have described lipid peroxidation caused by heavy metals.[24,55] Metal specific studies on lipid peroxidation (LPO) induced by heavy metals in fish are also available in the scientific literature. Maiti *et al.* reported that Pb exposure induced LPO in the catfish, *C. batrachus.*[56] Mercury induced LPO in the liver, kidney, brain and gills of fresh water catfish *I. Punctatus.*[57] A significant increase in malondialdehyde (MDA) was reported after 24 h exposure of the gilt-head (sea) bream *Sparus aurata* (Spariformes, Sparidae) to Cd.[58] MDA in the liver, kidney, gills and heart of *C. gariepinus* was determined after exposure to Zn, Cu, Cd, As and Pb as an indicator of water pollution in the Ogun River of Nigeria.[59] Bioaccumulation of heavy metals in fish leads to significant alterations in oxidative stress parameters that can be used as biomarkers of their toxicity.[60] A non-radical species, H_2O_2, has also been considered as a biomarker of aquatic pollution.[61] Its application in monitoring the health status of market (edible) fish has recently been discussed.[62]

Nanoparticles/nanomaterials are also known to induce toxicity in fish through reactive oxygen species (ROS).[63] The possibility of using them as biomarkers in ecotoxicological studies has been discussed. The suitability of various biomarkers in determining the toxicity of nanoparticles, especially in fishes, has been recently explored.[64] However, the impact of nanoparticles on aquatic fauna still remains largely unknown as for today.

2.8 Reduced Glutathione System

Reduced glutathione (GSH) is a ubiquitous tripeptide. The GSH/GSSG cycle plays a crucial role in two electron reduction of peroxides. GSH functions as a co-substrate for the enzyme GPx glutathione peroxidase(s) that scavenges

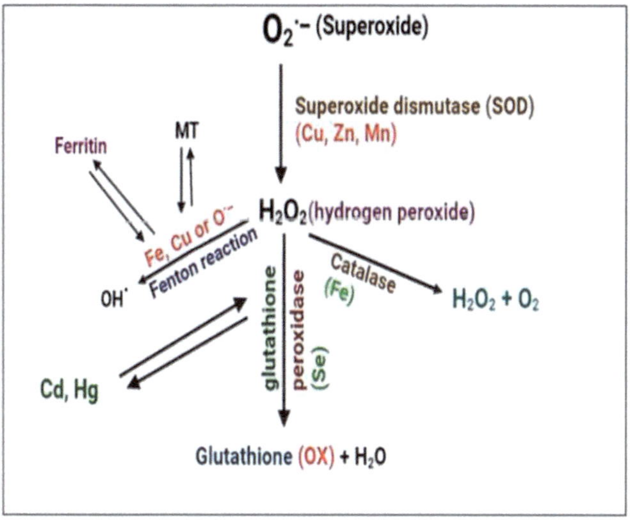

Figure 2.6 Schematic illustration of the antioxidant system that offers protection against trace element toxicity in fish.

H_2O_2. Two molecules of GSH donate two electrons to reduce H_2O_2 to water and are themselves oxidized to GSSG, which is recycled back to GSH by glutathione reductase (GR) using NADPH as the electron donor (see Figure 2.6). In eukaryotes, GSH is involved in many metabolic processes including ascorbate metabolism.[65] It prevents oxidation of SH groups and cross-linking and plays a major role as an antioxidant. It can react with $OH^•$, $HOCl^•$, $ONOO^•$, $RO^•$, $ROO^•$, $NO_2{}^•$ and carbon centred radicals. Rana *et al.* discussed its role in different pathological conditions and designated it as "inevitable glutathione".[66]

GSH is a preferred target molecule of heavy metals due to its strong affinity with sulfhydryl groups. Cadmium, Cu, Cr, Hg and As predominantly bind with the thiol groups and induce oxidative stress. Inhibition of GSH in liver and other tissues of fish on exposure to toxic metals has been considered as a biomarker of their toxicity. Copper sulphate inhibited GSH in the liver and muscles of the three-spined stickleback *Gasterosteus aculeatus* (Scorpaeniformes, Gasterosteidae).[67,68] GSH reduces trivalent chromium to Cr(VI) to participate in the Fenton reaction. Mieiro *et al.* opined that Hg directly binds with SH moiety of GSH.[69] It is implicated in the toxicity of Hg and Cd.[57,70] The affinity of inorganic As with GSH has also been reported.[71,72] A decline in GSH values is considered as an indicator of oxidative stress. Farombi *et al.* considered GSH as a biomarker of heavy metal pollution in the African catfish *C. gariepinus* inhabiting the Ogun River in Nigeria.[59] In zebra fish, *D. rerio*, GSH also was found to be a reliable marker of arsenic toxicity.[73] Recently, Ullah *et al.* used multiple biomarkers including glutathione to detect heavy metal accumulation and toxicity in an endangered fish species, the putitor mahseer *Tor putitora* (Cypriniformes, Cyprinidae)[74]. Thus, there exists considerable experimental evidence favouring GSH as a reliable biomarker in the assessment of heavy metal toxic effects in fish.

2.9 Antioxidant Enzymes

Defence against oxidative stress in all eukaryotes is offered by antioxidant enzymes. These include superoxide dismutases (SODs), catalase (CAT) and glutathione dependent enzymes. Since toxic elements manifest their toxicity through oxidative stress, their effects on antioxidant enzymes can be considered as biomarkers of effect. The activity of SOD and CAT against numerous heavy metals has been studied in bivalves and polychaetes.[75] However, as for today, insufficient information is available in fish.

2.9.1 Superoxide Dismutase

Irwin Fridovich and Joe McCord discovered this enzyme.[76] Superoxide dismutase (SOD) (EC.1.15.1.1) alternately catalyzes the dismutation (or partitioning) of the superoxide (O_2^-) radical into molecular oxygen (O_2) and hydrogen peroxide (H_2O_2). Superoxide is produced as the by-product of oxygen metabolism and, if not regulated, causes many types of cell damage. Employment of an antioxidant defence system as a biochemical marker of metal induced oxidative stress was suggested by Sevcikova *et al.*[55] Similar observations had also been made earlier.[24] A dose-dependent inhibition of SOD in the embryos of the medaka *O. latipes* was recorded after exposure to nano Fe and Hg.[78] A relationship between Fe contamination and increased erythrocyte SOD activity in a cichlid fish was reported.[79] Asagba *et al.* demonstrated increased SOD activity in the liver of African catfish *C. gariepinus* after exposure to Cd.[80] Results on antioxidant enzyme activity were recently treated as biomarkers of Hg induced oxidative damage in the liver and gills of the juveniles of the European sea bass *Dicentrarchus labrax* (Moroniformes, Moronidae).[81] Effects of Hg on multiple antioxidant parameters *viz.* SOD, CAT, MDA, total antioxidant capacity (TAOC), heat shock protein 70 (HSP70), metallothionein (MT) and gene expression changes were investigated in the grass carp *C. idella*.[82] A similar approach was made by Trivedi and co-authors while studying the effects of mercuric chloride in a fresh water fish *C. punctatus*.[83] Investigations on potential antioxidant enzyme biomarkers of Hg exposure in three Amazonian fish were recently made as well.[84] The study identified 21 metal binding proteins as possible biomarkers of Hg exposure in fish.

2.9.2 Catalase

Catalase (CAT) (EC 1.11.1.21) is the main enzyme involved in the enzymic degradation of H_2O_2 to molecular O_2 and H_2O. It is ubiquitous in nature. In many cases it is localized in subcellular organelles such as peroxisomes. Sufficient information is available on the role of catalase in human health and disease,[85] but as far as we know, pharmaco-toxicological studies on catalase in fish remain very limited, so far. Nevertheless, it has been considered as a reliable marker of Cd and Zn exposure in fish. A decrease in CAT activity was reported in the European sea bass, *D. labrax*, after exposure to Cd.[86] Similar observations were recorded in another fish, the gilt-head (sea) bream *S. aurata.*[87] Elia *et al.* registered a negative dose–response relationship between Cd exposure and CAT activity in black bull head catfish *Ameiurus melas* (Siluriformes, Ictaluridae).[88] However, CAT activity offered protection against Fe toxicity in *C.* gariepinus.[89] It served as a reliable marker of Pb and Ni toxicity in *C. punctatus.*[90] Its response to a combined exposure of Pb + Cd in the kidney of *O. niloticus* was also studied.[91] Sublethal effects of Zn, Cu and Ni on antioxidant enzymes were studied in *L. rohita.*[92] Recently, CAT activity in fresh water fishes has been treated as a potential biomarker of environmental pollution.[93] Further studies are needed in order to establish its true significance as a biomarker of metal pollution in aquatic ecosystems.

2.9.3 Glutathione Peroxidase

Glutathione peroxidase (GPx) (EC.1.11.1.9) is the general name given to an enzyme family with peroxidative activity whose main biological function is to protect the organism from oxidative damage. It was discovered by Gordon C. Mills in 1957. It was the first mammalian protein shown to incorporate Se in the form of selenocysteine-C into the catalytic site and was assumed to be associated with antioxidant activity of selenium.[94] At present, GPx is well known for catalyzing the reaction of H_2O_2 and organic hydroperoxides, thus protecting the cells from oxidative damage.

$$\text{ROOH} + 2\text{GSH} \longrightarrow \text{ROH} + \text{GSSG} + \text{H}_2\text{O}$$

The protein chemistry and kinetic mechanisms of glutathione peroxidase have been described in detail by Flohé.[95] Even when the relationship between GPx and metal/xenobiotic toxicity has also been studied by a number of researchers,[96,97] its role in metal induced toxicity in fish has been poorly mirrored. Glutathione dependent enzymes in the liver and kidney of Cd treated *C. punctatus* and *C. batrachus* were studied by Rana and Singh.[98] The study emphasized species' differences in the activity of conjugating enzymes. The GSH-dependent enzymes offered resistance to As-induced oxidative stress in fresh water fish *C. punctatus*.[99] The same group of workers reported protective effects of GPx against arsenic toxicity in *C. punctatus*.[100] Inhibition of enzyme activity has been registered after exposure of fish to Hg, Pb and Zn.[101,102] Antioxidant enzymes in target organs of fish like liver and kidney are now considered as reliable biomarkers of oxidative stress induced by heavy metals.[77]

2.10 Biomarkers of Genotoxicity

The potential risk of cancer is associated with exposure to some metals which can induce genotoxicity.[103] Among the several types of assays for detecting and evaluating genotoxicity, the most employed worldwide are, *e.g.*, induction of sister chromatid exchanges (SCEs), micronuclei (MNs), chromosomal aberrations (CAs) and all the variants of the comet or single cell gel electrophoresis (SCGE) assay, among others. DNA or protein adducts, or mutagenicity in urine, have not been successfully applied for monitoring the genotoxicity of metals, especially in fish species, yet.

Nevertheless, fresh water fish are often considered as ideal candidates for laboratory studies on genotoxicity. The SCGE assay is the most widely used sensitive and reliable marker of genotoxicity in aquatic species.[104] A few species, *viz.* the goldfish *C. auratus*, the carp *C. carpio* and the rainbow trout *O. mykiss*, have been extensively used as experimental models to study the genotoxic effects of mutagenic compounds.[105,106] Embryos of the zebrafish, *D. rerio,* have been used to screen the genotoxicity of Cu from Laguna Lake in the Philippines using the SCGE assay.[107] A comet assay demonstrated an enhancement of DNA damage in the Sacramento sucker *Catastomus occidentalis* (Cypriniformes, Catastomidae), which was linked to runoff events and supported by Ames' test data.[108] A limited number of genotoxic studies have been made in marine fish using the SCGE assay including the European eel *Anguilla anguilla* (Anguilliformes, Anguillidae).[109] As an example, Cd and Hg were found to damage DNA in the latter species.[110]

The micronucleus (MN) test is another widely used method to detect genotoxicity of a wide range of compounds, especially in fish.[111,112] The method was originally developed in mammalian cells[113] but is now widely applied in fish and other aquatic organisms, *i.e.*, sea urchins, mussels, oysters, crabs, as well as in other terrestrial species.[114] The MN test utilizes an interphase stage of proliferating cells irrespective of the species under study. Cells from gills, fins, kidney and liver as well as peripheral erythrocytes are the most commonly used cells to record the genotoxicity of heavy metals (see Figure 2.7). Singh and co-authors studied the genotoxicity of

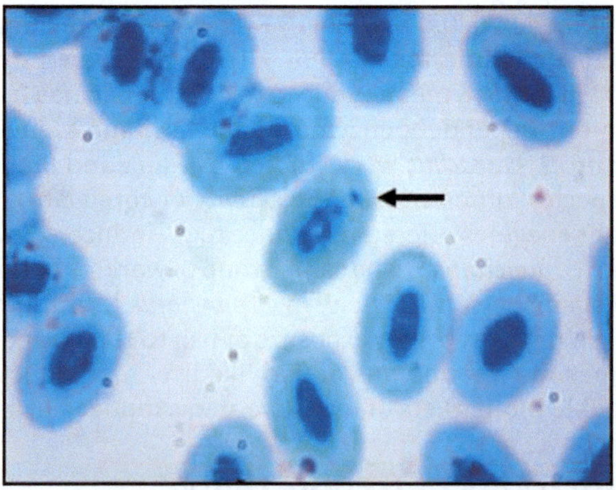

Figure 2.7 Erythrocytes of the spotted snakehead fish *Channa punctata* (Anabantiformes, Channidae) exhibiting the formation of an ovoid micronucleus (black arrow) after exposure to Cr for 30 days. Magnification 400×. Reproduced from ref. 115 with permission from Springer Nature, Copyright 2019.

Cd, Cr, Ni and As in the fresh water fish *C. punctatus*.[115] The frequency of nuclear abnormalities *viz.* blebbed, notched, lobed and vacuolated nuclei differed after exposure to these elements. Chromium exposure also led to the induction of MNs in *C. carpio*.[116] Sublethal exposure to heavy metals *viz.* Hg, As and Cu induced MNs formation in *C. punctatus*.[117] Genotoxic effects of $HgCl_2$ on the Neotropical fish, green terror *Andinoacara rivulatus* (Cichliformes, Cichlidae) were also studied employing the MN test.[118] Furthermore, MN frequency and other nuclear abnormalities have been recorded in the Western mosquito fish *Gambusia affinis* (Cyprinodontiformes, Poeciliidae) after Cd and Cu exposure.[119]

The genotoxicity of nanoparticles in fish has been poorly studied. However, an increased number of MNs has been reported after exposure to cathionic liposomes, dendrimer-modified silica and superparamagnetic Fe oxide (SPIO) particles.[120] The significance of *in vitro* assays in testing the genotoxicity of CdONP in different models was reported by Demir and co-workers.[121] It is worth mentioning that while results on the Ames test showed no genotoxicity, positive results were in fact obtained with the MN assay. Another study also showed that AgNPs did cause MN induction in the rohu, *L. rohita*.[122]

2.11 Molecular Markers

This is an emerging concept applied to record the effect of drugs/chemicals on the expression of thousands of genes either simultaneously or sequentially along regulatory pathways. Gene variations are treated as biomarkers of effect or susceptibility. High throughput omic technologies including genomics, proteomics and metabolomics have driven major efforts in improving environment health risk assessment. Recently, a few studies have been made on gene expression changes caused by heavy metals in fish. An earlier study considered metallothionein gene expression as a biomarker of heavy metal toxicity in the Sleek unicorn *Naso hexacanthus* (Acanthuriformes, Acanthuridae) found in the Jeddah coast, Saudi Arabia.[123] The results were also supported by histopathological observations on liver and gills. The effect of natural brominated chemicals on mRNA expression in the European perch *Perca fluviatilis* (Perciformes, Percidae) of the Baltic has been also investigated.[124] The study identified about 4000 genes with a significant difference in mRNA expression levels. A gene ontology enrichment analysis showed that differently expressed genes were involved in biological processes *viz.* complement activation, ion homeostasis, cholesterol biosynthesis and oxidative stress. Gene expression changes related to oxidative stress induced by Cu, Cr, Cd and Pb in the zebra fish *D. rerio* were studied too.[125] In addition, mRNA levels related to the antioxidant system (SOD1, SOD2 and Nrf2) were upregulated by non-lethal concentration of these elements. In brief, the importance of gene expression assays in environmental risk assessment was proposed. The effects of Ni, Cd and Cr on metallothionein and glutathione peroxidase gene expression along with other health biomarkers were studied in the African sharptooth catfish *C.*

gariepinus by Swaleh *et al.*[126] In the Nile tilapia, *O. niloticus*, alterations in GST gene expression were found to be sensitive indicators of aquatic pollution of Manzala Lake, Egypt.[127] The precise role of molecular markers as reliable indicators of heavy metal toxicity has been discussed as well.[128] It is envisaged that an integrated genomic, proteomic and metabolomic approach would be a future paradigm in the discipline of predictive ecotoxicology of heavy metals, especially in fish.

2.12 Conclusion and Perspectives

The information discussed in the above paragraphs highlights the fact that fish can be used as a suitable model to be implemented in a predictive toxicological evaluation of toxic elements in diverse aquatic ecosystems. From the zebra fish to the medaka as well as from catfish to carps, several other species can be employed to determine the water quality and ecosystem health applying *ad hoc* biomarkers (see Figures 2.8–2.10). Biomarkers can

Figure 2.8 Dorsal view of adults of the spotted snakehead fish *Channa punctata* (Anabantiformes, Channidae) used as an experimental model.

Figure 2.9 Dorsal view of adults of the Asian stinging catfish *Heteopneustes fossilis* (Siluriformes, Heteropneustidae) used as an experimental model.

Figure 2.10 Dorsal view of adults of the walking catfish *Clarias batrachus* (Siluriformes, Clariidae) used as an experimental model.

be further categorized as biomarkers of exposure, susceptibility and effect. These biomarkers may be morphological *viz.* scales, chromatophores, gills, otoliths and other tissues. Several biomarkers including genetic, biochemical and molecular traits can be treated as biomarkers of effect where a considerable number of studies have been performed on these markers too. Our study on chromatophores in the fresh water fish *C. punctatus* yielded useful information on As toxicity.[30] Similarly, histopathological and ultrastructural observations made in the gills of *C. punctatus* demonstrated toxic end points of Cd nanoparticles.[41,63] Earlier reports from our laboratory have affirmed that biochemical markers *viz.* metallothionein(s), stress proteins, free radicals, oxidative stress and antioxidant enzymes were useful in determining the toxicity of elements *viz.* Cd and Hg.[57,70,71,98] Our investigations showed that recently developed biomarkers *viz.* genetic and molecular markers are expected to be helpful in determining the environmental health risks posed by As, Cr, Cd and Ni.[115]

In spite of all the progress achieved, no single biomarker or individual fish species appears to be conclusive. Therefore, a rigorous effort is needed to apply an integrative holistic biomarker approach in an environmental health risk assessment programme. This effort should integrate conventional methods combined with recently developed tools *viz.* DNA and RNA sequencing, bioinformatics, *in silico* methods, genomics, proteomics and eco-genomics, as appropriate. This combinational approach is expected to accelerate the speed and precision of biomarker end point determinations and their application, especially in the management of aquatic environmental health risk assessment strategies.

Acknowledgements

The author sincerely acknowledges the excellent technical support provided by Dr Meenu Singh.

References

1. National Research Council (NRC), *Biological Markers in Reproductive Toxicology*, Committee on Biological markers, National Academy Press, Washington, DC, 1989.
2. H. L. Needleman and C. D. Bellinger, The epidemiology of low level lead exposure in childhood, *J. Am. Acad. Child Psychiatry*, 1981, **20**, 496–512.
3. P. A. Schulte and F. P. Perera, *Molecular Epidemiology: Principles and Practices*, Academic Press, London, UK, 1st edn, 1998.
4. S. V. S. Rana and S. Kumar, Liver function in rats treated individually and with a combination of xylene, toluene and methanol, *Toxicol. Ind. Health*, 1993, **9**, 479–483.
5. V. F. Bergerova and M. Ogata, *Biological Monitoring of Exposure to Industrial Chemicals*, Amer. Conf. Gov. Indust. Hyg. Inc., Cincinnati, Ohio, 1990.
6. M. Ikeda, in *Biological Monitoring of Exposure to Industrial Chemicals*, ed. V. F. Bergerova, and M. Ogata, Amer. Conf. Gov. Indust. Hyg. Inc., Cincinnati, Ohio, 1990, pp. 127–130.
7. S. V. S. Rana, Biological monitoring as a tool for assessing occupational health. An Indian perspective, *Res. J. Environ. Toxicol.*, 2018, **12**, 50–55.
8. C. L. Bolis, M. Piccolella, A. Z. Dalla Valle and J. C. Rankin, Fish as model in pharmacological and biological research, *Pharmacol. Res.*, 2001, **44**, 265–280.
9. M. Schaeck, W. V. D. Broeck, K. Hermans and A. Decostere, Fish as research tools: alternatives to *in vivo* experiments, *Altern. Lab. Anim.*, 2013, **41**, 219–229.
10. G. Pandey, A review of fish model in experimental toxicology, *Int. Res. J. Pharm.*, 2011, **2**, 33–36.
11. T. W. Kim, J. H. Che and J. W. Yun, Use of stem cells as alternate methods in animal experimentation in predictive toxicology, *Regul. Toxicol. Pharmacol.*, 2019, **105**, 15–29.
12. X. Ma, J. Liu, W. Zhu, M. Tang, N. Lawrence, C. Yu, M. Gou and S. Chen S, 3D bioprinting of functional tissue models for personalized drug screening and *in vitro* disease modelling, *Adv. Drug Delivery Rev.*, 2018, **132**, 235–251.
13. S. Kar and J. Leszczynski, Exploration of computational approaches to predict the toxicity of chemical mixtures, *Toxics*, 2019, **7**, 15.
14. T. P. Barros, W. K. Alderton, H. M. Reynolds, A. G. Roach and S. Berghmans, Zebra fish: an emerging technology for *in vitro* pharmacological assessment to identify potential safety liabilities in early drug delivery, *Br. J. Pharmacol.*, 2008, **154**, 1400–1413.
15. P. R. Hunt, The *C. elegans* model in toxicity testing, *J. Appl. Toxicol.*, 2017, **37**, 50–59.
16. C. Santoriello and L. I. Zon, Modelling human disease in zebra fish, *J. Clin. Invest.*, 2012, **122**, 2337–2343.

17. D. A. Meshalkina, E. V. Kysil, J. E. Warnick, K. A. Demin and A. V. Kalueff, Adult zebra fish on CNS disease modelling: a tank that is half full, not half empty, and still filling, *Lab. Anim.*, 2017, **46**, 378–387.
18. J. O. Nriagu and J. M. Pacyna, Quantitative assessment of worldwide contamination of air, water, and soils by trace metals, *Nature*, 1988, **133**, 134–139.
19. D. R. Livingstone, Oxidative stress in aquatic organisms in relation to pollution and agriculture, *Rev. Med. Vet.*, 2003, **154**, 427–430.
20. S. Tao, Y. Wen, A. Long, R. Dawson, J. Cao and F. Xu, Simulation of acid base condition and copper speciation in fish gill microenvironment, *Comput. Chem.*, 2001, **25**, 215–222.
21. L. S. Heir, B. Lien, A. E. Strømseng, M. Ljønes, B. O. Rosseland, K.-E. Tollefsen and B. Salbu, Speciation of lead, copper, zinc and antimony in water during a shooting range- time dependent metal accumulation and biomarker responses in brown trout (*Salmo trutta*), *Sci. Total Environ.*, 2009, **407**, 4047–4055.
22. R. K. Zalpus and J. Koropatnick, *Molecular Biology and Toxicology of Metals*, Taylor and Francis, London, 2000.
23. M. Valko, C. J. Rhodes, J. Moncol, M. Izakovic and M. Mazur, Free radicals, metals and antioxidants in oxidative stress induced cancer, *Chem.-Biol. Interact.*, 2006, **160**, 1–40.
24. F. W. Sunderman Jr, Metals and lipid peroxidation, *Acta Pharmacol. Toxicol.*, 1986, **59**, 48–55.
25. S. V. S. Rana, Metals and apoptosis-recent developments, *J. Trace Elem. Med. Biol.*, 2008, **22**, 262–284.
26. S. V. S. Rana, Role of molecular markers in environmental health and risk assessment, *Mol. Cell. Toxicol.*, 2007, **3**, 34.3.
27. S. J. Yoon, S. S. Han and S. V. S. Rana, Molecular markers of heavy metal toxicity - A new paradigm for health risk assessment, *J. Environ. Biol.*, 2008, **29**, 1–14.
28. A. Singh and J. S. Dutta Munshi, Effect of sublethal mercuric chloride exposure to melanophores in a cat fish, *Heteropneustes fossilis* (Bloch), *J. Environ. Biol.*, 1992, **13**, 303–308.
29. U. S. Chaudhary, V. Rathod and G. N. Vankhede, Effect of water extract of the bark of *Buchanania lanzan* (Linn) on behaviour and chromatophores of a fresh water fish *Labeo rohita*, *J. Environ. Biol.*, 2001, **22**, 229–231.
30. T. Allen, A. Awasthi and S. V. S. Rana, Fish chromatophores as biomarkers of arsenic exposure, *Environ. Biol. Fishes*, 2004, **71**, 7–11.
31. I. Nur, W. O. Erni and M. Idris, Yusnaini, Alterations in pigmentation and morphology of gold fish (*Carassius auratus*) exposed to sub lethal treatment with mercury, *AACL Bioflux.*, 2019, **12**, 2147–2156.
32. S. Ahmad, S. Shukla, A. Mishra, D. Kasherwani, V. P. Swami and S. Shukla, Effect of cadmium chloride on general body colouration and chromatophores of stinging cat fish *Heteropneustes fossilis* (Bloch), *J. Appl. Nat. Sci.*, 2018, **10**, 655–660.

33. R. Jindal and M. Kaur, Ultrastructural alterations in scales of *Ctenopharyngodon idellus (Cuvier and Valanciennes)* induced by chloropyrophos: a promising tool as bioindicator of pesticide pollution, *Int. J. Fish. Aquat. Stud.*, 2015, **2**, 58–62.
34. D. Hidayati, N. Sulaiman, S. Othman and B. S. Ismail, Fish scale deformation analysis using scanning electron microscope: new potential biomarker in aquatic environment monitoring of aluminium and iron contamination, *AIP Conf. Proc.*, 2014, **1571**, 563–568.
35. K. L. Foyle, S. Hess, M. D. Powell and N. A. Herbert, What is gill health and what is its role in marine fish aquaculture in the face of a changing climate, *Front. Mar. Sci.*, 2020, **7**, 400.
36. H. H. Q. Oliveira, J. A. Reis-Filho, J. A. C. C. Nunes, R. Melo Dos Santos, E. de F. Esteves Santiago, L. Aguilar, P. R. A. de Mello Affonso and A. L. da Cruz, Gill histopathological markers in fish exposed to trace metals in the Todos os Santos Bay, Brazil, *Biol. Trace Elem. Res.*, 2022, **200**, 3388–3399.
37. N. Shah, A. Khan, R. Ali, K. Marimuthu, M. N. Uddin, M. Rizwan, K. U. Rahman, M. Alam, M. Adnan, Muhammad, S. M. Jawad, S. Hussain and M. Khisroon, Monitoring bioaccumulation (in gills and muscle tissue), hematology and genotoxic alterations in *Ctenopharyngodon idella* exposed to selected heavy metals, *BioMed Res. Int.*, 2020, **2020**, 6185231.
38. K. Muthukumaravel and P. Rajaraman, A study on the toxicity of chromium on the histology of gill and liver of freshwater fish, *Labeo rohita*, *J. Pure Appl. Zool.*, 2013, **1**, 122–126.
39. T. Ostaszewska, M. Chojnacki, M. Kamaszewski and E. Sawosz-Chwalibóg, Histological effects of silver and copper nanoparticles in the epidermis, gills, and liver of Siberian sturgeon, *Environ. Sci. Pollut. Res.*, 2016, **23**, 1621–1623.
40. R. Marcirella and E. Brunelli, Morphological alterations in zebra fish (*Danio rerio*) gills after exposure to mercuric chloride, *Int. J. Mol. Sci.*, 2017, **18**, 824.
41. Y. Verma, V. Rani and S. V. S. Rana, Assessment of cadmium sulphide nanoparticle toxicity in the gills of a fresh water fish, *Environ. Nanotechnol., Monit. Manage.*, 2020, **13**, 100280.
42. A. M. El-Agri, M. A. Emam, H. S. Gaber, E. A. Hassan and S. M. Hamdy, Integrated use of biomarkers to assess the impact of heavy metal pollution on *Solea aegyptica* fish in Lake Qarun, *Environ. Sci. Eur.*, 2022, **34**, 74.
43. I. T. de Castro Sachi, M. Marques Bonomo, M. M. Sakuragui, P. Zaganin Modena, M. G. Paulino, R. M. Carlos, J. B. Fernandes and M. N. Fernandes, Biochemical and morphological biomarker responses in the gills of Neotropical fish exposed to new flavonoid metal insecticide, *Ecotoxicol. Environ. Saf.*, 2021, **208**, 111455.
44. M. Javed, M. I. Ahmad, N. Usmani and M. Ahmad, Multiple biomarker responses (serum biochemistry, oxidative stress, genotoxicity and

histopathology) in *Channa punctatus* exposed to heavy metal loaded waste water, *Sci. Rep.*, 2017, **7**, 1675.
45. G. Atli and M. Canli, Essential metal (Cu, Zn) exposures alter the activity of ATPases in gill, kidney and muscle of tilapia *Oreochromis niloticus*, *Ecotoxicology*, 2011, **20**, 1861–1869.
46. M. Margoshes and B. L. Vallee, A cadmium protein from equine kidney cortex, *J. Am. Chem. Soc.*, 1957, **79**, 4813–4814.
47. J. H. R. Kägi and Y. Kojima, Chemistry and biochemistry of metallothionein, *Exper. Suppl.*, 1987, **52**, 25–61.
48. L. Bargelloni, R. Scudiero, E. Parisi, V. Carginale, C. Capasso and T. Patarnello, Metallothioneins in antarctic fish: evidence for independent duplication and gene conversion, *Mol. Biol. Evol.*, 1999, **16**, 885–897.
49. R. van der Oost, J. Beyer and N. P. E. Vermeulen, Fish bioaccumulation and biomarkers in environmental risk assessment: a review, *Environ. Toxicol. Pharmacol.*, 2003, **13**, 57–149.
50. K. M. Chan, Metallothionein: potential biomarker for monitoring heavy metal pollution in fish around Hong Kong, *Mar. Pollut. Bull.*, 1995, **31**, 411–415.
51. D. Schlenk, M. Chelius, L. Wolford, S. Khan and K. M. Chan, Characterization of hepatic metallothionein expression in channel cat fish (*Ictalurus punctatus*) by reverse transcriptase polymerase chain reaction, *Biomarkers*, 1997, **2**, 161–167.
52. E. M'kandawire, A. Mierek-Adamska, S. R. Stürzenbaum, K. Choongo, J. Yabe, M. Mwase, N. Saasa and C. A. Blindauer, Metallothionein from wild populations of the African cat fish *Clarius gariepinus*: from sequence, protein expression and metal binding properties to transcriptional biomarker of metal pollution, *Int. J. Mol. Sci.*, 2017, **18**, 1548.
53. J. D. Martin, T.-L. L. Colson, V. S. Langlois and C. D. Metcalfe, Biomarkers of exposure to nanosilver and silver accumulation in yellow perch (*Perca flavescens*), *Environ. Toxicol. Chem.*, 2017, **36**, 1211–1220.
54. C. Hongstrand, P.-E. Olsson and C. Haux, A radioimmunoassay for metallothionein in fish, *Environ. Res.*, 1989, **28**, 183–186.
55. M. Sevcikova, H. Modra, A. Slaninova and Z. Svobodova, Metals as cause of oxidative stress in fish: a review, *Vet. Med.*, 2011, **56**, 537–546.
56. A. K. Maiti, N. C. Saha and G. Paul, Effect of lead on oxidative stress, $Na^{+}K^{+}$ ATPase activity and the mitochondrial electron transport chain activity of the brain of *Clarius batrachus*, *Bull. Environ. Contam. Toxicol.*, 2010, **84**, 672–676.
57. S. V. S. Rana, R. Singh and S. Verma, Mercury induced lipid peroxidation in liver, kidney, brain and gills of fresh water fish *Channa punctatus*, *Jpn. J. Ichthyol.*, 1995, **42**, 255–259.
58. G. Souid, N. Souayed, F. Yaktiti and K. Maaroufi, Effect of acute cadmium exposure on metal accumulation and oxidative stress markers in *Sparus aurata*, *Ecotoxicol. Environ. Saf.*, 2013, **89**, 1–7.
59. E. O. Farombi, O. A. Adelowo and Y. R. Ajimoko, Biomarkers of oxidative stress and heavy metal levels as indicators of environmental pollution

in African cat fish (*Clarius batrachus*) from Nigeria Ogun River, *Int. J. Environ. Res. Public Health*, 2007, **4**, 158–165.
60. P. Garai, P. Banerjee, P. Mondal and N. C. Saha, Effect of heavy metals on fishes: toxicity and bioaccumulation, *J. Clin. Toxicol.*, 2021, **11**, S18.
61. T. Asaeda, M. Rahman and H. D. L. Abeynayaka, Hydrogen peroxide can be a plausible biomarker in cyanobacterial bloom treatment, *Sci. Rep.*, 2022, **12**, 12.
62. N. Nagarani, G. Krishnaveni, K. Darshini, K. G. Manohari, M. A. Jeyapandi, P. V. Mariappan and R. Sangeetha, Biomarkers as ecological inducers in monitoring the status of market fish, *J. Basic Appl. Zool.*, 2023, **84**, 2.
63. S. V. S. Rana, Toxicity induced by nanoparticles in fresh water fish - A review of recent developments, *EC Pharmacol. Toxicol.*, 2021, **9**, 37–56.
64. A. R. Samim, M. Arshad and H. Vaseem, An insight into the various biomarkers to study toxicological impact of nanoparticles in fishes: explored and missing information, *Int. J. Environ. Sci. Technol.*, 2023, **20**, 10533–10552.
65. Sohini and S. V. S. Rana, Protective effects of ascorbic acid against oxidative stress induced by inorganic arsenic in the liver and kidney of rat, *Indian J. Exp. Biol.*, 2007, **45**, 371–375.
66. S. V. S. Rana, T. Allen and R. Singh, Inevitable glutathione, then and now, *Indian J. Exp. Biol.*, 2002, **40**, 706–716.
67. W. Sanchez, O. Palluel, L. Meunier, M. Coquery, J.-M. Porcher and S. Aït-Aïssa, Copper induced oxidative stress in three spined stickleback: relationship with hepatic metal levels, *Environ. Toxicol. Pharmacol.*, 2005, **19**, 177–183.
68. S. D. Jena, M. Behera, J. Dandapat and N. Mohanty, Nonenzymatic antioxidant status and modulation of lipid peroxidation in the muscles of *Labeo rohita* by sublethal exposure to copper sulphate, *Vet. Res. Commun.*, 2008, **33**, 421–429.
69. C. L. Mieiro, I. Ahmad, M. E. Pereira, A. C. Duarte and M. Pacheco, Antioxidant system breakdown in brain of feral golden grey mullet (*Liza aureta*) as an effect of mercury exposure, *Ecotoxicology*, 2010, **19**, 1034–1045.
70. S. V. S. Rana and R. Singh, Specie's differences in glutathione dependent enzymes in liver and kidney of two fresh water fishes and their implications in cadmium toxicity, *Ichthyol. Res.*, 1996, **43**, 223–229.
71. T. Allen, R. Singhal and S. V. S. Rana, Resistance to oxidative stress in a fresh water fish *Channa punctatus* after exposure to inorganic arsenic, *Biol. Trace Elem. Res.*, 2004, **98**, 63–72.
72. A. Bhattacharya and S. Bhattacharya, Induction of oxidative stress by arsenic in *Clarius batrachus*. Involvement of peroxisomes, *Ecotoxicol. Environ. Saf.*, 2007, **66**, 178–187.
73. Sunaina, A. K. Bhardwaj and B. A. Ansari, Oxidative stress biomarkers in assessing arsenic trioxide toxicity in the zebra fish, *Danio rerio*, *Int. J. Fish. Aquat. Stud.*, 2016, **4**, 8–13.

74. S. Ullah, L. Zhongqui, S. Hassan, S. Ahmad, X. Guo, K. Wanghe and G. Nabi, Heavy metal bioaccumulation and subsequent multiple biomarkers based appraisal of toxicity in the critically endangered *Tor putitora*, *Ecol. Environ. Saf.*, 2021, **228**, 113032.
75. R. Freitas, E. Costa, C. Velez, J. Santos, A. Lima, C. Oliveira, A. M. Rodrigues, V. Quintino and E. Figueira, Looking for suitable biomarkers in benthic macroinvertebrates inhabiting coastal areas with low metal contamination: comparison between the bivalve *Cerastoderma edule* and the polychaete *Diopatra neapolitina*, *Ecotoxicol. Environ. Saf.*, 2012, **75**, 109–118.
76. J. M. McCord and I. Fridovich, Superoxide dismutase. An enzymatic function of erythroprotein (hemocuprein), *J. Biol. Chem.*, 1969, **244**, 6049–6055.
77. M. K. Kadim and Y. Risjani, Biomarker for monitoring heavy metal pollution in aquatic environment: an overview towards molecular perspectives, *Emerging Contam.*, 2022, **8**, 195–205.
78. H. Li, J. Ruan and R. Durbin, Mapping short DNA sequencing reads and calling variants using mapping quality scores, *Genome Res.*, 2008, **18**, 1851–1858.
79. C. Bonafé Gaspar Ruas, C. dos Santos Carvalho, H. Sobreiro Selistre de Araújo, E. L. Gaeta Espíndola and M. N. Fernandes, Oxidative stress biomarkers of exposure in the blood of cichlid species from a metal contaminated river, *Ecotoxicol. Environ. Saf.*, 2008, **71**, 86–93.
80. S. O. Asagba, G. E. Eriyamremu and M. E. Igberaese, Bioaccumulation of cadmium and its biochemical effects on selected tissues of cat fish, *Clarius gariepinus*, *Fish Physiol. Biochem.*, 2008, **34**, 61–69.
81. L. G. A. Barboza, L. Russo Vieira, V. Branco, C. Carvalho and L. Guilhermino, Microplastics increase mercury concentration in gills and bioaccumulation in the liver and cause oxidative stress and damage in *Dicentrarchus labrax* juveniles, *Sci. Rep.*, 2018, **8**, 15655.
82. P. Li, Z.-H. Li and Y. Wu, Interactive effects of temperature and mercury exposure on the stress related responses in the fresh water fish *Ctenopharygodon idella*, *Aquacult. Res.*, 2021, **52**, 2070–2077.
83. S. P. Trivedi, S. Singh, A. Trivedi and M. Kumar, Mercuric chloride induced oxidative stress, genotoxicity, haematological changes and histopathological alterations in fish *Channa punctatus* (Bloch), *J. Fish Biol.*, 2022, **100**, 868–883.
84. J. Cavalcante Souza Vieira, C. Pereira Braga, J. V. de Queiroz, B. Cavecci-Mendonça, G. de Oliveira, N. Galvão de Freitas, A. A. Henrique Fernandes, M. da Silva Fernandes, M. Afonso Rabelo Buzalaf, J. Adamec, L. F. Zara and P. de Magalhães Padilha, The effects of mercury exposure on Amazonian fishes: An investigation of potential biomarkers, *Chemosphere*, 2023, **316**, 137779.
85. M. Ogata, Acatalasemia, *Hum. Genet.*, 1991, **86**, 331–340.
86. M. Roméo, N. Bennani, M. Gnassia-Barelli, M. Lafaurie and J. P. Girard, Cadmium and copper display different responses towards oxidative

stress in the kidney of sea bass *Dicentrarchus labrax*, *Aquat. Toxicol.*, 2000, **48**, 185–194.
87. A. Vaglio and C. Landriscina, Changes in liver enzyme activity in the teleost *Sparus aurata* in response to cadmium intoxication, *Ecotoxicol. Environ. Saf.*, 1999, **43**, 111–116.
88. A. C. Elia, A. J. Dörr and R. Galarini, Comparison of organochlorine pesticides, PCBs, and heavy metal contamination and detoxifying responses in tissues of *Ameiurus melas* from Corbara, Alviano, and Trasimeno Lakes, Italy, *Bull. Environ. Contam. Toxicol.*, 2007, **78**, 463–468.
89. I. I. Ujah, C. A. Nsude, I. O. Okpako and C. M. Ugwu, Induction of catalase response in *Clarias* exposed to graded concentration of iron as fish's body defence mechanism against environmental pollutants, *Acta Sci. Microbiol.*, 2019, **2**, 49–52.
90. R. Arshad, S. Abdullah, H. Naz and K. Abbas, Catalase activity as bio-indicator of lead + nickel toxicity in carnivorous fish, *Channa striata*, *Proc. Pak. Acad. Sci.*, 2018, **55**, 37–43.
91. T. Ahmed, S. Abdullah, K. Abbas and M. A. Zia, Catalase enzyme response to chronic Pb+Cd metal mixture exposure, its purification and partial characterization from kidney of fresh water fish, *Orechromis niloticus*, *Pak. J. Zool.*, 2016, **48**, 1733–1740.
92. Z. M. Anjum, A. Qurat-ul, S. Abdullah, K. H. Abbas and I. Ahmed, Sublethal impacts of heavy metals on antioxidant enzymes and biochemical parameters of rohu (*Labeo rohita*), *Iran. Fish. Sci. J.*, 2017, **16**, 668–683.
93. R. Creţu, Research on some fresh water catalase activity - as a potential biomarker for environmental pollution, *Ann. Univer. "Dunărea de Jos" Galaţi., Metall. Mater. Sci.*, 2021, **39**, 22–29.
94. U. Tinggi, Selenium: its role as antioxidant in human health, *Environ. Health Prev. Med.*, 2008, **13**, 102–108.
95. L. Flohé, Die Glutathionperoxidase: Enzymologie und biologische Aspekte, *Klin. Wochenschr.*, 1971, **49**, 669–683.
96. G. S. Shukla, T. Hussain, R. S. Srivastava and S. V. Chandra, Glutathione peroxidase and catalase in liver, kidney, testis and brain regions of rats following cadmium exposure and subsequent withdrawal, *Ind. Health*, 1989, **27**, 59–69.
97. S. V. Rana and S. Kumar, Antioxidative enzyme in the liver of rats after exposure to xylene, toluene and methyl alcohol separately and in combination, *Physiol. Chem. Phys. Med. NMR*, 1995, **27**, 25–29.
98. S. V. S. Rana and R. Singh, Species differences in glutathione-dependent enzymes in the liver and kidney of two fresh water fishes and their implications in cadmium toxicity, *Ichthyol. Res.*, 1996, **43**, 223–229.
99. T. Allen, R. Singhal and S. V. S. Rana, Resistance to oxidative stress in a fresh water fish *Channa punctatus* after exposure to inorganic mercury, *Biol. Trace Elem. Res.*, 2004, **98**, 63–72.
100. T. Allen and S. V. S. Rana, Effect of arsenic (AsIII) on glutathione-dependent enzymes in liver and kidney of fresh water fish *Channa punctatus*, *Biol. Trace Elem. Res.*, 2004, **109**, 39–48.

101. G. W. Winston and R. T. Di Giulio, Prooxidant and antioxidant mechanisms in aquatic organisms, *Aquat. Toxicol.*, 1991, **19**, 137–161.
102. J. L. Franco, T. Posser, J. J. Mattos, A. Sánchez-Chardi, R. Trevisan, C. S. Oliveira, P. S. M. Carvalho, R. B. Leal, M. R. F. Marques, A. C. D. Bainy and A. L. Dafre, Biochemical alterations in juvenile carp (*Cyprinus carpio*) exposed to zinc: glutathione reductase as a target, *Mar. Environ. Res.*, 2008, **66**, 88–89.
103. R. J. Albertini, D. Anderson, G. R. Douglas, L. Hagmar, K. Hemminki, F. Merlo, A. T. Natarajan, H. Norppa, D. E. Shuker, R. Tice, M. D. Waters and A. Aitio, ICPS Guidelines for monitoring of genotoxic effects of carcinogens in humans. International Program on Chemical Safety, *Mutat. Res.*, 2000, **163**, 111–172.
104. G. Frenzili, M. Nigro and B. P. Lyons, The comet assay for the evaluation of genotoxic impacts in aquatic environments, *Mutat. Res.*, 2009, **681**, 80–92.
105. B. Ateeq, M. A. Farah and W. Ahmad, Detection of DNA damage by alkaline single cell gel electrophoresis in 2,4-dichlorophenoxyacetic-acid and butachlor-exposed erythrocytes of *Clarius batrachus*, *Ecotoxicol. Environ. Saf.*, 2005, **62**, 348–354.
106. S. Pandey, N. S. Nagpure, R. Kumar, S. Sharma, S. K. Srivastava and M. S. Verma, Genotoxicity evaluation of acute doses of endosulfan to fresh water teleost *Channa punctatus* (Bloch) by alkaline single-cell gel electrophoresis, *Ecotoxicol. Environ. Saf.*, 2006, **65**, 56–61.
107. T. Kosmehl, A. V. Hallare, T. Braunbeck and H. Hollert, DNA damage induced by xenotoxicants in zebra fish (*Danio rerio*) embryos after contact exposure to freeze-dried sediment and sediment extracts from Laguna Lake (The Phillipines) as measured by comet assay, *Mutat. Res.*, 2008, **650**, 1–14.
108. A. Whitehead, K. M. Kuivila, J. L. Orlando, S. Kotelevtsev and S. L. Anderson, Genotoxicity in native fish associated with agricultural runoff events, *Environ. Toxicol. Chem.*, 2004, **23**, 2868–2877.
109. F. Regoli, G. W. Winston, S. Gorbi, G. Frenzilli, M. Nigro, I. Corsi and S. Focardi, Integrating enzymatic responses to organic chemical exposure with total oxyradical absorbing capacity and DNA damage in the European eel *Anguilla anguilla*, *Environ. Toxicol. Chem.*, 2003, **22**, 2120–2129.
110. S. Sanchez-Galan, A. R. Linde and E. Garcia-Vazquez, Brown trout and European minnow as target species for genotoxicity tests: differential sensitivity to heavy metals, *Ecotoxicol. Environ. Saf.*, 1999, **43**, 301–304.
111. J. A. Heddle, M. C. Cimino, M. Hayashi, F. Romagna, M. D. Shelby, J. D. Tucker, P. Vanparys and J. T. MacGregor, Micronuclei as an index of cytogenic damage: past, present, and future, *Environ. Mol. Mutagen.*, 1991, **18**, 277–291.
112. C. Russo, L. Rocco, M. A. Morescalchi and V. Stingo, Assessment of environmental stress by the micronucleus test and the comet assay on the genome of teleost population from two natural environments, *Ecotoxicol. Environ. Saf.*, 2004, **57**, 168–174.

113. J. A. Heddle, M. Hite, B. Kirkhart, J. T. MacGregor, G. W. Newell and M. F. Salamone, The induction of micronuclei as a measure of genotoxicity. A report of the U. S. Protection Agency Gene-Tox Program, *Mutat. Res.*, 1983, **123**, 61–118.
114. C. Bolognesi and M. Hayashi, Micronucleus assay in aquatic animals, *Mutagenesis*, 2011, **26**, 205–213.
115. M. Singh, H. Khan, Y. Verma and S. V. S. Rana, Distinctive fingerprints of genotoxicity induced by As, Cr, Cd and Ni in a fresh water fish, *Environ. Sci. Pollut. Res.*, 2019, **26**, 19445–19452.
116. Y. Zhu, J. Wang, Y. Bai and R. Zhang, Cadmium, chromium and copper induce polychromatocyte micronuclei in carp (*Cyprinus carpio*), *Bull. Environ. Contam. Toxicol.*, 2004, **72**, 78–86.
117. K. K. Yadav and S. P. Trivedi, Sublethal exposure to heavy metals induces micronuclei in fish, *Channa punctatus*, *Chemosphere*, 2009, **77**, 1495–1500.
118. M. Nirchio, O. C. Veintimilla, P. F. Q. Cordero, P. F. Quizhpe, J. G. Hernández and C. Oliveira, Genotoxic effects of mercuric chloride on the Neotropical fish *Andinoacara rivulatus* (Cichlidae: Cichlasomatini), *Rev. Biol. Trop.*, 2019, **67**, 745–754.
119. U. Güner and F. D. G. Muranli, Micronucleus test, nuclear abnormalities and accumulation of Cu and Cd in *Gambusia affinis* (Baird and Girard, 1853), *Turk. J. Fish. Aquat. Sci.*, 2011, **11**, 615–622.
120. V. Shah, O. Taratula, O. B. Garbuzenko, M. L. Patil, R. Savla, M. Zhang and T. Minko, Genotoxicity of different nanocarriers: possible modifications for the delivery of nucleic acids, *Curr. Drug Discovery Technol.*, 2013, **10**, 8–15.
121. E. Demir, T. Qin, Y. Li, Y. Zhang, X. Guo, T. Ingle, J. Yan, A. I. Orza, A. S. Biris, S. Ghorai, T. Zhou and T. Chen, Cytotoxicity and genotoxicity of cadmium oxide nanoparticles evaluated using *in vitro* assays, *Mutat. Res.*, 2020, **850–851**, 503149.
122. M. S. Khan, N. A. Qureshi and F. Jabeen, Assessment of toxicity in fresh water fish, *Labeo rohita* treated with silver nanoparticles, *Appl. Nanosci.*, 2017, **7**, 167–179.
123. M. Montaser, M. E. Mahfouz, S. A. M. El-Shazly, G. H. Abdel-Rahman and S. Bakry, Toxicity of heavy metals on fish at Jeddah Coast KSA: Metallothionein expression as a biomarker and histopathological study on liver and gills, *World J. Fish. Mar. Sci.*, 2010, **2**, 174–185.
124. L. Förlin, N. Asker, M. Töpel, T. Österlund, E. Kristiansson, J. Parkkonen, P. Haglund, S. Faxneld and J. Sturve, mRNA expression and biomarker responses in perch at a biomonitoring site in the Baltic Sea - Possible influence of natural brominated chemicals, *Front. Mar. Sci.*, 2019, **6**, 316.
125. J. Yin, A.-P. Wang, W.-F. Li, R. Shi, H.-T. Jin and J.-F. Wei, Time-response characteristic and potential biomarker identification of heavy metal induced toxicity in zebra fish, *Fish Shellfish Immunol.*, 2018, **72**, 309–317.

126. S. B. Swaleh, U. Z. Banday, M.-A. Asadi and N. Usmani, Biochemical profile and gene expression of *Clarius gariepinus* as a signature of heavy metal stress, *Environ. Pollut.*, 2020, **264**, 114693.
127. E. Hassan, G. El-sayed, A. Hassan and S. Abd Elsalam, Effect of pollution with lead, copper, cadmium on gene expression pattern of liver GST and serum lysozymes in Nile tilapia (*Orechromis niloticus*), *Mansoura Vet. Med. J.*, 2020, **21**, 53–60.
128. S. Yoon, S. S. Han and S. V. S. Rana, Molecular markers of heavy metal toxicity-a new paradigm for health risk assessment, *J. Environ. Biol.*, 2008, **29**, 1–14.

CHAPTER 3

Edible Fish Species as Bioaccumulators and Bioindicators of Heavy Metal Pollution in the Durban Basin, South Africa

HENRY JOSEPH ODUOR OGOLA[a,b*], SANJEEV DEBIPERSADH[b], KENNEDY ODHIAMBO ACHIENG[a], KALU CHIMDI MANG[b] AND VIRGINIA KIMANI WAMBUI[c]

[a]School of Agricultural and Food Sciences, Jaramogi Oginga Odinga University of Science and Technology, P.O. Box 210-40601, Bondo, Kenya; [b]Department of Environmental Science, University of South Africa, Florida Campus, Roodepoort, 1709, South Africa; [c]Industrial Microbiology and Biotechnology Research Centre (IMB-RC), Kenya Industrial Research and Development Institute (KIRDI), Popo Road off Mombasa Road, South C, Nairobi, Kenya
*E-mail: henryogola@gmail.com

3.1 Introduction

Coastal environments are dynamic and intricate ecological systems that are currently burdened by diverse anthropogenic activities. These activities encompass industrialization to agricultural practices and other human practices that have profound consequences on the health of marine ecosystems.[1,2]

Issues in Toxicology No. 50
Fish Species in Environmental Risk Assessment Strategies
Edited by Guillermo Eli Liwszyc and Marcelo L. Larramendy

Published by the Royal Society of Chemistry, www.rsc.org

The repercussions extend to the coastal communities that rely on these ecosystems as a primary source of livelihoods.[3] In recent years, concerns about seafood quality and its implications for public health have increased, primarily due to the presence of toxic metals and metalloids. This contamination can be attributed to the escalating levels of pollution in marine environments; a trend projected to triple over the next five decades.[4]

The Durban basin area, a key industrialized hotspot in the eastern seaboard of South Africa, plays a central role in this ecological narrative. This region is renowned for hosting a multitude of manufacturing and refining industries, establishing it as a critical industrial hub along the Indian Ocean coastline.[5–7] Notably, the Durban basin is home to two of the country's largest oil refineries, which collectively refine approximately 60 percent of South Africa's petroleum. Moreover, it houses wastewater treatment works, numerous toxic waste landfill sites, a paper manufacturing plant and a multitude of chemical process industries, including over 120 industries that generate substantial quantities of waste.[8] Among these pollutants, heavy metals, such as Cd, Ni, Pb, Cr and Hg have emerged as major concerns. The marine ecosystem serves as the ultimate repository for all forms of these land-based contaminants.[9,10] Consequently, the pervasive pollution of marine environments, including protected marine ecosystems, with chemical contaminants from industrial and agricultural effluents, particularly HM, ranks as one of the critical environmental problems in the region.[11–14]

The contamination of these marine environments with HMs presents a substantial challenge, particularly given the crucial role of fish in the human diet. Heavy metals in the aquatic environment are generally persistent, non-degradable and are considered hazardous to biological systems due to their oxidative and carcinogenic potential.[15] Under specific environmental conditions, HMs can accumulate to toxic levels, causing ecological damage. However, some HMs such as Cu, Zn and Mn are essential micro-nutrients which play an important role in biological systems at lower concentration and generally decay into harmless or less harmful forms during their metabolism. In contrast, non-essential and non-biodegradable metals like As, Cd, Hg, Cr, Ni and Pb and their organometals such as alkylated Pb, tributyl-, tri- and methylmercury [$(CH_3Hg)^+$], believed to have no known role, can be toxic even at lower concentrations.[16] The greatest concern related to these toxic metals are the ease of bioaccumulation by marine organisms and consequent biomagnification in aquatic food chains.[17] This may eventually be deleterious to human health primarily through consumption of contaminated fish, inducing renal failure, liver damage, cardiovascular disease and even death.[18–20]

Fish, being an essential part of the human diet along the South African coastline, play a pivotal role in this intricate ecological scenario. Numerous studies have reported significant HM contamination in marine fish caught within the Durban basin area of South Africa.[11,21–25] Moreover, there is a growing body of research linking the consumption of contaminated fish and other aquatic foods to various human health risks.[16,26,27] Some studies even suggest the use of fish species as potential indicators of pollution and for monitoring the accumulation of toxic HM in the marine environment.[16,28] The primary goal of bioindicator research is to identify species capable of reliably detecting environmental disturbances and illustrating how these disturbances impact other species and overall biodiversity. Within this context, fish emerge as valuable bioindicators of HM pollution in aquatic environments.[16,28–30] Their ability to accumulate, biomagnify and respond to environmental changes positions them as effective tools for evaluating the health of aquatic ecosystems and the potential risks associated with HM contamination.[29] However, it is important to note that the selection of a bioindicator species for assessing human health effects should adhere to specific criteria, including being commonly consumed in the area, having a wide geographic distribution, the potential to accumulate high metal concentrations and sufficient tissue mass for residue analyses.

This chapter explores the multifaceted issue of HM pollution in the Eastern Seaboard of the Indian Ocean, focusing on its impact on common edible and bait fish species. Specifically, we delve into the bioaccumulation patterns of HM and the bioindicative potential of these fish species for monitoring of their accumulation in the Durban basin. Through this exploration, we aim to provide a comprehensive understanding of the intricate interplay between HM pollution, marine ecosystems and human well-being in this vital region.

3.2 Heavy Metal Pollution in the Durban Basin's Marine Ecosystem

The increase in anthropogenic disturbances, common in large coastal cities worldwide, can lead to a continuous discharge of pollutants into the ocean, potentially raising contamination levels, which are a significant concern for marine environments. Durban, located along the Indian Ocean coastline, is a prominent industrial and commercial centre in South Africa and its marine ecosystem faces pollution challenges.[7,8,31] This pollution arises from various sources, including industrial runoff, wastewater discharge, river mouths carrying metal pollutants from agricultural and industrial areas and harbours with increased metal concentrations due to oil spills and shipping-related pollution.[6] Anthropogenic activities in the area encompass a variety of industries, petrochemical processing, pulp and paper production and vehicular traffic, leading to the accumulation of toxins in sediments, effectively creating an extended toxin sink.[11,14] The heavy metals and chemical toxins from these wastewater sources can become bioavailable due to remobilization events, such as floods, dredging or bioturbation, potentially introducing them to benthic and related organisms and adversely affecting both human and natural exposure.[32] As depicted in Figure 3.1, non-biogenic HMs and other chemical pollutants tend to persist in the environment due to their non-biodegradable and recalcitrant nature. They can undergo trophic transfer and bioaccumulation within aquatic food webs, posing a significant risk to both the environment and human health.

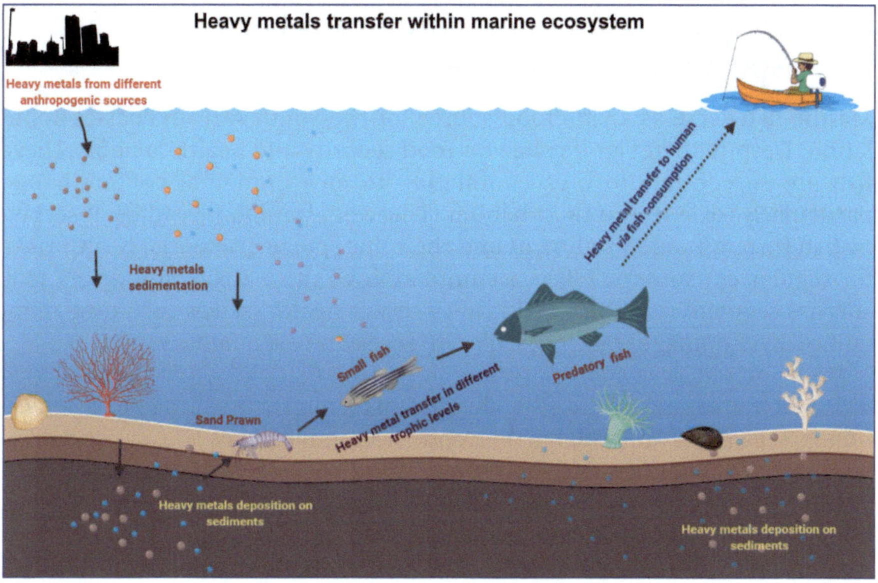

Figure 3.1 The transfer of toxic heavy metal along aquatic food webs in a marine ecosystem. Adapted from ref. 34 with permission from the author.

It is believed that deleterious trace elements and compounds associated with sewage outfalls are primarily localized at specific point sources along the Durban coastline. It has been estimated that there are over 67 discharge points contributing approximately 1.3 million m^3 of wastewater into the ocean daily within the basin.[33] Interestingly, the pollution profile of HM and other toxicants typically displays both spatial and temporal variations, likely linked to the distance from pollution sources and the physical conditions of the surroundings. For Instance, Debipersadh reported elevated heavy metal levels above the background upper continental crust values.[34] The values reported generally exhibited spatial variability, with samples from Durban South and harbour showing higher values of Al, As, Mn, Cr, Cu and Ni compared to other sites. Conversely, Durban harbour had significantly higher average values for Cu, Pb and Zn in the sediment samples compared with all other sites. Similarly, Vetrimurugan *et al.* also reported the enrichment of HM concentration levels in five different beach sediments along the Durban South coastline, exceeding background reference values.[13] The presence of a clear spatial separation of heavy metal levels suggests that there may be multiple metal sources or factors influencing metal concentrations in this region. Collectively, these findings also illustrated the urgent need to continuously monitor the levels of these pollutants in various components of the marine environment. In the subsequent subsection, therefore, we focus on the heavy metal's bioaccumulation in both edible and bait fish species common within this basin, taking into consideration localized hotspots for chronic metal pollution affecting the region.

3.3 Heavy Metal Bioaccumulation Patterns in Fish Species in the Durban Basin

Marine fish serve as a vital food source for the South African coastal population. Despite their significance for food security and health benefits, these fish are susceptible to bioaccumulation to environmental contaminants, particularly toxic HM in their habitat. These contaminants can be absorbed by fish through water, sediment and their diet, posing potential health risks for regular consumers.[26] The accumulation of these toxic substances may surpass regulatory limits[35] and lead to severe health issues, including renal failure, liver damage, cardiovascular diseases and even fatality.[18,19,36]

Heavy metals are widely pervasive in marine fish and can accumulate *via* different routes such as the gills, absorption through the skin as well as ingestion of seawater and food. Furthermore, accumulation of HM in fishes may be correlated to fat content, food sources, the physiological state of the individual and the toxicological dynamics of the specific metal.[37] Consequently, a high variability in HM concentrations may exist among fishes and their habitats. Numerous studies have highlighted the HM pollution in coastal areas[7,13,14,31,38] and contamination in marine fish caught within the Durban basin.[11,12,24,39,40]

Debipersadh *et al.* conducted a series of studies on four common edible fish species, namely the bronze bream *Pachymetopon grande* (Spariformes, Sparidae), the catface rockcod *Epinephelus andersoni* (Perciformes, Serranidae), the slinger seabream *Chrysoblephus puniceus* (Spariformes, Sparidae) and the soldier, also known as the Santer seabream *Cheimerius nufar* (Perciformes, Sparidae), and two bait fish, namely the Maasbanker or Atlantic horse mackerel *Trachurus trachurus* (Carangiformes, Carangidae) and the Pacific chub mackerel *Scomber japonicas* (Scombriformes, Scombridae), illustrated in Figure 3.2.[34,39,41]

These studies revealed distinct patterns of heavy metal accumulation, with the patterns being notably influenced by the fish species, the type of fish organ and the specific sampling locations within the Durban basin. Importantly, some of these species exhibited elevated levels of As, Cr, Ni, Pb and Zn in their muscles, surpassing the international regulations established for human consumption, as illustrated in Table 3.1 and Figure 3.3. In the case of bait fishes, significantly higher levels of Al and Cr were observed in Durban Central, while the average accumulation of As was higher in Durban South, as depicted in Figure 3.3. Furthermore, Moodley *et al.*[12] reported similar findings, specifically higher accumulation of As and Cr in the muscle tissues above the maximum permissible limits defined by the World Health Organization (WHO)/FAO Codex Alimentarius.[42] Their research focused on various fish species, including the African pompano *Trachinotus africanus* (Carangiformes, Carangidae), the karanteen *Crenidens crenidens* (Spariformes, Sparidae), the flathead grey mullet *Mugil cephalus* (Mugiliformes, Mugilidae), the sargo or white seabream *Diplodus capensis* (Spariformes, Sparidae) and the St Paul's fingerfin *Acantholatris*

Common name (Scientific name)		Lifestyle	Trophic level
Catface rockcod (*Epinephelus andersoni*) (Spariformes, Serranidae),		Demersal	Carnivore (feeds on crabs, other crustaceans and small fish)
Slinger-Seabream (*Chrysoblephus puniceus*) (Spariformes, Sparidae),		Demersal	Carnivore (feeds on bivalve mollusks, crustaceans and small fish)
Santer-Seabream/Soldier (*Cheimerius nufar*) (Perciformes, Sparidae),		Demersal	Carnivore (feeds on small fish, squid, and crustaceans)
Bronze/Copper-Seabream (*Pachymetopon grande*) (Spariformes, Sparidae),		Benthic	Omnivore (feeds on algae and small invertebrates)
Spotted Grunter (*Pomadasys commersonni*) (Perciformes, Haemulidae),		Benthic	Carnivore (feeds on crustaceans, shrimps, sand prawns and mollusks)
Maasbanker (*Trachurus trachurus* (Carangiformes, Carangidae),		Benthic/Pelagic	Carnivore (feeds on zooplanktons, crustaceans, such as copepods, amphipods and euphasiids)
Atlantic Chub Mackerel (*Scomber japonicas*) Scombriformes, Scombridae),		Pelagic	Carnivore (feeds on small fish, squids, prawns, and shrimp species)

Figure 3.2 Common examples of edible fish species in the Durban basin, their lifestyle and trophic groups.

monodactylus (Perciformes, Cheilodactylidae), which were caught from Cuttings, Isipingo and Amanzimtoti beaches in the South Durban industrial basin. The study attributed the high As and Cr levels in seawater and fish tissues to dissolved and particulate forms of the metal and their metalloids entering the ocean through rivers enriched by overland flow or waste discharges as well as from land-based discharges into coastal waters. Generally, As is highly mobile in aquatic environments with the ocean serving as its primary sink. Its soluble form, gaseous arsine, is extremely toxic to aquatic organisms. On the other hand, elevated Cr levels in the water could possibly be attributed to anthropogenic sources such as the production of refractory steel. It is used as electroplating cleaning agents, in catalyst production, in cooling towers of water treatment plants and in wood preservation, particularly in the heavily industrialized South Durban basin.[12]

Table 3.1 Summary of the mean concentrations (SE) of heavy metals measured in the muscle tissues of common fish species caught in the Durban Basin and the maximum permissible limits for fish (mg per kg ww).

	Al	As	Cr	Cu	Mn	Fe	Ni	Pb	Se	Zn
Fish species										
Catface rockcod *Epinephelus andersoni*[34,39,41]	18 (1.98)	1.1 (0.22)	1.3 (0.25)	2.7 (0.27)	1.2 (0.27)	—	1.2 (0.18)	**2.8 (0.35)**[a]	—	18 (1.59)
Soider/Santer seabream *Cheimerius nufar*[34,39,41]	21 (1.92)	1.1 (0.18)	1.5 (0.20)	2.2 (0.22)	1.6 (0.21)	—	1.0 (0.14)	**2.8 (0.28)**	—	26 (1.75)
Bronze bream[34,39,41] Pachymetopon grande	27 (2.51)	1.6 (0.20)	2.0 (0.23)	3.5 (0.25)	2.3 (0.24)	—	1.6 (0.16)	3.2 (0.32)	—	29 (1.79)
Slinger seabream[34,39,41] *Chrysoblephus puniceus*	35 (2.92)	**2.2 (0.19)**	**2.7 (0.22)**	3.6 (0.24)	3.3 (0.23)	—	**1.9 (0.15)**	**4.1 (0.31)**	—	30 (2.00)
Smallspotted grunter[34,39,41] *Pomadasys commersonnii*	24 (19.1)	1.5 (1.47)	1.7 (1.42)	3.9 (2.46)	1.9 (1.97)	—	**1.4 (0.97)**	**2.4 (0.90)**	—	24 (9.86)
Maasbanker *Trachurus trachurus*[34,39,41]	25 (3.91)	1.1 (0.03)	1.8 (0.30)	5.6 (0.64)	1.2 (0.10)	—	**1.6 (0.22)**	**4.7 (0.61)**	—	**52 (3.1)**
Pacific chub mackerel *Scomber japonicas*[34,39,41]	7.7 (0.90)	1.4 (0.19)	1.8 (0.22)	5.5 (0.43)	1.0 (0.07)	—	1.0 (0.11)	**3.1 (0.37)**	—	**34 (2.36)**
African pompano *Trachinotus africanus*[12]	142 (6)	**8.7 (0.76)**	0.9 (0.06)	0.9 (0.06)	3.7 (0.10)	23 (1.9)	—	0.2 (0.02)	1.4 (0.12)	16 (0.40)
Karanteen *Crenidens crenidens*[12]	89–123	**5.9–6.3**	0.76–0.83	1.12–1.14	0.4–1.89	18–19.8	—	0.21–0.36	1.23–1.53	13.4–21.4
Flathead grey mullet *Mugil cephalus*[12]	98.9 (1)	**6.3 (0.66)**	0.8 (0.03)	1.0 (0.03)	0.9 (0.11)	15 (1.3)	—	0.1 (0.01)	1.2 (0.15)	11.9 (0.1)

White seabream *Diplodus capensis*[12]	83–101	**4.2–8.9**	071–0.32	0.89–1.88	1.2–2.60	9.7–12.8	—	0.09–0.89	1.02–1.37	12.6–13.2
St. Paul's finger-fin *Acantholatris monodactylus*[12]	88 (10)	7.5 (1.14)	0.62 (0.06)	0.75 (0.06)	ND[b]	12 (1.1)	—	1.09 (0.13)	1.15 (0.11)	12.3 (0.9)
Maximum permissible limits (MAL)										
FAO[42]	—	—	—	30	—		—	—		30
EC[47]	—	—	—	—	—		—	0.3		—
Codex Alimentarius[48]	—	—	—	—	—		—	0.3		—
China[49]	—	0.1	2	—	—		—	0.5		—
Australia and New Zealand[50]	—	2	—	—	—		—	0.5		—
South Africa[51]	—	3	—	—	—		—	0.5		—
USA[52]	—	—	—	10	11		1	—		40

[a]Values higher than maximum allowable limits (MAL) are shown in bold.
[b]ND: Not determined.

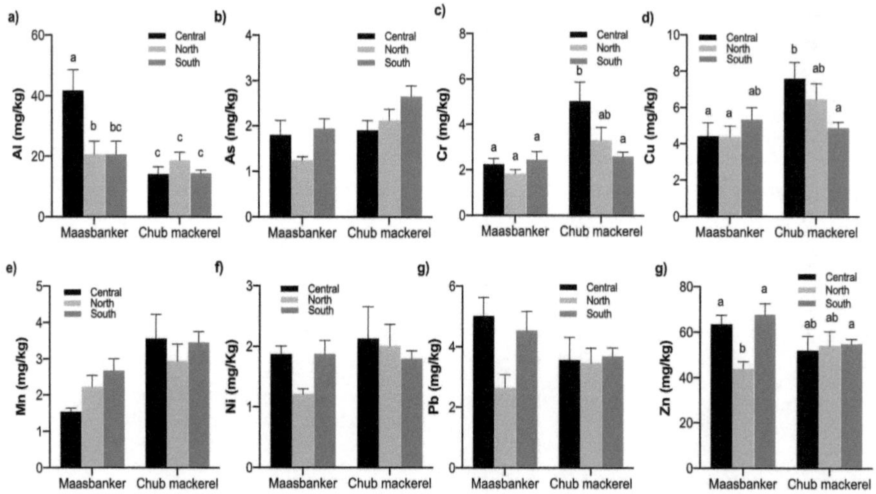

Figure 3.3 Average heavy metal concentrations comparison by baitfish species and location. Mean values followed with different letters within each species are statistically different (ANOVA; Tukey's test, $p > 0.05$). Adapted from ref. 34 with permission from the author.

Debipersadh also reported an inter-element relationship of HM in baitfishes, suggesting that they are external as the concentration patterns were higher due to localized anthropogenic activities.[34] For example, there were strong positive correlations ($r > 0.78$) between Al, Cu and Mn, indicating their origin from similar pollution point sources, especially within the Durban Central area. A similar trend has been observed also in cracker shrimp and grunter samples caught within Durban harbour. Since Durban harbour drains into Durban Central, the observed higher levels and the associated inter-element relationships in fish is highly dependent on the nature of the ports and human activities in the vicinity of the port area as has been reported in other seaports elsewhere.[43–46] Since Durban Central region is also home to the Central Works Outfall that discharges 135 mega L per day of sanitary wastewater, coupled with a prominent vessel transportation industry in the region, Debipersadh posited that these activities may account for the observed higher HM levels in fish species caught in the area.[34]

In contrast to baitfishes, among the edible fish species, the impact of species on HM accumulation patterns was highly significant, regardless of the sampling location.[34] It's worth noting that various studies have highlighted that factors such as dietary preferences (carnivorous/omnivorous), trophic level, fish size, age, fish physiology (storage or elimination of HM) and the lifestyle of fish species (migratory, demersal, pelagic or benthic) play pivotal roles in shaping observed variations in HM accumulation, even in similar polluted environments.[47,48] Comparatively, the catface rockcod *E. andersoni* and the Santer seabream *C. nufar* occupy a higher trophic level than bronze bream *P. grande* and slinger seabream *C. puniceus*, exhibiting distinct feeding behaviours.[34,39,41] Interestingly, the lower trophic level species, such as slinger *C. puniceus* and bronze bream *P. grande*, primarily feed on benthic organisms and demonstrated higher heavy metal accumulation patterns (see Table 3.1).[34,39,41] Specifically, the slinger seabream exhibited higher accumulation of As, Cr, Cu, Mn, Ni, Pb and Zn whereas the bronze bream had comparable values for As, Cu and Zn. Except for Mn, all the metal accumulations were higher in the liver compared to the fish muscles and gills, emphasizing the importance of prey type.[34]

In terms of lifestyle, the maasbanker *T. trachurus* is a pelagic-benthic species, mainly carnivorous feeding on a mixed diet of zooplankton, benthos (such as polychaeta, crustacea) and small fishes, with krill (malacostraca, euphausiacea) as the dominant prey.[49,50] In contrast, the Pacific chub mackerel *S. japonicas* are exclusively pelagic and planktivorous during the initial stages of development before incorporating piscivory at adult stage, but mainly exhibit filter- and/or particulate feeding behaviour.[51,52] Although pelagic fish have been reported to accumulate lower HM compared to the benthic fish,[53] the filter feeding on suspended particulates and organic matter rich in metals such as Al, As, Cu and Mn in the seawater column in addition to algae and small zooplankton, which have limited capacity to regulate metal levels, implies that Pacific chub mackerel *S. japonicas* may have greater exposure to heavy metals through its diet than the maasbanker *T. trachurus*.[54] Such filter feeding behaviour has also been implicated for the accumulation of microplastics in the same species.[55–57] However, it was observed that the slinger seabream *C. puniceus* and bronze bream *P. grande*, as the lowest trophic level species among the four, did not align with the general consensus that HM and pollutant levels increase in higher trophic species.[52,58,59] It is worth mentioning that the catface rockcod *E. andersoni* and the soldier *C. nufar* are piscivorous demersal species, primarily preying on smaller fish. In contrast, the demersal slinger seabream mainly feeds on bivalves, molluscs, crustaceans and small fish species, while the omnivorous and benthic bronze bream primarily consumes algae and small invertebrates. Besides differences in prey type and feeding behaviour, the observed variation in this study could be attributed to factors like fish physiology (some species may possess more efficient mechanisms for storing and detoxifying heavy metals), or variations in exposure levels (linked to fish size, age and migratory behaviour), which were not explored in the study by Debipersadh.[34]

Another key observation made by the above studies was the distinct bioaccumulation patterns in the fish species dependent on the organ or tissue type. Debipersadh reported the mean metal concentrations observed in four edible fish species in the descending order as follows: liver > gills > muscles.[34] Notwithstanding, the subtle differences could be ascribed to specific fish species and specific metals. However, HM accumulation in the gills and liver exhibited a spatial variability that can be attributed to different localized pollution. Among the fish organs, gills are the first target of water-borne metals due to their large surface area and direct contact with the aquatic environment. Gill surfaces consist of an epithelial membrane rich in negatively charged phospholipids and a mucous layer that provides a potential site for gill–metal interaction for positively charged metals.[60] Thus, in heavily contaminated marine aquatic environments, gills will be the main route for the contaminants to enter the fish body.[61] The digestive system is the main route for HM contamination through feeding, where the absorbed metals mainly accumulate and/or are detoxified by the liver before excretion by kidneys.[62] Thus, there is general consensus that liver acts as an environmental indicator of water pollution due to its ability to accumulate various kinds of pollutants at higher levels than the surrounding environment, playing an important role in contaminant storage, detoxification, redistribution and transformation while acting as an active pathological site.[59] In contrast, muscle tissue does not actively accumulate metals and seems to have a very fast decontamination rate, and hence has lower HM contamination compared to other fish organs. Nevertheless, interspecific variations in muscle of these pollutant concentrations, that were metal-specific, were recorded in both bait and edible fish species caught within the same area, pointing to several factors in play.

Within species, the differences in fish sharing the same habitat indicate that they do not necessarily share the same levels of metal accumulation. Conversely, differences in location and the environment from which the fish are sampled could also cause variation in metal accumulation within and between fish species. It should be noted that the sampling sites in this study were limited to approximately a 50 km stretch along the Durban coast with no clear boundaries. While adult slinger seabream *C. puniceus*, bronze bream *P. grande* and catface rockcod *E. andersoni* specimens are fairly residential, soldier *C. nufar* specimens are known to undertake extensive migrations with a mean distance of over 29 km.[63] Given the relatively small sampling area, easy dispersion of HM by concentration gradients, wind/waves with possible roaming of the fish within the sampling sites cannot be overruled. Overall, the diet and lifestyle of fish play important roles in influencing interspecies variability of HM accumulation patterns in the Durban basin.

To assess HM toxicity in fish, researchers often examine histopathological changes in fish tissues; for instance, alterations in liver and kidney tissues, including cytoplasmic disorganization, vacuolization, sinusoidal dilation, hepatic cellular necrosis, blood cell congestion and oedema.[64,65] These changes are commonly associated in fish with higher accumulations of metals such as Pb, Cu and Ni. Additionally, Pb toxicity has been linked to physical manifestations like darkening of the caudal region (black tails) and spinal curvatures.[26] However, a study by Moodley *et al.* revealed an intriguing finding.[12] Despite the accumulation of Pb exceeding the maximum permissible limits in the edible muscles of the fish, none of the fish species displayed visible signs of Pb and As toxicity. Similarly, no histopathological analyses were undertaken for fish species.[12] This absence of observable toxicity signs can be attributed to several factors, including, *e.g.*, species-specific tolerance, sublethal effects, detoxification mechanisms, storage of metals in other organs and interactions with other environmental factors.[66] These complex interactions may mask the immediate effects of HM accumulation. Nevertheless, it is essential to emphasize that even when apparent toxicity signs are absent, sublethal effects and metal accumulation beyond regulatory limits can have long-term implications for both ecological and human health. This underscores the ongoing need for monitoring and further research to fully understand the potential consequences of HM contamination in aquatic environments.

3.4 Fish as Bioindicators in Environmental Risk Assessment

It is accepted worldwide that bioindicators are species which are utilized to assess the health of the environment and are also an important tool for detecting changes in the environment, either positive or negative, and their subsequent effects on human society.[67] Fish species have gained significant attention in studies evaluating biological and biochemical responses to environmental contaminants.[68] Due to their aquatic habitat and feeding behaviours, fish are particularly susceptible and heavily exposed to pollution, unable to evade the detrimental effects of pollutants.[69] Therefore, fish species are frequently employed as contamination indicators in aquatic environments, especially when compared to invertebrates. Fish are more sensitive to various toxic substances, making them a convenient test subject for assessing ecosystem health.

However, it's important to note that numerous intrinsic species-specific factors complicate the comparison of HM concentration results and the selection of bioindicator species for assessing human health effects. Ideally, a bioindicator species should: (1) be commonly consumed in the area, (2) have a wide geographic distribution, and (3) possess the potential to accumulate high metal concentrations while providing sufficient tissue mass for residue analysis.[29] Some fish species may serve as better bioindicators of specific heavy metal contamination compared to others such as the yellowfin tuna *Thunnus albacores* (Perciformes, Scombridae), the narrow-barred Spanish mackerel *Scomberomorus commerson* (Perciformes, Scombridae), the tigertooth croaker *Otolithes ruber* (Perciformes, Sciaenidae), the silver pomfret *Pampus argenteus* (Scombriformes, Stromateidae) and the orange-spotted grouper *Epinephelus coioides* (Perciformes, Serranidae), among others.[29,70]

Nevertheless, the use of fish as environmental biomonitors also entails the greatest technical, personnel and time-related requirements and it carries a big load of uncertainty. Many fish species exhibit migratory behaviour for reasons such as spawning, feeding or territorial exploration. While some fish undertake extensive spawning migrations, covering thousands of km, as seen in diadromous species like eels or salmon, others restrict their migrations to riverine systems.[71,72] Migration affects the accuracy of adult fish sampling. These uncertainties of fish origin could lead to a biased evaluation of given localities or misrepresent the source of pollution identification. Another crucial point in biomonitoring programmes is the choice of reference fish species.

Concentrations of pollutants, as HMs, may differ between fish groups because of their different trophic position.[73] It is necessary to choose a reference species which can be caught at all investigated sites. From a practical point of view, this can cause complications in different habitat types samplings; a particular species might be absent in some localities. Regardless, catching sufficient numbers of individuals of the same species at each locality requires a considerable effort, which will require more personnel, technical and financial resources. Concentrations of certain pollutants differ not only between species, but individuals of the same species too.[74,75] This fact is caused mostly by the different age of the sampled animals. However, it should be taken into consideration that migration can play a role in these differences as well. For this reason, a sufficient number of individuals must be sampled to obtain relevant results about the contamination at a given locality. On the other hand, sampling of high numbers of adult fish can cause a problem. As these fish are, in most of the cases, within their reproductive age, their sampling could significantly affect the broodstock at the locality in some fish species. Pollutants may differ greatly between the tissues of the body within one individual. Where a wide range of different contaminants are to be analysed, different organs and tissues must be verified to obtain relevant data. Different chemicals accumulate in specific parts in the fish body,[76] thus, sampling of only muscle or liver is not sufficient, as some contaminants are found in fat or in the brain in higher concentrations. Sampling and analysing several tissues or organs can be time-consuming and expensive.

3.5 The Maasbanker *Trachurus trachurus* and the Slinger Seabream *Chrysoblephus puniceus* as Potential Biomonitors of Heavy Metal Pollution

Recently, a study conducted by Debipersadh *et al.* highlighted the potential of using two fish species, namely the maasbanker *T. trachurus* and slinger seabream *C. puniceus* as bioindicators for HM pollution in the Durban basin.[39] This study is notable as it is the first report on higher HM accumulation in the slinger seabream *C. puniceus* compared to other species, rendering it as a promising bioindicator candidate for pollution in the Durban coast. The research aimed to compare the HM bioaccumulation patterns in these two fish species across two distinct habitats along the eastern seaboard of South Africa: Cape Vidal and the Durban basin. Cape Vidal, a marine protected area, is generally considered pristine and unaffected by pollution. In contrast, the Durban basin faces substantial contamination due to industrial and agricultural activities in its catchment area. The study sought to assess how these differing environments influenced HM accumulation in maasbanker *T. trachurus* and slinger seabream *C. puniceus*, making them valuable bioindicators for detecting pollution exerted by these contaminants. We have analysed element concentrations of HM in the muscle tissues of these two fish species from the two habitats and the results are presented in Table 3.2.

The study employed a metal pollution index (MPI) to compare the total metal accumulation level in muscle tissues of these two fish species.[39] The MPI values indicated a significant increase in cumulative metal accumulation in fish collected from the Durban basin compared to those from Cape Vidal (see Table 3.2). The study also revealed that liver and gill HM concentration directly reflected those of the surrounding habitat.[39] This pointed to the pronounced species-specific HM accumulation patterns, which were influenced by the distinct environments – polluted and anthropogenically disturbed in contrast to pristine ones. It is worth noting that the MPI values observed in this study were relatively lower than those reported for freshwater fish species from both the Nansi and Taihu lake (China), indicating that marine fishes tend to show higher MPI values than their freshwater counterparts, especially in polluted marine and estuarine environments.[77] For example, higher MPI values were reported for the spotted snakehead *Channa puntata* (Anabantiformes, Channidae) exposed to thermal power plant effluent in various tissues.[78] However, the values in this study were comparatively lower, partly due to the dilution effect of pollutants by seawater.[78] These values are higher than those reported in this study, which is partly attributed to the same dilution effect. Consistent with these findings, other studies have reported higher HM accumulation in species closely related to the maasbanker, such as the Mediterranean horse mackerel *T. mediterraneus*,[79] with time-dependent accumulation of Zn and Pb in *T. trachurus* in polluted environments being also reported,[30,48] suggesting its suitability as a biomonitor for metal pollution. This study is the first report on higher heavy metal accumulation in the slinger seabream *C. puniceus* compared to other species, making it a promising candidate as a bioindicator for pollution in the Durban coast. Furthermore, different patterns in both species collectively show that the fish muscle is not an active tissue in accumulating HM, whereas liver and gills are good monitors of water pollution with metals because their concentrations are proportional to those present in the environment.

In summary, the study of Debipersadh *et al.* has successfully demonstrated the effectiveness of native biota, including maasbanker *T. trachurus* and slinger seabream *C. puniceus*, as bioindicators for monitoring moderate levels of HM pollution in marine ecosystems across different locations.[39] However, it is important to note that species-specific variations in metal levels, as observed in the aforementioned geographical differences, may not be sufficient for effectively monitoring subtle differences, such as time trends, without a more comprehensive analysis of monitoring data. Therefore, future studies should consider comparing the temporal trends in HM accumulation patterns in both water and fish species across the two geographical locations to fully showcase the monitoring capabilities of maasbanker *T. trachurus* and slinger seabream *C. puniceus*. Additionally, there is a need for future biomonitoring studies that assess the responses of native species and their ability to indicate environmental degradation accurately.

Table 3.2 Mean (±SE) comparison of heavy metal concentration (mg per kg dw) and metal pollution index (MPI) value of the total metal accumulation level in different body parts of the maasbanker *Trachurus trachurus* and the slinger seabream *Chrysoblephus puniceus* in Cape Vidal and Durban basin. Adapted from ref. 34 and 39 with permission from the authors.

Species/parts	Location	Al	As	Cr	Cu	Mn	Ni	Pb	Zn	MPI[a]
Maasbanker										
Gut	Durban	51 ± 6.37	1.82 ± 0.21	2.69 ± 0.33	4.56 ± 0.74	2.84 ± 0.34	1.43 ± 0.11	4.96 ± 0.70	65 ± 6.04	7.62
	Cape Vidal	10 ± 0.91	1.44 ± 0.07	1.74 ± 0.07	4.38 ± 0.73	3.08 ± 0.18	1.17 ± 0.06	4.18 ± 0.78	43 ± 4.57	4.96
	t-Value[b]	−6.27***	−1.73	−2.85 *	−0.18	0.622	−2.022	−0.750	−2.938**	
Slinger seabream										
Fleshy tissue	Durban	34 ± 2.85	2.25 ± 0.28	2.75 ± 0.27	3.57 ± 0.28	3.41 ± 0.31	1.96 ± 0.18	4.25 ± 0.38	26 ± 1.79	6.62
	Cape Vidal	12 ± 3.26	0.88 ± 0.11	1.78 ± 0.30	3.40 ± 0.74	1.29 ± 0.35	1.45 ± 0.29	1.81 ± 0.34	23 ± 4.25	3.40
	t-Value	−4.975***	−4.536***	−2.414*	−0.221	−4.540***	−1.494	−4.765***	−0.687	

[a]MPI values were calculated using the equation by Usero *et al.*: $MPI = (Cf_1 \times Cf_2 \times \ldots \times Cf_n)^{1/n}$, where Cf_n is the contents for the metal n in the sample.
[b]Two-tailed two sample Student's *t*-test ($*p < 0.05$; $**p < 0.01$; $***p < 0.001$).

3.6 Conclusion and Perspectives

The findings presented in this chapter have helped us in our comprehension of HM accumulation in fish species, particularly within the Durban basin. It accentuates the intricate web of factors that influence the distribution of pollutants, underscoring the significance of species-specific disparities encompassing dietary preferences, trophic levels as well as fish size, age, physiology and lifestyle. Furthermore, it sheds light on the distinct functions that gills, liver and muscles play in the accumulation and processing of HM in various fish species. As evidenced by some species exhibiting higher concentrations of As, Cr, Ni, Pb and Zn in their muscles compared to South African and international regulations, this review offers a holistic perspective on the ecological and physiological aspects influencing their bioaccumulation. The presence of anthropogenic HM exceeding regulatory limits for seafood, particularly As, Cr, Ni, Pb and Zn, underscores the immediate need for comprehensive spatial and temporal monitoring and the development of national and regional strategies to address and manage HM pollution in the Durban basin effectively.

Moreover, novel insights into the potential of native fish species, specifically the maasbanker *T. trachurus* and the slinger seabream *C. puniceus*, demonstrate their usefulness as effective bioindicators for assessing moderate levels of HM pollution in marine ecosystems. The research reveals significant disparities in HM accumulation patterns between these species across two distinct geographical locales, Cape Vidal and the Durban basin. These findings underscore the importance of accounting for species-specific variations when employing bioindicators for pollution monitoring.

Looking ahead, several promising avenues for future research emerge. First and foremost, longitudinal studies will be essential to gain a more comprehensive understanding of the capabilities of these bioindicator species. Such studies can track how HM accumulation patterns change over time, thus establishing these species as effective monitors of temporal trends in pollution levels. Additionally, in this case, a comparative analysis of maasbanker and slinger seabream, in conjunction with other potential bioindicator species within the same regions, can offer valuable insights. This approach could help identify the most suitable species for different monitoring scenarios, enhancing the precision of pollution assessments. Moreover, expanding the research to encompass broader ecosystem health evaluations will be critical. Understanding how HM pollution impacts not only the chosen bioindicator species but also other biota and environmental parameters can provide a more holistic view of ecological consequences. Furthermore, unravelling the intricate mechanisms underlying HM accumulation in this fish species is a priority. Investigating the species' feeding habits, physiological adaptations and interactions with their environment will also enrich our comprehension of their suitability as bioindicators. Given that some of these bioindicator species are part of human diets, assessing the potential transfer of HM to consumers is also crucial. Particularly in regions where these fish are commonly consumed, studying the associated public health risks is essential. Lastly, the insights generated by this research offer valuable guidance for environmental management and policy decisions. Future studies should concentrate on how data provided by bioindicator species can be effectively translated into actionable policies to mitigate HM pollution and safeguard marine ecosystems. This proactive approach will ensure that research findings can lead to meaningful environmental conservation measures.

References

1. P. K. Gusso-Choueri, R. B. Choueri, G. S. Santos, G. S. de Araújo, A. C. F. Cruz, T. Stremel, S. X. de Campos, M. M. Cestari, C. A. O. Ribeiro and D. M. de Sousa Abessa, Assessing genotoxic effects in fish from a marine protected area influenced by former mining activities and other stressors, *Mar. Pollut. Bull.*, 2016, **104**, 229–239.
2. D. M. S. Abessa, H. C. Albuquerque, L. G. Morais, G. S. Araújo, T. G. Fonseca, A. C. F. Cruz, B. G. Campos, J. B. D. A. Camargo, P. K. Gusso-Choueri, F. C. Perina, R. B. Choueri and L. M. Buruaem, Pollution status of marine protected areas worldwide and the consequent toxic effects are unknown, *Environ. Pollut.*, 2018, **243**, 1450–1459.
3. N. V. Nair and P. K. Nayak, Exploring water quality as a determinant of small-scale fisheries vulnerability, *Sustainability*, 2023, **15**, 13238.
4. N. Kovaleva, Ecology: Life in the "Unstable Biosphere", in *Reconsidering the Limits to Growth*, ed. A. Sadovnichy, A. Akaev, I. Ilyin, S. Malkov, L. Grinin, and A. Korotayev, World-Systems Evolution and Global Futures, Springer, Cham, Rome, 5th edn, 2023, pp. 71–95.
5. V. Wepener and N. Degger, Status of marine pollution research in South Africa (1960–present), *Mar. Pollut. Bull.*, 2012, **64**, 1508–1512.
6. eThekwini Municipality, *Economic Development and Growth in eThekwini*, Durban, 2014.
7. E. Vetrimurugan, V. C. Shruti, M. P. Jonathan, P. D. Roy, N. W. Kunene, L. Elizabeth and C. Villegas, Metal concentration in the tourist beaches of South Durban: An industrial hub of South Africa, *Mar. Pollut. Bull.*, 2017, **117**, 538–546.
8. B. J. Goble, R. P. van der Elst and L. K. Oellermann, *Ugu Lwethu–Our Coast, A Profile of Coastal KwaZulu-Natal*, KwaZulu-Natal Department of Agriculture and Environmental Affairs and the Oceanographic Research Institute, Cedara, 2014.
9. A. P. W. Barrows, S. E. Cathey and C. W. Petersen, Marine environment microfiber contamination: Global patterns and the diversity of microparticle origins, *Environ. Pollut.*, 2018, **237**, 275–284.
10. C. Kroeze, S. Gabbert, N. Hofstra, A. A. Koelmans, A. Li, A. Löhr, F. Ludwig, M. Strokal, C. Verburg, L. Vermeulen, M. T. van Vliet, W. de Vries, M. Wang and J. van Wijnen, Global modelling of surface water quality: a multi-pollutant approach, *Curr. Opin. Environ. Sustainability*, 2016, **23**, 35–45.

11. S. Debipersadh, T. Sibanda, R. Selvarajan and R. Naidoo, Investigating toxic metal levels in popular edible fishes from the South Durban basin: implications for public health and food security, *Environ. Monit. Assess.*, 2018, **190**, 476.
12. R. Moodley, N. T. Mahlangeni and P. Reddy, Determination of heavy metals in selected fish species and seawater from the South Durban Industrial Basin, KwaZulu-Natal, South Africa, *Environ. Monit. Assess.*, 2021, **193**, 206.
13. E. Vetrimurugan, V. C. Shruti, M. P. Jonathan, P. D. Roy, S. K. Sarkar, B. K. Rawlins and L. E. C. Villegas, Comprehensive study on metal contents and their ecological risks in beach sediments of KwaZulu-Natal province, South Africa, *Mar. Pollut. Bull.*, 2019, **149**, 110555.
14. E. Vetrimurugan, V. C. Shruti, M. P. Jonathan, P. D. Roy, B. K. Rawlins and D. M. Rivera-Rivera, Metals and their ecological impact on beach sediments near the marine protected sites of Sodwana Bay and St. Lucia, South Africa, *Mar. Pollut. Bull.*, 2018, **127**, 568–575.
15. K. Moodley, S. Pillay, K. Pather and H. Ballabh, Heavy metal contamination of the Palmiet River: KwaZulu Natal South Africa, *Int. J. Sci. Res. Environ. Sci.*, 2014, **2**, 397–409.
16. M. Canli and G. Atli, The relationships between heavy metal (Cd, Cr, Cu, Fe, Pb, Zn) levels and the size of six Mediterranean fish species, *Environ. Pollut.*, 2003, **121**, 129–136.
17. M. K. Ahmed, M. A. Baki, G. K. Kundu, M. S. Islam, M. M. Islam and M. M. Hossain, Human health risks from heavy metals in fish of Buriganga river, Bangladesh, *Springerplus*, 2016, **5**, 1697.
18. E. M. Alissa and G. A. Ferns, Heavy metal poisoning and cardiovascular disease, *J. Toxicol.*, 2011, **2011**, 870125.
19. P. Lentini, L. Zanoli, M. de Cal, A. Granata and R. Dell'Aquila, in *Critical Care in Nephrology*, ed. C. Ronco, R. Bellomo, J. Kellum, and Z. Ricci, Elsevier Inc., 3rd edn, 2017, pp. 1324–1330.
20. S. Rajeshkumar and X. Li, Bioaccumulation of heavy metals in fish species from the Meiliang Bay, Taihu Lake, China, *Toxicol. Rep.*, 2018, **5**, 288–295.
21. L. A. M. Guastella, A quantitative assessment of recreational angling in Durban Harbour, South Africa, *S. Afr. J. Mar. Sci.*, 1994, **14**, 187–203.
22. S. Thawley, S. Morris and A. Vosloo, Zn and Cd accumulation in *Potamonautes warreni* from sites in the North – West Province of South Africa, *Int. Congr. Ser.*, 2004, **1275**, 180–188.
23. R. C. Newell, D. W. Maughan, L. J. Seiderer, M. W. Trett and C. G. Nicholson, Assessing the impact of liquid wastes discharge in the sea. A case study off Amanzimtoti, near Durban, Natal, South Africa, *S. Afr. J. Mar. Sci.*, 1991, **10**, 368–396.
24. A. O. Olaniran, K. Naicker and B. Pillay, Assessment of physico-chemical qualities and heavy metal concentrations of Umgeni and Umdloti Rivers in Durban, South Africa, *Environ. Monit. Assess.*, 2014, **186**, 2629–2639.

25. J. Salinas, V. Carpena, M. R. Martínez-Gallardo, M. Segado, M. J. Estrella-González, A. J. Toribio, M. M. Jurado, J. A. López-González, F. Suárez-Estrella and M. J. López, Development of plastic-degrading microbial consortia by induced selection in microcosms, *Front. Microbiol.*, 2023, **14**, 1143769.
26. A. C. Bosch, B. O'Neill, G. O. Sigge, S. E. Kerwath and L. C. Hoffman, Heavy metals in marine fish meat and consumer health: A review, *J. Sci. Food Agric.*, 2016, **96**, 32–48.
27. J. Kearns and A. Turner, An evaluation of the toxicity and bioaccumulation of bismuth in the coastal environment using three species of macroalga, *Environ. Pollut.*, 2016, **208**, 435–441.
28. M. Jayaprakash, R. Senthil Kumar, L. Giridharan, S. B. Sujitha, S. K. Sarkar and M. P. Jonathan, Bioaccumulation of metals in fish species from water and sediments in macrotidal Ennore creek, Chennai, SE coast of India: A metropolitan city effect, *Ecotoxicol. Environ. Saf.*, 2015, **120**, 243–255.
29. P. A. Cunningham, E. E. Sullivan, K. H. Everett, S. S. Kovach, A. Rajan and M. C. Barber, Assessment of metal contamination in Arabian/Persian Gulf fish: A review, *Mar. Pollut. Bull.*, 2019, **143**, 264–283.
30. O. Jitar, C. Teodosiu, M. Nicoara and G. Plavan, Study of heavy metal pollution and bioaccumulation in the Black Sea living environment, *Environ. Eng. Manage. J.*, 2013, **12**, 271–276.
31. E. Vetrimurugan, M. P. Jonathan, V. C. Shruti and B. K. Rawlins, in *Coastal Management: Global Challenges and Innovations*, ed. R. R. Krishnamurthy, M. P. Jonathan, S. Srinivasalu, and B. Glaeser, Elsevier Inc., 2018, pp. 121–139.
32. T. Vogt, R. Pieters, J. Giesy and B. K. Newman, Biological toxicity estimates show involvement of a wider range of toxic compounds in sediments from Durban, South Africa than indicated from instrumental analyses, *Mar. Pollut. Bull.*, 2019, **138**, 49–57.
33. P. Reopanichkul, T. A. Schlacher, R. W. Carter and S. Worachananant, Sewage impacts coral reefs at multiple levels of ecological organization, *Mar. Pollut. Bull.*, 2009, **58**, 1356–1362.
34. S. Debipersadh, Dynamics of heavy metals in seawater, sediment, mudprawns and fish tissues caught of Durban basin in South Africa: impacts and implications, Ph.D. Dissertation, University of South Africa, 2020.
35. G. Vandermeersch, H. M. Lourenço, D. Alvarez-Muñoz, S. Cunha, G. Cano-Sancho, J. J. Sloth, C. Kwadijk, D. Barcelo, W. Allegaert, K. Bekaert, J. O. Fernandes, A. Marques and J. Robbens, Environmental contaminants of emerging concern in seafood – European database on contaminant levels, *Environ. Res.*, 2015, **143**, 29–45.
36. N. H. Kim, Y. Y. Hyun, K. B. Lee, Y. Chang, S. Rhu, K. H. Oh and C. Ahn, Environmental heavy metal exposure and chronic kidney disease in the general population, *J. Korean Med. Sci.*, 2015, **30**, 272–277.
37. B. Sea, L. Bat, H. C. Öztekin and F. Üstün, Heavy metal levels in four commercial fishes caught in Sinop Coasts of Black Sea, Turkey, *Turk. J. Fish. Aqua. Sci.*, 2015, **15**, 399–405.

38. E. Vetrimurugan, M. P. Jonathan, P. D. Roy, V. C. Shruti and O. M. Ndwandwe, Bioavailable metals in tourist beaches of Richards Bay, Kwazulu-Natal, South Africa, *Mar. Pollut. Bull.*, 2016, **105**, 430–436.
39. S. Debipersadh, H. J. O. Ogola, K. Mearns and R. Selvarajan, Bioaccumulation, bioindication and health risk assessment of heavy metals in Cape horse mackerel (*Trachurus trachurus*) and slinger seabream (*Chrysoblephus puniceus*) in the Durban Basin and Cape Vidal, South Africa, *Arch. Environ. Contam. Toxicol.*, 2023, **85**, 165–180.
40. S. Debipersadh, R. Selvarajan, T. Sibanda and R. Naidoo, Assessing toxic elemental concentrations in marine fish *Trachurus capensis* (Cape horse mackerel) and implications for public health, *Pol. J. Environ. Stud.*, 2018, **27**, 1395–1400.
41. B. Austin, The effects of pollution on fish health, *J. Appl. Microbiol.*, 1999, **85**, 234S–242S.
42. FAO, Food and Agriculture Organization - Compilation of legal limits for hazardous substances in fish and fishery products, FAO Fishery Circular No. 464, Rome, Italy, 1983.
43. J. Verdouw, Heavy metal contamination in muscle tissue of four key recreational fish species from the Derwent estuary, Ph.D. Dissertation, University of Tasmania, 2008.
44. H. M. Mzimela, V. Wepener and D. P. Cyrus, Spatial and temporal variations in selected heavy metals in water and sediment from the Mhlathuze Estuary, Richards Bay, *Afr. J. Environ. Sci. Technol.*, 2014, **8**, 670–683.
45. B. P. Jupp, S. W. Fowler, S. Dobretsov, H. van der Wiele and A. Al-Ghafri, Assessment of heavy metal and petroleum hydrocarbon contamination in the Sultanate of Oman with emphasis on harbours, marinas, terminals and ports, *Mar. Pollut. Bull.*, 2017, **121**, 260–273.
46. S. Jahan and V. Strezov, Assessment of trace elements pollution in the sea ports of New South Wales (NSW), Australia using oysters as bioindicators, *Sci. Rep.*, 2019, **9**, 1416.
47. M. Mendoza-Carranza, A. Sepúlveda-Lozada, C. Dias-Ferreira and V. Geissen, Distribution and bioconcentration of heavy metals in a tropical aquatic food web: A case study of a tropical estuarine lagoon in SE Mexico, *Environ. Pollut.*, 2016, **210**, 155–165.
48. E. Lozano-Bilbao, G. Lozano, S. Jiménez, A. Jurado-Ruzafa, A. Hardisson, C. Rubio, D. G. Weller, S. Paz and Á. J. Gutiérrez, Ontogenic and seasonal variations of metal content in a small pelagic fish (*Trachurus picturatus*) in northwestern African waters, *Mar. Pollut. Bull.*, 2020, **156**, 111251.
49. Y. G. Georgieva, G. M. Daskalov, S. L. Klayn, K. B. Stefanova and E. S. Stefanova, Seasonal diet and feeding strategy of horse mackerel *Trachurus mediterraneus* (Steindachner, 1868) (Perciformes: Carangidae) in the South-Western Black Sea, *Acta Zool. Bulg.*, 2019, **71**, 201–210.
50. H. K. Kadila, D. N. Nakwaya, M. Butler and J. A. Iitembu, Insights into feeding interactions of shallow water cape hake (*Merluccius capensis*) and cape horse mackerel (*Trachurus capensis*) from the Northern Benguela (Namibia), *Reg. Stud. Mar. Sci.*, 2020, **34**, 101071.

51. P. Cury, A. Bakun, R. J. M. Crawford, A. Jarre, R. A. Quinones, L. J. Shannon and H. M. Verheye, Small pelagics in upwelling systems: patterns of interaction and structural changes in "wasp-waist" ecosystems, *ICES J. Mar. Sci.*, 2000, **57**, 603–618.
52. S. Garrido, A. Silva, J. Pastor, R. Dominguez, A. V. Silva and A. M. Santos, Trophic ecology of pelagic fish species off the Iberian coast: Diet overlap, cannibalism and intraguild predation, *Mar. Ecol.: Prog. Ser.*, 2015, **539**, 271–285.
53. A. Gundogdu, S. T. Culha, F. Kocbas and M. Culha, Heavy metal accummulation in muscles and total bodies of *Mullus barbatus, Trachurus trachurus* and *Engraulis encrasicolus* captured from the coast of sinop, Black Sea, *Pak. J. Zool.*, 2016, **48**, 25–34.
54. G. W. Bryan and A. Darracott, Bioaccumulation of marine pollutants, *Philos. Trans. R. Soc., B*, 1979, **286**, 483–505.
55. H. Maaghloud, R. Houssa, S. Ouansafi, F. Bellali, K. El Bouqdaoui, N. Charouki and A. Fahde, Ingestion of microplastics by pelagic fish from the Moroccan Central Atlantic coast, *Environ. Pollut.*, 2020, **261**, 114194.
56. A. Herrera, A. Ŝtindlová, I. Martínez, J. Rapp, V. Romero-Kutzner, M. D. Samper, T. Montoto, B. Aguiar-González, T. Packard and M. Gómez, Microplastic ingestion by Atlantic chub mackerel (*Scomber colias*) in the Canary Islands coast, *Mar. Pollut. Bull.*, 2019, **139**, 127–135.
57. C. Lopes, J. Raimundo, M. Caetano and S. Garrido, Microplastic ingestion and diet composition of planktivorous fish, *Limnol. Oceanogr. Lett.*, 2020, **5**, 103–112.
58. P. Hernández-Almaraz, L. Méndez-Rodríguez, T. Zenteno-Savín, F. García-Domínguez, A. Vázquez-Botello and E. Serviere-Zaragoza, Metal mobility and bioaccumulation differences at lower trophic levels in marine ecosystems dominated by Sargassum species, *J. Mar. Biol. Assoc. U. K.*, 2014, **94**, 435–442.
59. F. O. Vilches, M. A. Bobinac, A. C. Labudía, M. N. Paso Viola, J. E. Marcovecchio, H. L. Cappozzo and M. V. Panebianco, Metals concentration and bioaccumulation in the marine-coastal trophic web from Buenos Aires Province southern coast, Argentina, *J. Chem. Ecol.*, 2019, **35**, 501–523.
60. D. Crafford and A. Avenant-Oldewage, Bioaccumulation of non-essential trace metals in tissues and organs of *Clarias gariepinus* (sharptooth catfish) from the Vaal River system - Strontium, aluminium, lead and nickel, *Water SA*, 2010, **36**, 621–640.
61. A. Aich, A. R. Goswami, U. S. Roy and S. K. Mukhopadhyay, Ecotoxicological assessment of tannery effluent using guppy fish (*Poecilia reticulata*) as an experimental model: A biomarker study, *J. Toxicol. Environ. Health, Part A*, 2015, **78**, 278–286.
62. V. Yancheva, S. Stoyanova, I. Velcheva, S. Petrova and E. Georgieva, Metal bioaccumulation in common carp and rudd from the Topolnitsa reservoir, Bulgaria, *Arh. Hig. Rada. Toksikol.*, 2014, **65**, 57–66.

63. B. Q. Mann, Southern African marine linefish species profiles, *Oceanographic Research Institute Special Publication No. 9*, Oceanographic Research Institute, Durban, 2013.
64. S. O. Ayoola, Heavy metals concentration and histopathological profile of some commercial fish species at Makoko Slum neighbourhood environment, Lagos, Nigeria, *Aceh. J. Anim. Sci.*, 2019, **4**, 1–10.
65. A. Iqbal, A. Tabinda, A. Yasar and Y. Mahfooz, Heavy metal uptake and toxicity in tissues of commercially important freshwater fish (*Labeo rohita* and *Wallago attu*) from the Indus River, Pakistan, *Pol. J. Environ. Stud.*, 2017, **26**, 627–633.
66. A. C. Bosch, B. O'Neill, G. O. Sigge, S. E. Kerwath and L. C. Hoffman, Heavy metal accumulation and toxicity in smoothhound (*Mustelus mustelus*) shark from Langebaan Lagoon, South Africa, *Food Chem.*, 2016, **190**, 871–878.
67. E. A. Holt and S. W. Miller, Bioindicators : Using organisms to measure environmental impacts, *Nat. Educ. Knowl.*, 2010, **3**, 8.
68. D. Oost, J. Beyer and N. P. E. Vermeulen, Fish bioaccumulation and biomarkers in environmental risk assessment : A review, *Environ. Toxicol. Pharmacol.*, 2003, **13**, 57–149.
69. E. Yarsan and M. Yipel, The Important Terms of Marine Pollution "Biomarkers and Biomonitoring, Bioaccumulation, Bioconcentration, Biomagnification", *J. Mol. Biomarkers Diagn.*, 2013, **S1**, 003.
70. J. Burger and M. Gochfeld, Mercury and selenium levels in 19 species of saltwater fish from New Jersey as a function of species, size, and season, *Sci. Total Environ.*, 2011, **409**, 1418–1429.
71. J. Kotusz, A. Witkowski, M. Baran and J. Błachuta, Fish migrations in a large lowland river (Odra River, Poland) – based on fish pass observations, *Folia Zool.*, 2006, **55**, 386–398.
72. M. Prchalová, P. Horký, O. Slavík, L. Vetešník and K. Halačka, Fish occurrence in the fishpass on the lowland section of the River Elbe, Czech Republic, with respect to water temperature, water flow and fish size, *Folia Zool.*, 2011, **60**, 104–114.
73. C. Miege, A. Roy, P. Labadie, H. Budzinski, B. Le Bizec, K. Vorkamp, H. Persat, M. Coquery, M. Babut, C. Miege, A. Roy, P. Labadie, H. Budzinski and B. Le Bizec, Occurrence of priority and emerging organic compounds in fishes from the Rhone River (France), *Anal. Bioanal. Chem.*, 2012, **404**, 2721–2735.
74. M. Pandelova, Levels of PCDD/F and dioxin-like PCB in Baltic fish of different age and gender, *Chemosphere*, 2007, **71**, 369–378.
75. J. J. Verdouw, C. K. Macleod, B. F. Nowak and J. M. Lyle, Implications of age, size and region on mercury contamination in estuarine fish species, *Water, Air, Soil Pollut.*, 2011, **214**, 297–306.
76. N. Yamaguchi, D. Gazzard, G. Scholey and D. W. Macdonald, Concentrations and hazard assessment of PCBs, organochlorine pesticides and mercury in fish species from the upper Thames : River pollution and its potential effects on top predators, *Chemosphere*, 2003, **50**, 265–273.

77. P. Li, J. Zhang, H. Xie, C. Liu, S. Liang, Y. Ren and W. Wang, Heavy metal bioaccumulation and health hazard assessment for three fish species from Nansi Lake, China, *Bull. Environ. Contam. Toxicol.*, 2015, **94**, 431–436.
78. M. Javed, I. Ahmad, N. Usmani and M. Ahmad, Bioaccumulation, oxidative stress and genotoxicity in fish (*Channa punctatus*) exposed to a thermal power plant effluent, *Ecotoxicol. Environ. Saf.*, 2016, **127**, 163–169.
79. B. Rios-Fuster, C. Alomar, M. Compa, B. Guijarro and S. Deudero, Anthropogenic particles ingestion in fish species from two areas of the western Mediterranean Sea, *Mar. Pollut. Bull.*, 2019, **144**, 325–333.

CHAPTER 4

Fish Microbiota Disruption by Ecotoxicology Agents: A Bioindicator of Health and Pollution

ESTEFANÍA GARIBAY-VALDEZ[a†], DIANA MEDINA-FÉLIX[b†], FRANCISCO VARGAS-ALBORES[a], EDILMAR CORTÉS-JACINTO[c] AND MARCEL MARTÍNEZ-PORCHAS[a*]

[a]Centro de Investigación en Alimentación y Desarrollo, A.C. Biología de Organismos Acuáticos, Hermosillo, Sonora, Mexico; [b]Universidad Estatal de Sonora, Hermosillo, Sonora, Mexico; [c]Centro de Investigaciones Biológicas del Noroeste, La Paz, Baja California Sur, Mexico
*E-mail: marcel@ciad.mx

4.1 Introduction

Aquatic ecosystems are unfortunately often considered as disposal and recycling sites for sewage and toxic wastes, increasing the amounts of toxic compounds and affecting the biological communities.[1] They are also regarded as the principal contaminants' final destination, and their sediments are the primary sink for their storage.[2] Various toxic pollutants and the accumulation of contaminants affect fish health; these, in turn, induce genetic effects and

†These authors contributed equally.

Issues in Toxicology No. 50
Fish Species in Environmental Risk Assessment Strategies
Edited by Guillermo Eli Liwszyc and Marcelo L. Larramendy

Published by the Royal Society of Chemistry, www.rsc.org

may lead to mortality. Thus, special attention should be given to the aquatic ecosystems and how aquatic organisms, from bacteria to fish, are affected by pollutants. In the host-environment relationship, the microbiome is a relevant, yet understudied compartment in ecotoxicology research. Notably, the gut microbiota regulates the overall health of fish, including a bi-directional interaction with the immune system. This chapter describes some effects on the fish gut microbiome and their interactions with contaminants.

4.2 Ecotoxicology Agents and Fish Health

4.2.1 Importance of Ecotoxicology in Ecosystems

Ecosystems are mainly under threat from contaminants released through cumulative anthropogenic activities. In this sense, the importance of an ecotoxicological evaluation for environmental health has increased.[3] Ecotoxicology focuses on toxic chemicals' effects on aquatic and terrestrial ecosystems and their biological populations and communities (*i.e.*, microorganisms, plants, invertebrates, vertebrates).[4,5] The Ecotoxicology (ECOTOX) Knowledgebase Resource Hub (https://www.epa.gov/comptox-tools/ecotoxicology-ecotox-knowledgebase-resource-hub) provides useful information about the Toxic Substances Control Act Chemical Substance Inventory (TSCA Inventory) and is regulated by the United States Environmental Protection Agency (U.S. EPA).[6] The TSCA Inventory defines a "chemical substance" as any organic or inorganic substance of a particular molecular identity, including any combination of these substances occurring in whole or in part due to a chemical reaction or occurring in nature, and any element or uncombined radical.[6] The ECOTOX database registers 86 685 chemicals to date. Nevertheless, only 12 830 are reported as ecotoxic compounds affecting ecosystems.[7]

Terrestrial and aquatic ecosystems are at a constant risk of ecotoxicological effects,[8] which occur at all biological organization levels, from the molecular to the ecosystem.[9] These ecotoxicological effects include, among others, alterations in growth rate and reproduction in terrestrial plants and animals.[10–12] However, aquatic ecosystems are also influenced by showing poor physiological, growth and reproductive performance of marine and freshwater fish, shellfish and even zooplankton as a result of collective biological disorders in the animal community.[13–15] Furthermore, hormone disruption is also reported to be an ecotoxicological effect in freshwater fish, promoting deregulation in hormone syntheses.[16]

4.2.2 Aquatic Ecotoxicological Agents

Chemical compounds are known to affect more than 13 872 living species on Earth, 56% of which belong to terrestrial environments, whereas the remaining 45% are aquatic. Interestingly, from the total of the reported ecotoxic chemical compounds, 40% of these affect terrestrial species, while 60% impact their aquatic counterpart.[6] Thus, special attention should be given to water bodies and how habitant organisms, from bacteria to fish, are affected by pollutants. The occurrence of ecotoxicological agents in aquatic ecosystems increases due to the accelerated production of anthropogenic contaminants, turning the water bodies into the main contaminants' final destination, and their sediments are the sink for their storage.[2] These compounds are constantly released into the environment through direct discharge, sewage treatment plant (STP) effluent and runoff of manure.[17] The role of fertilizer wash-off should be considered as well. In addition, urban stormwater runoff degrades water quality and ecosystem function by transporting anthropogenic chemical mixtures to receiving water bodies.[18]

Fresh and marine water ecosystems have been widely affected by ecotoxicological agents such as drugs, disinfection by-products and pesticides, among others.[16,19,20] The aquatic ecotoxicological chemical agents reported by the ECOTOX database include several and diverse xenobiotics. Among them can be included conazoles, cyanide, cyanotoxins, DDT and its metabolites, dibenzofurans, explosives, glycol ethers, major ions, neonicotinoids, nitrosamines, perchlorates, per- and polyfluoroalkyl substances (PFAS), phthalate esters, polyaromatic hydrocarbons (PAHs), polybrominated diphenyl ethers (PBDEs), polychlorinated biphenyls (PCBs), pharmaceutical personal care products (PPCPs), strobins, organotin and several elements including Al, Sb, As, Ba, Be, Cd, Cr, Co, Cu, Fe, Ld, Mn, Hg, Ni, Ag, Se, Tl, V and Zn (see Figure 4.1).

Conversely, the most abundant ecotoxicological compounds in the aquatic environment are those derived from pharmaceuticals and personal care

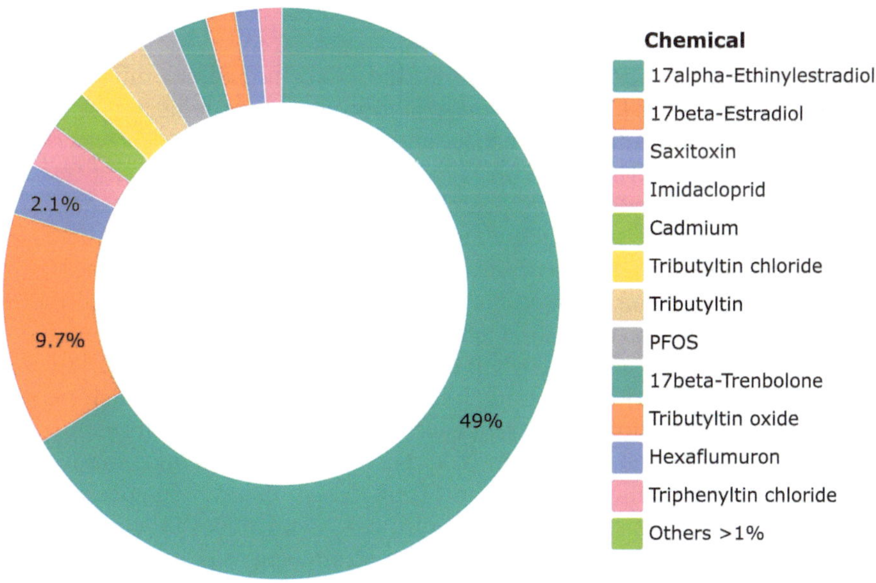

Figure 4.1 Ecotoxicological chemical agents reported in aquatic environments. Graphical information was retrieved from the ECOTOX database.[6]

products (PPCPs).[21,22] This category encompasses antibiotics, hormones, drugs, antimicrobial agents, insect repellants and sunscreen UV filters.[23] Precisely, in the ecotoxic database, 17-α-ethinylestradiol is placed as the main ecotoxicological agent in the aquatic environment with 49% of the reported cases and in second place, 17-β-estradiol with a 9.7% incidence.[24] Both steroid oestrogens are classified as endocrine-disrupting chemicals (EDCs), mainly affecting growth, development and reproduction.[25] On the other hand, EDCs are used worldwide for human therapy and in the livestock rearing and aquaculture industries.[26]

In third place we find saxitoxin, which, although not considered a xenobiotic *per se*, has a relevant role in ecotoxicology, increasing and intensifying harmful algal blooms (HABs) and their subsequent toxin production. HABs are, in turn, favoured by eutrophication, which is induced by climate change and the over-development of industrial sewage discharged to water bodies from increased anthropogenic activities.[27,28] Saxitoxin is a toxin notoriously implied in paralytic shellfish poisoning, produced by several dinoflagellates and cyanobacteria, which are then released into aquatic environments.[29] The influx of ecotoxicological agents in marine ecosystems constantly accumulates in water, sediments and aquatic species. Thus, humans indirectly consume saxitoxin through the food chain, resulting in multiple effects at several levels.[30]

4.2.3 Ecotoxicology Implications in Fish Health

According to the U.S. EPA,[6] there are 2944 studies at the time of publication of this book, on 71 fish species related to ecotoxicological exposure, in which the zebrafish *Danio rerio* (Cypriniformes, Cyprinidae) is the most used experimental model species, with 777 studies, followed by the fathead minnow *Pimephales promelas* (Cypriniformes, Cyprinidae), the rainbow trout *Oncorhynchus mykiss* (Salmoniformes, Salmonidae), the medaka *Oryzias latipes* (Beloniformes, Adrianichthyidae) and the common carp *Cyprinus carpio* (Cypriniformes, Cyprinidae) with 517, 444, 202 and 136 studies, respectively. Thus, the remaining 868 ecotoxicological studies were done on 66 non-conventional model fish species (see Figure 4.2).

Ecotoxicological compounds in aquatic ecosystems, both sediment and water column, contaminate and affect all lifeforms.[31,32] However, sediments are considered to be a sink of ecotoxicological compounds since they present very high concentrations when compared to the column water.[33] In general, ecotoxicological compounds have several implications for fish health; genetics (18%), mortality (16%) and accumulation (16%) constitute the principal effects observed (see Figure 4.3). For instance, herbicides induce several consequences at cytogenetic and DNA level in fish. When specimens of the goldfish *Carassius auratus* (Cypriniformes, Cyprinidae) were experimentally exposed to a commercial herbicide, micronuclei, nuclear abnormalities and DNA strand breaks were observed in the fish peripheral erythrocytes.[34] Similar toxicogenic erythrocytic nuclear abnormalities were reported in the presence of metals (Al, Ba, Cr, Cu, Fe, Mn, Ti and Zn) in the pearl cichlid

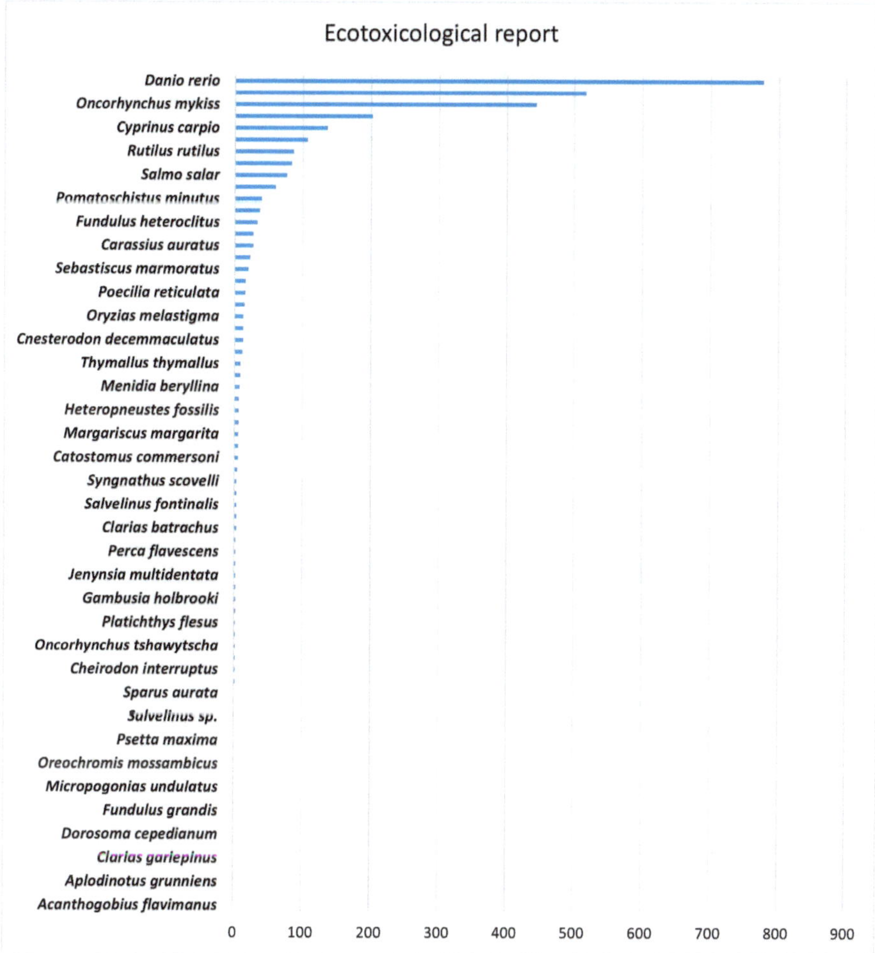

Figure 4.2 Conventional and non-conventional fish species used as a model study for ecotoxicological analyses.

Geophagus brasiliensis (Cichliformes, Cichlidae) sampled from a river in a zone with high agricultural activities.[35]

Acute mortality is another significant effect of the ecotoxicological compounds. In particular, a chemical derived from tyre rubber, 6PPD-quinone, is reported to be lethally toxic in two different salmonid species.[36,37] Additionally, mortality is a collateral effect of 17β-estradiol (E2) and diethylstilbestrol (DES), both synthetic hormones for medical and off-label purposes, which induce growth retardation, cumulative mortality and malformations in zebrafish *D. rerio* embryos.[38]

The third effect observed in fish tissues is the accumulation of ecotoxicological compounds and heavy metals. After exposure to Pb, high concentrations of this metal were detected in the Northern snakehead *Channa argus*

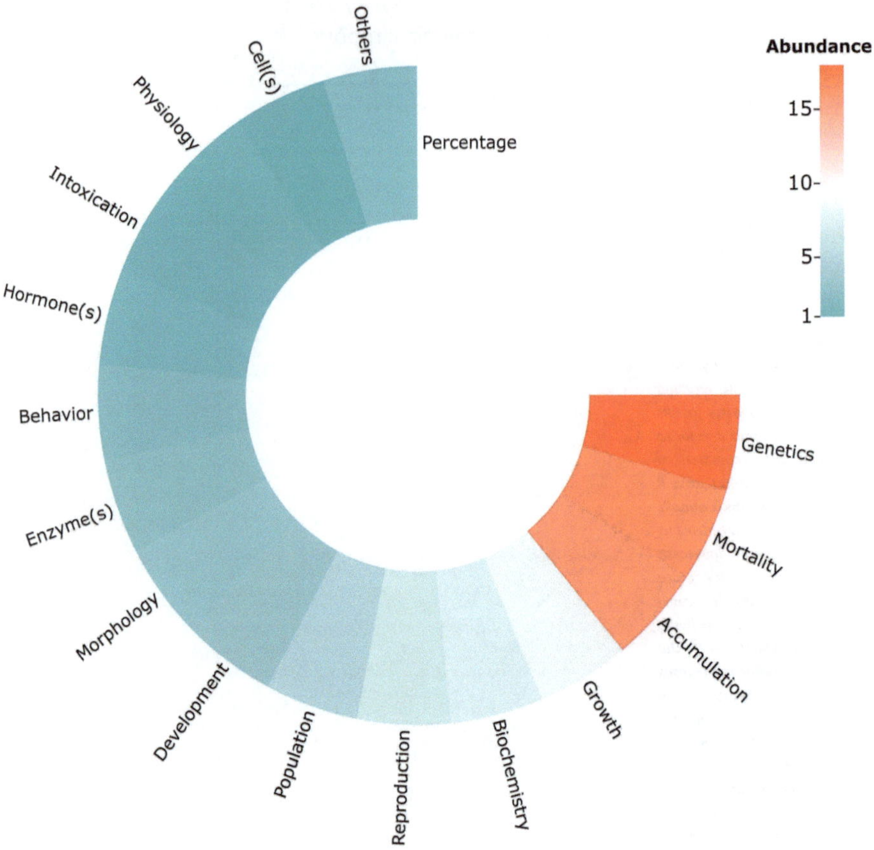

Figure 4.3 Ecotoxicological effects in fish, retrieved from the ECOTOX database.

(Anabantiformes, Channidae) gills, intestines, liver and muscles.[38] Similarly, the accumulation of Cd and other metals in the common carp *C. carpio* tissues has also been recorded.[39,40] Aquatic organisms accumulate contaminants by a bioconcentration and bioaccumulation process. Bioaccumulation refers to the process of chemical compound accumulation in an organism through various sources like water, diet and sediments, a process influenced by physical and chemical factors that affect the xenobiotics' adsorption. Additionally, biological factors like feeding habits and their trophic interaction also play a role in the process, as they affect the bioavailability of the chemicals in the environment. Thus, their concentration in organisms exceeds levels present in the water body. In the first case, fish incorporate contaminants by direct water uptake and exposure. Bioaccumulation results from chemical uptake through all possible ways of chemical exposure, *i.e.*, *via* water or foodborne, and elimination occurs from all possible routes through time.[41,42] The relative importance of food or water as a source of contaminant accumulation is highly dependent on the contaminant and organism in question.[41]

4.2.4 Fish in the Trophic Networks in the Aquatic Ecosystem

The microorganisms that inhabit the ecosystem are in constant interaction, forming a trophic network among species. Ecotoxicological adverse effects influence species' abundance patterns, diversity, community composition and species' interactions across different trophic levels.[43] In the dynamic aquatic ecosystem, fish consume plants, invertebrates or other vertebrates. At the same time, fish occupy many trophic niches, acting as prey and hosts for various species.[44,45] In the trophic network, in addition to being of great significance as food for human consumption,[46] fish have vital importance in the functionality of marine ecosystems because bioaccumulated ecotoxic compounds are transferred through biomagnification. This process involves the movement of toxic compounds from a lower to a higher trophic level, increasing chemical concentration with each trophic level step transformation in the food web. Consequently, the highest concentrations of these compounds are found in the upper trophic levels.[41,42,47]

Ecotoxicological agents such as microplastics and nanoplastics can affect the lower trophic level organisms in aquatic systems.[48] For instance, a trophic network was monitored to evaluate the trophic transfer in marine ecosystems among the microalgae *Dunaliella salina*, small crustaceans such as the branchiopod *Artemia franciscana* and small yellow croakersfish *Larimichthys polyactis* (Acanthuriformes, Sciaenidae). The results revealed that nanoplastics adsorbed in microalgae cell walls were gradually transferred to the higher trophic-level organisms (fish), and digestive α-amylase activity was inhibited.[49] Similar trophic transfers have also been demonstrated in freshwater ecosystems. For example, polyethylene (PE) microparticles were found in the gammarid Crustacea (*Echinogammarus veneris*) gut content after consuming the aquatic duckweed *Spirodela polyrhiza*.[50] Since the food web finds its direct or indirect way into the human diet, the ingestion of microplastics and their bioaccumulation in the aquatic ecosystem is a worrying issue which has only recently started to emerge.

4.2.5 Disorders in the Fish Physiology, Immune System and Gastrointestinal Microbiota

Protection against pathogens is an essential function of the immune system, in order to maintain a constant balance with the associated microbiota. The fish immune system has an innate response and an adaptive system which, acting in a coordinate fashion, is responsible for defending the host from pathogens. The innate immune system is the first defence barrier, initiated by the pattern recognition receptors, antimicrobial peptides, lectins, cytokines and phagocytes interacting to avoid pathogen proliferation in the host. Meanwhile, immunoglobulins, immunological memory and cellular reactions mediate the adaptive immune system.[51] As in all vertebrates, the fish immune system includes membrane and intracellular receptors cross-linked through diverse signalling pathways. The physiological function of this signalling network is to orchestrate immune cell differentiation and functioning as well as to integrate the immune system with other physiological systems of the organism.[52]

Recently, the aryl-hydrocarbon receptor (AhR) has been described in the fish immune system, which can mediate the effects of xenobiotics, such as halogenated aromatic and polyaromatic hydrocarbons.[52] Furthermore, the fish gut microbiota communicates with the host through small molecules and metabolites regulating immune system responses and physiological activities.[53] Unfortunately, the ecotoxic compounds alter the innate immune system, as demonstrated with benzo[*a*]pyrene in an *in vivo* assay in the zebrafish *D. rerio* larvae, in which the exposure to benzo[*a*]pyrene decreased reactive oxygen species (ROS) production and increased macrophage but not neutrophil numbers.[54] Similarly, pesticides produce ROS and induce oxidative damage in fish tissue, such as abnormalities in the liver and gills.[55]

Additionally, xenobiotics exert toxic effects on the reproductive and nervous systems. Nano and microplastics, such as polystyrene, could pass through the gonad blood barrier and accumulate in gonad tissues. These events could further affect reproductive activity by inducing apoptosis of germ cells in fish gonad tissues,[56] resulting in oxidative stress and tissue damage that disrupts the reproductive endocrine system.[57] Furthermore, xenobiotic exposure can also induce thrombosis in fish. Chemicals such as triclosan, perfluorooctanoate and triphenyl phosphate can disrupt the homeostasis of cardiolipins (CLs) and related metabolites, typical indicators of thrombosis.[58] Moreover, exposure to polycyclic aromatic hydrocarbons is involved in fish thermal behaviour. In agreement with this concept, it has been observed that tilapia *O. niloticus* exposed to benzo[*a*]pyrene showed a deficiency in identifying its optimal temperature, probably due to neurotoxicity triggered by exposure to this compound.[59]

Xenobiotics may also compromise the gastrointestinal system. In this regard, they cause intestine barrier damage and trigger pathological states associated with dysbiosis, characterized by alterations in the community structure of host-associated microbiota.[60] Subsequently, this affects the host's homeostatic interaction with the microbiota, causing inflammatory, metabolic, neurological, as well as autoimmune diseases and disorders.[61,62] Even so, xenobiotic toxicity can be reduced by the transformation carried out by the gut microbiota, as occurs in the modification of plastic additives (bisphenols and triphenyl phosphate), and has been recently reported.[63]

Overall, we may ascertain that fish exposed to microplastics, pesticides and various nanoparticles experience ROS generation, which results in oxidative stress, inflammation, immunotoxicity, genotoxicity, DNA damage and changes in gut microbiota.[64]

4.2.6 Cytotoxicity and Genotoxicity (Bioaccumulation Pathways)

Monitoring the process of absorption, distribution, biotransformation and excretion is fundamental to determining which ecotoxicological compounds are bioaccumulated and/or bioconcentrated by fish.[65] In the case of different fish species exposed to antibiotics, the highest bioaccumulation factor for ciprofloxacin was observed in the common carp *C. carpio*, while it was significantly lower in the climbing perch *Anabas testudineus* (Anabantiformes, Anabantidae) and other omnivorous fish.[66] Similarly, a chemical's affinity for specific organs and excretion rate contributes to bioaccumulation.[67] As revealed by Njinga, heavy metals vary in different organs of both fish species evaluated, the African native tilapia *T. brevimanus* (Cichliformes, Cichlidae) and the false albacore *Euthynnus alletteratus* (Scombriformes, Scombridae), with contents of As, Cd, Cr and Pb registering higher concentrations in the gills than in the muscle.[68] This organ affinity has also been reported with Cu, which presented a strong affinity for the liver, whereas Zn concentrated mostly in the gonads, all due to their primary metabolic functions in these organs.[69] Indeed, in this bioaccumulation process, it goes without saying that several cytotoxic and genotoxic effects occur in fish.

Fish possess various cytotoxic cell lineages, including, *i.e.,* nonspecific cytotoxic cells (NCC), neutrophilic granulocytes, monocyte/macrophages, cytotoxic T lymphocytes (CTL) and natural killer cells (NK). Additionally, cellular immune-type receptors (NITR) and cichlid killer leukocyte receptors (cKLR) are structurally related to mammalian NK receptors and are likely to play a role in cytotoxic function in fish.[70] A pathway for the cytotoxic effect of antibiotics has been described in the grass carp *Ctenopharyngodon idella* (Cypriniformes, Cyprinidae), where enrofloxacin produces ROS *via* apoptosis and exerts its cytotoxic effect, affecting the hepatic mitochondria.[71] Macrophages from the head kidney of the walking catfish *Clarias batrachus* (Siluriformes, Clariidae) were utilized to establish cytotoxicity characterized by head kidney hypertrophy, reduction in melano-macrophage population, increased hemosiderin accumulation and inhibition of macrophage phagocytic potential as well.[72]

On the other hand, genotoxicity is a deleterious action that affects cellular genetic material and integrity.[73] Genotoxic pollution includes several chemical compounds, *e.g.*, heavy metals,[74] herbicides[75] and polycyclic aromatic hydrocarbons (PAHs),[76] among other relevant xenobiotic compounds. These genotoxicants have been reported to cause damage not only at the chromosomal level but also directly to DNA. Micronucleus in fish erythrocytes are biomarkers of sublethal exposure to various xenobiotics, including metal and metalloid elements in aquatic organisms. Furthermore, the micronucleus assay is considered worldwide as one of the most standard methods for assessing these effects in fish blood cells.[77] For instance, DNA methylation was presented in the euryhaline fish, the inland silverside *Menidia beryllina* (Atheriniformes, Atherinopsidae), after exposure to endocrine-disrupting chemicals, and also showed transgenerational effects by the strict inheritance of DNA methylation alterations.[78] Additionally, bisphenol A (BPA) exposure in the viviparous blackfin goodea fish *Goodea atripinnis* (Cyprinodontiformes, Goodeidae) resulted in *Foxl2* overexpression, which participates in the differentiation of the ovary and maintenance of ovarian functions in mammals and fish.[79] Multiple studies assess these xenobiotics' genotoxic effects. However, the consequences and implications of genotoxicant exposure in fish need further research.

4.2.7 Dysbiosis Mechanism Caused by Xenobiotics

It has become clear that exposing fish to xenobiotic agents such as pesticides, heavy metals, microplastics and antibiotics disrupts their gut microbiota diversity. The degree of impact on the microbiota is proportional to the developmental toxicity of the agent.[60]

The interaction between xenobiotics and microbiota occurs mainly in the host's gastrointestinal tract. This tract provides different habitats for microorganisms, which are influenced by the physiology of epithelial cells, the pH level and the nutrient content as well. These variations affect the metabolic processes that microorganisms offer to the host. As a result, the interaction between microbiota and xenobiotics is highly dependent on the microbial composition and the metabolites they secrete.[80,81]

The gut microbiota can interact with xenobiotics before or after the host absorbs them. When consumed by the host, xenobiotics are processed by digestive enzymes before being absorbed. After this step, they pass through the intestinal epithelial cells, where digestive enzymes can modify them before being transported to the liver. In these circumstances, xenobiotics and their metabolites enter the systemic circulation, potentially damaging other tissues and affecting distal organs.[82] However, microorganisms in the gut microbiota perform various metabolic reactions such as xenobiotic reductions, hydrolysis, glucuronidation, lyase reactions and nucleophilic substitutions.[60]

In bacteria, xenobiotics can either increase or decrease growth, alter the metabolism of native bacteria, get metabolized, accumulate or affect its virulence.[81] These interactions are unique to each individual, the specific xenobiotic and the host's health status. Finally, other factors that can influence these interactions include the time and dosage of exposure to the xenobiotic and other extrinsic factors.

4.3 Implications of Xenobiotics in the Fish Gut Microbiota

4.3.1 Microbiota in Fish Development

Aquatic animals depend on the surrounding environment to fulfil their vital functions; this includes interaction with a high abundance of microorganisms such as viruses and bacteria.[83] These interactions could be mutualistic, commensal, symbiotic or pathogenic.[84] The mechanism of microorganism colonization in the fish intestine starts even before the fish begins feeding.[85] Other studies suggest that microorganisms colonize their host after hatching, and the microbiota composition is determined by genetics, environment, nutrition and trophic level.[86]

A healthy microbiota community is commonly represented by high taxonomic diversity, high microbial gene richness and stable core microbiota.[87] Mainly, the fish gut microbiota includes specimens of different species belonging to the phyla Proteobacteria (major phylum of Gram-negative bacteria), Firmicutes (with most species having Gram-positive cell wall structure), Fusobacteria (obligately anaerobic non-spore forming Gram-negative bacilli), Cyanobacteria (a phylum of Gram-negative bacteria that obtain energy *via* photosynthesis), Planctomycetes (which play a considerable role in global C and N cycles, with many species of this phylum capable of anaerobic ammonium oxidation, also known as anammox), Verrucomicrobia (phylum of Gram-negative bacteria), Bacteroidetes (composed of three large classes of Gram-negative, non-spore-forming, anaerobic or aerobic spp.), Actinobacteria (diverse phylum of Gram-positive bacteria with high G + C content), Crenarchaeota (which indeed are prokaryotes that have been classified as a phylum of the Archaea domain) and, finally, Acidobacteria (a phylum of Gram-negative bacteria).[86]

The gut microbiota and their host engage in a crucial mutualistic interaction. This interaction involves the production of essential metabolites, vitamins, amino and fatty acids that help regulate the host's homeostasis and, consequently, develop the immune system. In turn, the host's immune system also influences the microbiota composition. Furthermore, fish gut microbiota play a crucial role in nutrition and health.[88] Nutrition (or diet composition) is perhaps the most influential extrinsic factor in the gut microbiota of fish. For instance, predatory fish tend to have a higher abundance of halophilic bacteria, Gram-negative *Halomonas* (Proteobacteria) and the Gram-negative *Cetobacterium* (Fusobacteria), while herbivorous fish have a higher abundance of the Gram-positive bacteria *Clostridium* (Bacillota), Gram-negative coliform bacteria *Citrobacter* (Pseudomonadota) and the Gram-negative, non-spore-forming and non-motile bacteria *Leptotrichia* (Firmicutes).[86]

The studies conducted in gnotobiotic zebrafish *D. rerio* models have concluded that the gut microbiota plays a crucial role in the developmental process of the host. These findings were described using gnotobiotic zebrafish larvae, revealing that the microbiota plays a vital role in the whole-body host gene expression related to the signalling of the microbial pattern recognition by the immune system. Besides, the fish microbiota can modulate the expression of six cytochrome P450 monooxygenase genes associated with xenobiotic metabolism.[89] Moreover, fish health is highly linked to the gut's endogenous bacteria. For example, when rainbow trout *O. mykiss* larvae were infected with the fish pathogen *Flavobacterium columnare* (Bacteroidetes), a thin Gram-negative bacteria, compared with the germ-free rainbow trout larvae, those with endogenous microbiota were less susceptible to the infection.[90]

The availability of omics sciences sequencing technology and bioinformatics tools has increased the number of studies carried out and comprehension of the fish microbiota composition, since less than 0.1% of the microorganisms can be cultured and cannot be mass detected through traditional Gram test morphological approaches, their dependence on oxygen or any other conventional non-molecular method. In this regard, novel information about the network interaction between the gut microbiota and fish development is still expected to be revealed in the coming years.[91]

4.3.2 Microbial Disruption Causing Dysbiosis

The microbiota in the gastrointestinal tracts of fish plays a significant role in their disease resistance. Any disruption of this microbial structure could cause pathogenic bacteria to proliferate, leading to tissue damage and weakened epithelial barriers, consequently enabling pathogens and toxic agents to infiltrate the blood stream, causing further harm.[85] The gut microbiota of fish is a dynamic and complex ecosystem, usually in a state of equilibrium known as eubiosis, while dysbiosis is considered when this equilibrium is disrupted.[51] However, dysbiosis is commonly used, even in healthy individuals. A healthy microbiota is complex to define, mainly because of the high variability among healthy individuals.[87] Thus, dysbiosis is defined as a microbial community state that shows statistical association with disease and contributes functionally to its etiology, diagnosis and/or treatment.[92]

Three main conditions could induce dysbiosis of the gut microbiota. The first one is the bloom of pathogenic bacteria or pathobionts; the second is the loss of commensals or the regular members of the microbiota, and finally, the loss of diversity that can be measured through alpha diversity.[92] Nevertheless, pathogenic microorganisms have developed strategies to evade the immune system and exploit microbiota metabolites. This leads to dysbiosis as they change the environmental conditions in the gastrointestinal tract.[93]

In the same way, a state of dysbiosis could be induced by the action of anthropogenic factors, *e.g.*, the use of chemical additives, antibiotics or environmental pollution. The increase in population and urbanization has led to more significant water contamination from human activities such as agriculture, aquaculture, sewage, municipal waste and industry, thus increasing the amount of toxicants transferred to aquatic environments.[1]

4.3.3 Ecotoxicological Agents Influence Microbiota of Conventional and Non-conventional Fish Models

Fish are the most diverse and extensive group of vertebrates, surpassing any terrestrial animal. They have adapted to a wide range of environments, making them ideal for research in embryology, neurobiology, endocrinology, environmental biology and many other disciplines.[94] Conventional models include those species capable of quickly adapting to laboratory conditions. On the other hand, non-conventional model organisms are typically defined as any model that deviates from the typical laboratory species; some of these non-conventional models are adapted to unique environments.[95]

For instance, to assess water pollution induced by the presence of polycyclic aromatic hydrocarbons in freshwater, marine and estuary environments, the fish "tambaqui" *Colossoma macropomum* (Characiformes, Serrasalmidae) (see Figure 4.4) and the long-snouted seahorse *Hippocampus reidi* (Syngnathiformes, Sygnathidae) (see Figure 4.5) can be considered as valid non-conventional experimental models for the Orinoco and Amazon River basins.[96] Studies address the possibility that both fish species can be considered as valid bioindicators for exposure to polycyclic aromatic hydrocarbons.[96]

Figure 4.4 Non-conventional model fish "tambaqui" *Colossoma macropomum* (Characiformes, Serrasalmidae) proposed to assess water pollution induced by the presence of polycyclic aromatic hydrocarbons in the Orinoco and Amazon River basins.

Figure 4.5 Non-conventional model fish long-snouted seahorse *Hippocampus reidi* (Syngnathiformes, Sygnathidae) proposed to assess water pollution induced by the presence of polycyclic aromatic hydrocarbons in subtropical marine environments.

However, during xenobiotic exposure, the viability and function of the microbiota are disrupted.[97] Antibiotics, pesticides, heavy metals, nanoparticles, microplastics, polychlorinated biphenyls (PCBs), polybromodiphenylethers (PBDEs), pharmaceutical and personal care products (PPCPs), among others, constitute a significant concern for gut microbiota because they tend to accumulate and cause damage to the health of aquatic organisms.[98] Unfortunately, microbiota studies from non-conventional fish models after pollutant exposure are not fully understood yet.

As previously mentioned, the toxicity of pharmaceuticals and personal care products (PPCPs) is related to the stage and type of organisms, time of exposure, temperature and contaminant concentration.[99,100]

Another group of agents to be included in the top-ranking toxicants are antibiotics. They are among the most studied agents proven in several mammal and fish models to induce dysbiosis.[101] Antibiotics can be directly discharged into the aquatic environments due to inappropriate aquaculture and farming practices. This group of substances are used in aquaculture to control bacterial diseases, but also, at low concentrations, as growth promoters and in animal husbandry as well.[102] However, antibiotics target not only pathogenic bacteria but also other symbiotic microbes[103] and can be applied through the feed or directly into the water of ponds to decrease the impact of bacterial infection. Thus, antibiotics can find their way to aquatic environments through water exchange of ponds, simultaneously affecting the local ecosystems, including wild fish.[101,102,104,105] In this regard, antibiotics that have been used for decades, such as amoxicillin, ciprofloxacin, erythromycin, clarithromycin and azithromycin were tagged as potential polluters by the European Commission of the European Union (EU), highlighting the absence of regulations and the danger posed to ecosystems.[106] Quantifying antibiotic residues is crucial for assessing antibiotics' biological effects and ecotoxicological impact, which are considered to be emerging pollutants in aquatic environments. This is particularly important because certain antibiotics with lower bioavailability can remain in the gastrointestinal tract for more extended periods, resulting in a more significant impact on the microbiota (see Table 4.1).[107]

Conversely, microplastics are of high concern because they are not easily degraded and accumulate in animal tissues, causing cytotoxic and neurotoxic effects.[108,109] Plastics are classified into two groups: primary microplastics, which are small particles produced by personal care products and cleaning products that get transported through sewage systems, and secondary microplastics, which are products of the breakdown or degradation of larger plastic pieces that degrade in the coastal or marine ecosystems.[110] Plastic particles that measure less than 5 mm are considered microplastics making them easy for fish to consume.

Pesticides are commonly used in agricultural and domestic activities. Herbicides, insecticides and fungicides can be detected in food, water and aquatic environments, affecting the host's gut microbiota.[97] Due to the ecotoxic effect of pesticides on the aquatic environment, the safety and quality

Table 4.1 Ecotoxicology agents affecting the fish gut microbiota.

Ecotoxicology agents	Fish species	Microbiota effect	Author
Antibiotics			
Doxycycline (DH), oxytetracycline (OTCC) and florfenicol (FF)	Zebrafish *Danio rerio*	Exposure to antibiotics caused decrease of diversity of the gut microbiota and dysbiosis	Qian *et al.*[105]
Oxytetracycline (OTC) and amoxicillin (AMX)	Olive flounder *Paralichthys olivaceus*	Reduction of diversity and dysbiosis.	Kim *et al.*[101]
Growth promoter antibiotic, olaquindox (OLA)	Zebrafish *Danio rerio*	Gut microbiota dysbiosis	He *et al.*[102]
Pesticides			
Chlorpyrifos (CPF)	Zebrafish *Danio rerio*	Hepatic lipid metabolism disorders, associated with gut oxidative stress and microbiota dysbiosis	Wang *et al.*[113]
Chlorantraniliprole (CAP)	Crucian carp *Carassius carassius*	Oxidative stress and gut microbiota dysbiosis	Yin *et al.*[111]
Diazon	Prussian carp *Carassius gibelio*	Histopathological damage in gill and liver, oxidative stress in liver, and gut microbiota dysbiosis	Tang *et al.*[112]
Heavy metals			
Cr, Cd and Cu mixture	Common carp *Cyprinus carpio*	Toxicity and gut microbiota dysbiosis	Kakade[137]
Cd, Pb and Zn	Marine medaka *Oryzias melastigma*	Perturbation of the gut microbiota homeostasis and reproductive disturbance	Yan *et al.*[115]
Class 1 integron (intI-1): Zn, Pd, Cu and Cd	Eastern stripped grunter *Pelates sexlineatus*	The ability of fish to resist contaminated environments may be influenced by long-term exposure to chronic metal contamination in their gut microbiome	Suzzi *et al.*[116]
PCBs and PBDEs			
Twenty-eight persistent organic pollutants	European eels *Anguilla anguilla*	Accumulation of xenobiotics in fish muscle and gut microbiota dysbiosis	Bertucci[138]
Microplastics			
Polyethylene plastic powder	Indian mackerel *Rastrelliger kanagurta*	Accumulation of microplastics (0.01 mm) in the fish gut	Rajendran *et al.*[108]
Polystyrene microplastics	Grass carp *Ctenopharyn godonidella*	Growth decrease and intestinal damage	Hao *et al.*[109]
Polystyrene	Marine medaka *Oryzias melastigma*	Liver injury and gut microbiota dysbiosis	Zhang *et al.*[110]

of products from such environments used for foodstuffs must be evaluated carefully. Several pesticides are directly used in aquatic systems, like rice-fish-culture systems, reaching higher toxic effects on fish. Therefore, exposure to pesticides induces gut microbiota dysbiosis, while at the same time triggering oxidative stress responses, a bioindicator of the toxic effects that can be measured through the levels of oxidative products such as Cu/Zu-SOD, GST, CYP1A1 and CYP2B19.[111–113]

Heavy metals are non-biodegradable and tend to accumulate, making them major contaminants in aquatic environments and, in most cases, are a consequence of anthropogenic activities.[114] Additionally, heavy metals have been associated with gut microbiota disruption, altering metabolic and physiological processes.[115,116] Several heavy metals like Pb, Hg, Cd, As and Cr(VI) are listed as carcinogens by the IARC (International Agency for Research of Cancer, WHO, UN).[117] For instance, exposure of the common carp *C. carpio* to 1 mg Pb per mL caused histopathological changes in intestinal tissues, including epithelial cell necrosis. Additionally, changes in the microbiota structure were reported, with a significant decrease in the phyla Proteobacteria, Actinobacteria and Planctomycetes, whereas an increase in Fusobacteria, Firmicutes and Bacteroidetes were detected when compared to the non-exposed control group.[118] On the contrary, Cu exposure caused a decrease in Firmicutes and Bacteroidetes phyla in the common carp *C. carpio* gut microbiota.[119] Likewise, a significant reduction of Firmicutes phylum in the Nile tilapia *O. niloticus* gut microbiota is observed after Cu exposure in water. At the same time, Cu bioaccumulation was observed in the fish liver.[120]

Polychlorinated biphenyls (PCBs) are industrial chemicals that contain chlorine and are commonly used in electrical equipment, hydraulic fluids, heat transfer fluids, lubricants, plasticizers, surface coatings and inks.[121] Polychlorinated biphenyls accumulate in adipose tissue and induce toxicity due to their high lipophilicity and persistence.[122] Fish exposure to PCBs most possibly results in impairment of physiology, reproductive dysfunction, genetic disorders, abnormal development and neurotoxicity;[123,124] particularly, congestion of red blood cells, lymphocyte infiltration and increase of the mast cells in the gills and intestines of the gilt-head seabream *Sparus aurata* (Spariformes, Sparidae) have been reported.[125] Additionally, the effects of PCB chemicals on the zebrafish *D. rerio* microbiota varied considerably only with increasing exposure time but not with rising PCBs' concentrations, which resulted in a decrease in the population of Bacillota bacteria.[126]

4.4 Non-conventional Fish Gut Microbiota as a Biomarker for Environmental Contamination

Microbial biomarkers may also prove helpful for identifying xenobiotic toxicity. In the case of the identification of polycyclic aromatic hydrocarbon (PAH), specific bacterial taxa can be used as indicators to monitor environmental and fish health status. This is the case of the fish barramundi *Lates calcarifer* (Perciformes, Latidae), which was exposed to crude oil, among other toxicants, and the results showed that the genus *Photobacterium*, belonging to the phylum Pseudomonadota, proliferated proportionally to the PHA concentration.[127] Similarly, the microbiota signatures of the Atlantic cod *Gadus morhua* (Gadiformes, Gadidae) are associated with PAH contaminants and certain bacterial taxa can be used as indicators to monitor environmental and fish health status.[128] Moreover, the viviparous blackfin goodea fish *Goodea atripinnis* (Cyprinodontiformes, Goodeidae), endemic to the Central Mexican plateau, was proposed as a potential non-conventional model for assessing pollutant effects, specifically bisphenol (BPA) exposure.[79] In spite of the fact that non-conventional models are being mainly employed as surrogates for wildlife, or as indicators of ecosystem status in effects-oriented studies, there are still scarce studies available and further research is required for microbial composition practical application as a biomarker of the effects of ecotoxic agents.[129]

4.4.1 Loss in Microbial Diversity and Affected Taxa by Ecotoxicological Factors

It is a well-known fact that antibiotics inhibit bacterial growth and pathogen proliferation, selecting non-resistant strains. Simultaneously, they also reduce the diversity of non-target bacteria, which can lead to the expansion of opportunistic pathogens.[130] Antibiotics like erythromycin and ampicillin, administered to the Korean native ricefish, also known as medaka, *O. latipes*, increased the proportion of phylum Proteobacteria, especially *Aeromonas veronii*, while the phylum Fusobacteria decreased.[131] In the same way, the phylum Proteobacteria is also affected by fish exposure to the pesticide chlorpyrifos.[113] The antibiotic enrofloxacin is commonly detected in aquatic environments, causing an imbalance of *Vibrio* and *Flavobacteria* genera, resulting in a negative physiological impact for the marine medaka fish *O. melastigma*.[132]

Fish exposed to Cu showed a significant decrease in bacteria that produce short fatty acids, such as several belonging to the phylum Bacillota. Among them, Allobaculum, Blautia, Coprococcus, Faecali bacterium, Roseburia and Ruminococcus can be included. On the other hand, the study found an increase in pathogen-related bacteria from the phylum Pseudomonadota, like *Pseudomonas*, a genus of Gram-negative bacteria belonging to the family Pseudomonadaceae and *Acinetobacter*, a genus of Gram-negative bacteria belonging to the wider class of Gammaproteobacteria.[119] Among aquatic organisms, exposure to contaminants affects the phylum Proteobacteria the most, while the phyla Bacteroidetes and Firmicutes may respond differently.[98]

4.4.2 Discussion and Further Steps in the Implementation and Use of a Dysbiosis Index as a Reference in Fish Microbial Disruptions

Having defined dysbiosis, the magnitude of this imbalance can be measured by estimating the loss of richness and/or diversity applying the Microbiota Dysbiosis index (MD-index). However, because of individual significant variation of the gut microbiota among individuals, the definition of a "normal" or "healthy" microbiota is still inconclusive and challenging.[133]

The MD-index intends to measure the degree of dysbiosis in a microbiome. It is calculated as the logarithm of the ratio of the total abundance of organisms that have increased in the infected group over the total abundance of organisms that have decreased in the infected group when compared to the control group. The resulting value provides an empirical estimation of the degree of dysbiosis.[134,135] It is also a valuable tool to measure the healthy balance of the gut intestinal microbiota. An interesting example of the use of the MD-index as a dysbiosis indicator was observed during an experimental challenge in tilapia performed by our research group. Fish were inoculated with the bacterial pathogen *Nocardia* sp.; therefore, when comparing stomach bacterial communities between infected and non-infected tilapia, the MD-index showed a value of 2.63, which is related to the fact that the phylum Proteobacteria represented 94.8% of the stomach bacterial abundance in the infected tilapia, meanwhile, the non-infected group reported 58% of phylum abundance.[136]

The *caveat* in the definition of a "healthy" gut microbiota state is still prone to be controversial and will not be easily defined, since not all forms of bacterial imbalance cause evidential diseases. In order to determine whether the variation of the gut microbiota is a pertinent dysbiotic change or just a mere fluctuation within the normal range, it is necessary to complement this index with other appropriate biological and physiological indicators in order to have a realistic and holistic approach to the matter. It certainly is an interesting and promising avenue to be explored. Much work remains to be done in this sense.

Acknowledgements

The authors appreciate the laboratory members and their invaluable contributions, especially Azucena Santacruz.

References

1. D. Bukola, A. Zaid, E. Olalekan and A. Falilu, Consequences of anthropogenic activities on fish and the aquatic environment, *Poult. Fish. Wildl. Sci.*, 2015, **3**, 1–12.
2. C. Amiard-Triquet, in *Aquatic Ecotoxicology, Introduction*, ed. C. Amiard-Triquet, J. C. Amiard, and C. Mouneyrac, Academic Press, U.S.A., 2015, pp. 1–23.

3. D. Gouveia, C. Almunia, Y. Cogne, O. Pible, D. Degli-Esposti, A. Salvador, S. Cristobal, D. Sheehan, A. Chaumot, O. Geffard and J. Armengaud, Ecotoxicoproteomics: A decade of progress in our understanding of anthropogenic impact on the environment, *J. Proteomics*, 2019, **198**, 66–77.
4. K. Kobetičová and R. Černý, Ecotoxicology of building materials: A critical review of recent studies, *J. Cleaner Prod.*, 2017, **165**, 500–508.
5. W. H. Tolleson, Mechanistic ecotoxicology and environmental toxicology, *J. Environ. Sci. Health, Part C: Environ. Carcinog. Ecotoxicol. Rev.*, 2018, **36**, 164–166.
6. U. S. EPA, United States Environmental Protection Agency, Toxic substances control act (tsca) chemical substance inventory, Non-Confidential TSCA inventory, 2023, available from: https://www.epa.gov/tsca-inventory, accessed December 12, 2023.
7. J. H. Olker, C. M. Elonen, A. Pilli, A. Anderson, B. Kinziger, S. Erickson, M. Skopinski, A. Pomplun, C. A. LaLone, C. L. Russom and D. Hoff, The ECOTOXicology knowledgebase: A curated database of ecologically relevant toxicity tests to support environmental research and risk assessment, *Environ. Toxicol. Chem.*, 2022, **41**, 1520–1539.
8. T. P. Traas and C. J. van Leeuwen, in *Risk Assessment of Chemicals: An Introduction*, ed. C. J. V. Leeuwen, and T. G. Vermeire, Springer, The Netherlands, Dordrecht, 2007, pp. 281–356.
9. K. Fent, Ecotoxicological effects at contaminated sites, *Toxicology*, 2004, **205**, 223–240.
10. T. M. Melo, M. Schauerte, A. Bluhm, M. Slaný, M. Paller, N. Bolan, J. Bosch, A. Fritzsche and J. Rinklebe, Ecotoxicological effects of per- and polyfluoroalkyl substances (PFAS) and of a new PFAS adsorbing organoclay to immobilize PFAS in soils on earthworms and plants, *J. Hazard. Mater.*, 2022, **433**, 128771.
11. F. Liu, G.-G. Ying, L.-H. Yang and Q.-X. Zhou, Terrestrial ecotoxicological effects of the antimicrobial agent triclosan, *Ecotoxicol. Environ. Saf.*, 2009, **72**, 86–92.
12. Y. Zhou, X. Liu and J. Wang, Ecotoxicological effects of microplastics and cadmium on the earthworm *Eisenia foetida*, *J. Hazard. Mater.*, 2020, **392**, 122273.
13. A. A. Al-Badran, M. Fujiwara and M. A. Mora, Effects of insecticides, fipronil and imidacloprid, on the growth, survival, and behavior of brown shrimp *Farfantepenaeus aztecus*, *PLoS One*, 2019, **14**, e0223641.
14. F. Bhunia, N. C. Saha and A. Kaviraj, Effects of Aniline–An Aromatic amine to some freshwater organisms, *Ecotoxicology*, 2003, **12**, 397–404.
15. M. B. Hussain, M. Laabir and M. N. Daly Yahia, A novel index based on planktonic copepod reproductive traits as a tool for marine ecotoxicology studies, *Sci. Total Environ.*, 2020, **727**, 138621.
16. Y. Hu, D. Li, X. Ma, R. Liu, Y. Qi, C. Yuan and D. Huang, Effects of 2,4-dichlorophenol exposure on zebrafish: Implications for the sex hormone synthesis, *Aquat. Toxicol.*, 2021, **236**, 105868.
17. B. Huang, B. Wang, D. Ren, W. Jin, J. Liu, J. Peng and X. Pan, Occurrence, removal and bioaccumulation of steroid estrogens in Dianchi Lake catchment, China, *Environ. Int.*, 2013, **59**, 262–273.

18. K. T. Peter, Z. Tian, C. Wu, P. Lin, S. White, B. Du, J. K. McIntyre, N. L. Scholz and E. P. Kolodziej, Using high-resolution mass spectrometry to identify organic contaminants linked to urban stormwater mortality syndrome in coho salmon, *Environ. Sci. Technol.*, 2018, **52**, 10317–10327.
19. H. Cui, X. Zhu, Y. Zhu, Y. Huang and B. Chen, Ecotoxicological effects of DBPs on freshwater phytoplankton communities in co-culture systems, *J. Hazard. Mater.*, 2022, **421**, 126679.
20. F. T. Weiss, C. Ruepert, S. Echeverría-Sáenz, R. I. L. Eggen and C. Stamm, Agricultural pesticides pose a continuous ecotoxicological risk to aquatic organisms in a tropical horticulture catchment, *Environ. Adv.*, 2023, **11**, 100339.
21. L. Jiang, Y. Li, Y. Chen, B. Yao, X. Chen, Y. Yu, J. Yang and Y. Zhou, Pharmaceuticals and personal care products (PPCPs) in the aquatic environment: Biotoxicity, determination and electrochemical treatment, *J. Cleaner Prod.*, 2023, **388**, 135923.
22. R. Beiras and A. M. Schönemann, Water quality criteria for selected pharmaceuticals and personal care products for the protection of marine ecosystems, *Sci. Total Environ.*, 2021, **758**, 143589.
23. J.-L. Liu and M.-H. Wong, Pharmaceuticals and personal care products (PPCPs): A review on environmental contamination in China, *Environ. Int.*, 2013, **59**, 208–224.
24. A. S. Mortensen and A. Arukwe, Effects of 17α-ethynylestradiol on hormonal responses and xenobiotic biotransformation system of Atlantic salmon (*Salmo salar*), *Aquat. Toxicol.*, 2007, **85**, 113–123.
25. R. F. Salla, F. U. Gamero, R. Z. Rissoli, S. E. Dal-Medico, L. M. Castanho, C. d. S. Carvalho, E. C. M. Silva-Zacarin, A. L. Kalinin, F. C. Abdalla and M. J. Costa, Impact of an environmental relevant concentration of 17α-ethinylestradiol on the cardiac function of bullfrog tadpoles, *Chemosphere*, 2016, **144**, 1862–1868.
26. A. M. Razekenari, A. E. Fereidouni, A. Movahedinia and E. Z. Neyshabouri, Impacts of sublethal concentrations of 17 α-ethinylestradiol (EE2) on growth, reproductive performance, and survival in red cherry shrimp *Neocaridina davidi* (Crustacea, Atyidae) during consecutive spawnings, *Aquat. Toxicol.*, 2023, **259**, 106519.
27. A. W. Griffith and C. J. Gobler, Harmful algal blooms: A climate change co-stressor in marine and freshwater ecosystems, *Harmful Algae*, 2020, **91**, 101590.
28. J. Sha, H. Xiong, C. Li, Z. Lu, J. Zhang, H. Zhong, W. Zhang and B. Yan, Harmful algal blooms and their eco-environmental indication, *Chemosphere*, 2021, **274**, 129912.
29. M. N. Haque, S.-E. Nam, Y.-S. Han, H. S. Park and J.-S. Rhee, Chronic exposure to sublethal concentrations of saxitoxin reduces antioxidant activity and immunity in zebrafish but does not affect reproductive parameters, *Aquat. Toxicol.*, 2022, **243**, 106070.

30. V. G. Christensen and E. Khan, Freshwater neurotoxins and concerns for human, animal, and ecosystem health: A review of anatoxin-a and saxitoxin, *Sci. Total Environ.*, 2020, **736**, 139515.
31. B. Yüksel, F. Ustaoğlu, C. Tokatli and M. S. Islam, Ecotoxicological risk assessment for sediments of Çavuşlu stream in Giresun, Turkey: association between garbage disposal facility and metallic accumulation, *Environ. Sci. Pollut. Res.*, 2022, **29**, 17223–17240.
32. D. H. Youssef and F. T. Tayel, Metal accumulation by three *Tilapia* spp. from some Egyptian inland waters, *Chem. Ecol.*, 2004, **20**, 61–71.
33. A. Ndhlovu, L. R. D. Human, J. B. Adams, G. M. Rishworth, C. Olisah and T. G. Bornman, Ecological risk assessment of metal pollutants in two agriculturally impacted estuaries, *Mar. Pollut. Bull.*, 2023, **195**, 115572.
34. T. Çavaş and S. Könen, Detection of cytogenetic and DNA damage in peripheral erythrocytes of goldfish (*Carassius auratus*) exposed to a glyphosate formulation using the micronucleus test and the comet assay, *Mutagenesis*, 2007, **22**, 263–268.
35. D. Lacerda, C. d. S. Vergilio, T. da Silva Souza, L. H. Viana Costa, T. P. Rangel, B. C. Vaz de Oliveira, D. Q. Ribeiro de Almeida, I. A. Pestana, M. Gomes de Almeida and C. E. de Rezende, Comparative metal accumulation and toxicogenetic damage induction in three neotropical fish species with distinct foraging habits and feeding preferences, *Ecotoxicol. Environ. Saf.*, 2020, **195**, 110449.
36. K. Hiki and H. Yamamoto, The tire-derived chemical 6PPD-quinone is lethally toxic to the white-spotted char salvelinus leucomaenis pluvius but not to two other salmonid species, *Environ. Sci. Technol. Lett.*, 2022, **9**, 1050–1055.
37. Z. Tian, H. Zhao, K. T. Peter, M. Gonzalez, J. Wetzel, C. Wu, X. Hu, J. Prat, E. Mudrock, R. Hettinger, A. E. Cortina, R. G. Biswas, F. V. C. Kock, R. Soong, A. Jenne, B. Du, F. Hou, H. He, R. Lundeen, A. Gilbreath, R. Sutton, N. L. Scholz, J. W. Davis, M. C. Dodd, A. Simpson, J. K. McIntyre and E. P. Kolodziej, A ubiquitous tire rubber–derived chemical induces acute mortality in coho salmon, *Science*, 2021, **371**, 185–189.
38. A. H. B. Adam, L. H. J. de Haan, J. Louisse, I. M. C. M. Rietjens and L. Kamelia, Assessment of the in vitro developmental toxicity of diethylstilbestrol and estradiol in the zebrafish embryotoxicity test, *Toxicol. In Vitro*, 2021, **72**, 105088.
39. L. Zhao, Y.-G. Zheng, Y.-H. Feng, M.-Y. Li, G.-Q. Wang and Y.-F. Ma, Toxic effects of waterborne lead (Pb) on bioaccumulation, serum biochemistry, oxidative stress and heat shock protein-related genes expression in *Channa argus*, *Chemosphere*, 2020, **261**, 127714.
40. L. Bervoets, K. Van Campenhout, H. Reynders, D. Knapen, A. Covaci and R. Blust, Bioaccumulation of micropollutants and biomarker responses in caged carp (*Cyprinus carpio*), *Ecotoxicol. Environ. Saf.*, 2009, **72**, 720–728.

41. M. J. Kainz and A. T. Fisk, in *Lipids in Aquatic Ecosystems*, ed. M. Kainz, M. T. Brett, and M. T. Arts, Springer New York, New York, NY, 2009, pp. 93–114.
42. F. Gobas and H. A. Morrison, in *Handbook of Property Estimation Methods for Chemicals*, ed. D. Mackay, and R. S. Boethling, CRC Press, Boca Raton, FL, USA, 2000, pp. 189–231.
43. W. H. Clements and J. R. Rohr, Community responses to contaminants: Using basic ecological principles to predict ecotoxicological effects, *Environ. Toxicol. Chem.*, 2009, **28**, 1789–1800.
44. M. Traugott, B. Thalinger, C. Wallinger and D. Sint, Fish as predators and prey: DNA-based assessment of their role in food webs, *J. Fish Biol.*, 2021, **98**, 367–382.
45. A. Ramírez-García, E. Jeppesen, R. Moncayo-Estrada, N. Mercado-Silva and O. Domínguez-Domínguez, Diet and trophic structure of the fish community in a small sub-tropical lake in central Mexico, *Water*, 2023, **15**, 1301.
46. D. C. Dioses-Salinas, B. L. Pérez-Baca and G. E. De-la-Torre, Ecotoxicological effects of microplastics and adsorbed contaminants on aquatic organisms, *Manglar*, 2019, **16**, 173–182.
47. L. Córdoba-Tovar, J. Marrugo-Negrete, P. R. Barón and S. Díez, Drivers of biomagnification of Hg, As and Se in aquatic food webs: A review, *Environ. Res.*, 2022, **204**, 112226.
48. A. Bour, J. Sturve, J. Höjesjö and B. C. Almroth, Microplastic vector effects: are fish at risk when exposed via the trophic chain?, *Front. Environ. Sci.*, 2020, **8**, 90.
49. L. Kim, R. Cui, J. Il Kwak and Y.-J. An, Trophic transfer of nanoplastics through a microalgae–crustacean–small yellow croaker food chain: Inhibition of digestive enzyme activity in fish, *J. Hazard. Mater.*, 2022, **440**, 129715.
50. V. Iannilli, L. Passatore, S. Carloni, F. Lecce, G. Sciacca, M. Zacchini and F. Pietrini, Microplastic toxicity and trophic transfer in freshwater organisms: ecotoxicological and genotoxic assessment in *Spirodela polyrhiza* (L.) Schleid. and *Echinogammarus veneris* (Heller, 1865) treated with polyethylene microparticles, *Water*, 2023, **15**, 921.
51. D. Medina-Félix, E. Garibay-Valdez, F. Vargas-Albores and M. Martínez-Porchas, Fish disease and intestinal microbiota: A close and indivisible relationship, *Rev. Aquacult.*, 2023, **15**, 820–839.
52. H. Segner, C. Bailey, C. Tafalla and J. Bo, Immunotoxicity of xenobiotics in fish: A role for the aryl hydrocarbon receptor (AhR)?, *Int. J. Mol. Sci.*, 2021, **22**, 9460.
53. G. Wang, S. Huang, Y. Wang, S. Cai, H. Yu, H. Liu, X. Zeng, G. Zhang and S. Qiao, Bridging intestinal immunity and gut microbiota by metabolites, *Cell. Mol. Life Sci.*, 2019, **76**, 3917–3937.
54. D. W. Phelps, A. A. Fletcher, I. Rodriguez-Nunez, M. R. Balik-Meisner, D. A. Tokarz, D. M. Reif, D. R. Germolec and J. A. Yoder, *In vivo* assessment of respiratory burst inhibition by xenobiotic exposure using larval zebrafish, *J. Immunotoxicol.*, 2020, **17**, 94–104.

55. U. Subaramaniyam, R. S. Allimuthu, S. Vappu, D. Ramalingam, R. Balan, B. Paital, N. Panda, P. K. Rath, N. Ramalingam and D. K. Sahoo, Effects of microplastics, pesticides and nano-materials on fish health, oxidative stress and antioxidant defense mechanism, *Front. Physiol.*, 2023, **14**, 1217666.
56. S. Sarasamma, G. Audira, P. Siregar, N. Malhotra, Y.-H. Lai, S.-T. Liang, J.-R. Chen, K. H.-C. Chen and C.-D. Hsiao, Nanoplastics cause neurobehavioral impairments, reproductive and oxidative damages, and biomarker responses in zebrafish: Throwing up alarms of wide spread health risk of exposure, *Int. J. Mol. Sci.*, 2020, **21**, 1410.
57. J. Wang, Y. Li, L. Lu, M. Zheng, X. Zhang, H. Tian, W. Wang and S. Ru, Polystyrene microplastics cause tissue damages, sex-specific reproductive disruption and transgenerational effects in marine medaka (*Oryzias melastigma*), *Environ. Pollut.*, 2019, **254**, 113024.
58. H. Liu, H. Cui, Y. Huang, S. Gao, S. Tao, J. Hu and Y. Wan, Xenobiotics targeting cardiolipin metabolism to promote thrombosis in zebrafish, *Environ. Sci. Technol.*, 2021, **55**, 3855–3866.
59. M. Martinez-Porchas, M. Henández-Rodríguez, J. Dávila-Ortiz, V. Villa-Cruz and J. R. Ramos-Enríquez, A preliminary study about the effect of benzo[a] pyrene (BaP) injection on the thermal behavior and plasmatic parameters of Nile tilapia (*Oreochromis niloticus* L.) acclimated to different temperatures, *PanamJAS*, 2011, **6**, 76–85.
60. L. B. Bertotto, T. R. Catron and T. Tal, Exploring interactions between xenobiotics, microbiota, and neurotoxicity in zebrafish, *Neurotoxicology*, 2020, **76**, 235–244.
61. H. Wu, J. Gao, M. Xie, J. Wu, R. Song, X. Yuan, Y. Wu and D. Ou, Chronic exposure to deltamethrin disrupts intestinal health and intestinal microbiota in juvenile crucian carp, *Ecotoxicol. Environ. Saf.*, 2022, **241**, 113732.
62. H. Zhang, C. Ran, T. Teame, Q. Ding, S. H. Hoseinifar, M. Xie, Z. Zhang, Y. Yang, R. E. Olsen, D. M. Gatlin, E. Ringø, M. Duan and Z. Zhou, Research progress on gut health of farmers teleost fish: a viewpoint concerning the intestinal mucosal barrier and the impact of its damage, *Rev. Fish Biol. Fish.*, 2020, **30**, 569–586.
63. R. Hou, S. Zhang, Q. Huang, L. Lin, H. Li, J. Li, S. Liu, C. Sun and X. Xu, Role of gastrointestinal microbiota from crucian carp in microbial transformation and estrogenicity modification of novel plastic additives, *Environ. Sci. Technol.*, 2023, **57**, 11476–11488.
64. B. Ibarra-Mendoza, B. Gomez-Gil, M. Betancourt-Lozano, L. Raggi and B. Yáñez-Rivera, Microbial gut dysbiosis induced by xenobiotics in model organisms and the relevance of experimental criteria: A minireview, *Gut Microbiome*, 2023, **4**, e7.
65. J.-W. Lee, H. Choi, U.-K. Hwang, J.-C. Kang, Y. J. Kang, K. I. Kim and J.-H. Kim, Toxic effects of lead exposure on bioaccumulation, oxidative stress, neurotoxicity, and immune responses in fish: A review, *Environ. Toxicol. Pharmacol.*, 2019, **68**, 101–108.

66. J. Tang, J. Zhang, L. Su, Y. Jia and Y. Yang, Bioavailability and trophic magnification of antibiotics in aquatic food webs of Pearl River, China: Influence of physicochemical characteristics and biotransformation, *Sci. Total Environ.*, 2022, **820**, 153285.
67. M. Bonsignore, D. Salvagio Manta, S. Mirto, E. M. Quinci, F. Ape, V. Montalto, M. Gristina, A. Traina and M. Sprovieri, Bioaccumulation of heavy metals in fish, crustaceans, molluscs and echinoderms from the Tuscany coast, *Ecotoxicol. Environ. Saf.*, 2018, **162**, 554–562.
68. R. L. Njinga, A. S. Adebayo and A. P. Olufemi, Bioaccumulation of heavy metals in water and organs of *Tilapia brevimanus* and *Euthynnus alletteratus* from a coaster water in southwestern Nigeria, *Environ. Sci. Eur.*, 2023, **35**, 85.
69. B. Jezierska and M. Witeska, *The Metal Uptake and Accumulation in Fish Living in Polluted Waters*, Springer, Dordrecht, The Netherlands, 2006.
70. J. A. Yoder, Investigating the morphology, function and genetics of cytotoxic cells in bony fish, *Comp. Biochem. Physiol. C Toxicol. Pharmacol.*, 2004, **138**, 271–280.
71. B. Liu, Y. Cui, P. B. Brown, X. Ge, J. Xie and P. Xu, Cytotoxic effects and apoptosis induction of enrofloxacin in hepatic cell line of grass carp (*Ctenopharyngodon idellus*), *Fish Shellfish Immunol.*, 2015, **47**, 639–644.
72. S. Datta, D. Ghosh, D. R. Saha, S. Bhattacharaya and S. Mazumder, Chronic exposure to low concentration of arsenic is immunotoxic to fish: Role of head kidney macrophages as biomarkers of arsenic toxicity to *Clarias batrachus*, *Aquat. Toxicol.*, 2009, **92**, 86–94.
73. M. Mir, S. Khan, S. Bhat, A. A. Reshi, F. Shah, M. Balki and R. Manzoor, Scenario of genotoxicity in fishes and its impact on fish industry, *IOSR J. Environ. Sci., Toxicol. Food Technol.*, 2014, **8**, 2319–2402.
74. M. Fatima, N. Usmani, F. Firdaus, M. F. Zafeer, S. Ahmad, K. Akhtar, S. M. Dawar Husain, M. H. Ahmad, E. Anis and M. Mobarak Hossain, *In vivo* induction of antioxidant response and oxidative stress associated with genotoxicity and histopathological alteration in two commercial fish species due to heavy metals exposure in northern India (Kali) river, *Comp. Biochem. Physiol. C Toxicol. Pharmacol.*, 2015, **176–177**, 17–30.
75. P. Gupta and S. K. Verma, Evaluation of genotoxicity induced by herbicide pendimethalin in fresh water fish *Clarias batrachus* (Linn.) and possible role of oxidative stress in induced DNA damage, *Drug Chem. Toxicol.*, 2022, **45**, 750–759.
76. S. Khan, Z. Qamar, A. Khan, M. Waqas, J. Nawab, M. Khisroon and A. Khan, Genotoxic effects of polycyclic aromatic hydrocarbons (PAHs) present in vehicle-wash wastewater on grass carp (*Ctenopharyngodon idella*) and freshwater mussels (*Anodonta cygnea*), *Environ. Pollut.*, 2023, **327**, 121513.
77. Á. Cruz-Esquivel, S. Díez and J. L. Marrugo-Negrete, Genotoxicity effects in freshwater fish species associated with gold mining activities in tropical aquatic ecosystems, *Ecotoxicol. Environ. Saf.*, 2023, **253**, 114670.
78. K. M. Major, B. M. DeCourten, J. Li, M. Britton, M. L. Settles, A. C. Mehinto, R. E. Connon and S. M. Brander, Early life exposure to environmentally

relevant levels of endocrine disruptors drive multigenerational and transgenerational epigenetic changes in a fish model, *Front. Mar. Sci.*, 2020, **7**, 471.

79. I. Cervantes-Camacho, S. M. Guerrero-Estévez, M. F. López, E. Alarcón-Hernández and E. López-López, Effects of Bisphenol A on Foxl2 gene expression and DNA damage in adult viviparous fish *Goodea atripinnis*, *J. Toxicol. Environ. Health, Part A*, 2020, **83**, 95–112.
80. N. Koppel, V. Maini Rekdal and E. P. Balskus, Chemical transformation of xenobiotics by the human gut microbiota, *Science*, 2017, **356**, eaag2770.
81. A. E. Lindell, M. Zimmermann-Kogadeeva and K. R. Patil, Multimodal interactions of drugs, natural compounds and pollutants with the gut microbiota, *Nat. Rev. Microbiol.*, 2022, **20**, 431–443.
82. B. Wang, L. Hu and T. J. Siahaan, *Drug Delivery: Principles and Applications*, John Wiley & Sons Inc., Hoboken, New Jersey, 2016.
83. J. L. Clasen, S. M. Brigden, J. P. Payet and C. A. Suttle, Evidence that viral abundance across oceans and lakes is driven by different biological factors, *Freshwater Biol.*, 2008, **53**, 1090–1100.
84. S. Egerton, S. Culloty, J. Whooley, C. Stanton and R. P. Ross, The gut microbiota of marine fish, *Front. Microbiol.*, 2018, **9**, 873.
85. A. R. Wang, C. Ran, E. Ringø and Z. G. Zhou, Progress in fish gastrointestinal microbiota research, *Rev. Aquac.*, 2018, **10**, 626–640.
86. H. Liu, X. Guo, R. Gooneratne, R. Lai, C. Zeng, F. Zhan and W. Wang, The gut microbiome and degradation enzyme activity of wild freshwater fishes influenced by their trophic levels, *Sci. Rep.*, 2016, **6**, 24340.
87. J. Mougin and A. Joyce, Fish disease prevention via microbial dysbiosis-associated biomarkers in aquaculture, *Rev. Aquac.*, 2023, **15**, 579–594.
88. G. D. Gómez and J. L. Balcázar, A review on the interactions between gut microbiota and innate immunity of fish, *FEMS Immunol. Med. Microbiol.*, 2008, **52**, 145–154.
89. J. F. Rawls, B. S. Samuel and J. I. Gordon, Gnotobiotic zebrafish reveal evolutionarily conserved responses to the gut microbiota, *Proc. Natl. Acad. Sci. U. S. A.*, 2004, **101**, 4596–4601.
90. D. Pérez-Pascual, S. Vendrell-Fernández, B. Audrain, J. Bernal-Bayard, R. Patiño-Navarrete, V. Petit, D. Rigaudeau and J.-M. Ghigo, Gnotobiotic rainbow trout (*Oncorhynchus mykiss*) model reveals endogenous bacteria that protect against Flavobacterium columnare infection, *PLoS Pathog.*, 2021, **17**, e1009302.
91. Z. Zhou, B. Yao, J. Romero, P. Waines, E. Ringø, M. Emery, M. R. Liles and D. L. Merrifield, in *Aquaculture Nutrition: Gut Health, Probiotics and Prebiotics*, ed. D. Merrifield, and R. Ringø, Wiley Online Library, 2014, pp. 101–127.
92. M. Levy, A. A. Kolodziejczyk, C. A. Thaiss and E. Elinav, Dysbiosis and the immune system, *Nat. Rev. Immunol.*, 2017, **17**, 219–232.
93. E. J. Stevens, K. A. Bates and K. C. King, Host microbiota can facilitate pathogen infection, *PLoS Pathog.*, 2021, **17**, e1009514.

94. D. A. Powers, Fish as model systems, *Science*, 1989, **246**, 352–358.
95. A. Shepard and J. L. Kissil, The use of non-traditional models in the study of cancer resistance - The case of the naked mole rat, *Oncogene*, 2020, **39**, 5083–5097.
96. M. L. Larramendy, *Ecotoxicology and Genotoxicology: Non-traditional Aquatic Models*, Royal Society of Chemistry, Cambridge, UK, 2017.
97. R. Matsuzaki, E. Gunnigle, V. Geissen, G. Clarke, J. Nagpal and J. F. Cryan, Pesticide exposure and the microbiota-gut-brain axis, *ISME J.*, 2023, 1–14.
98. L. Evariste, M. Barret, A. Mottier, F. Mouchet, L. Gauthier and E. Pinelli, Gut microbiota of aquatic organisms: a key endpoint for ecotoxicological studies, *Environ. Pollut.*, 2019, **248**, 989–999.
99. J. Wang and L. Chu, Irradiation treatment of pharmaceutical and personal care products (PPCPs) in water and wastewater: An overview, *Radiat. Phys. Chem.*, 2016, **125**, 56–64.
100. H. K. Khan, M. Y. A. Rehman and R. N. Malik, Fate and toxicity of pharmaceuticals in water environment: An insight on their occurrence in South Asia, *J. Environ. Manage.*, 2020, **271**, 111030.
101. A. Kim, N. Kim, H. J. Roh, W.-K. Chun, D. T. Ho, Y. Lee and D.-H. Kim, Administration of antibiotics can cause dysbiosis in fish gut, *Aquaculture*, 2019, **512**, 734330.
102. S. He, Q. Wang, S. Li, C. Ran, X. Guo, Z. Zhang and Z. Zhou, Antibiotic growth promoter olaquindox increases pathogen susceptibility in fish by inducing gut microbiota dysbiosis, *Sci. China: Life Sci.*, 2017, **60**, 1260–1270.
103. S. M. Limbu, L. Q. Chen, M. L. Zhang and Z. Y. Du, A global analysis on the systemic effects of antibiotics in cultured fish and their potential human health risk: A review, *Rev. Aquac.*, 2021, **13**, 1015–1059.
104. T. P. Legrand, S. R. Catalano, M. L. Wos-Oxley, J. W. Wynne, L. S. Weyrich and A. Oxley, Antibiotic-induced alterations and repopulation dynamics of yellowtail kingfish microbiota, *Anim. Microbiome*, 2020, **2**, 1–16.
105. M. Qian, J. Wang, X. Ji, H. Yang, B. Tang, H. Zhang, G. Yang, Z. Bao and Y. Jin, Sub-chronic exposure to antibiotics doxycycline, oxytetracycline or florfenicol impacts gut barrier and induces gut microbiota dysbiosis in adult zebrafish (*Daino rerio*), *Ecotoxicol. Environ. Saf.*, 2021, **221**, 112464.
106. S. I. Polianciuc, A. E. Gurzău, B. Kiss, M. G. Ştefan and F. Loghin, Antibiotics in the environment: causes and consequences, *Med. Pharm. Rep.*, 2020, **93**, 231.
107. P.-L. Toutain, A. A. Ferran, A. Bousquet-Melou, L. Pelligand and P. Lees, Veterinary medicine needs new green antimicrobial drugs, *Front. Microbiol.*, 2016, **7**, 1196.
108. K. Rajendran, R. Rajendiran, R. Ravichandran and R. K. Velu, Investigation of microplastic accumulation in *Rastrelliger kanagurta* fish gut and microplastic degradation behaviour of existing gut bacteria *Pseudomonas* sp., *Arch. Microbiol.*, 2022, **204**, 626.

109. Y. Hao, Y. Sun, M. Li, X. Fang, Z. Wang, J. Zuo and C. Zhang, Adverse effects of polystyrene microplastics in the freshwater commercial fish, grass carp (*Ctenopharyngodon idella*): Emphasis on physiological response and intestinal microbiome, *Sci. Total Environ.*, 2023, **856**, 159270.
110. X. Zhang, K. Wen, D. Ding, J. Liu, Z. Lei, X. Chen, G. Ye, J. Zhang, H. Shen and C. Yan, Size-dependent adverse effects of microplastics on intestinal microbiota and metabolic homeostasis in the marine medaka (*Oryzias melastigma*), *Environ. Int.*, 2021, **151**, 106452.
111. H. Yin, Y. Huang, G. Yan, Q. Huang, Y. Wang, H. Liu, Z. Huang and Y. Hong, Effects of chlorantraniliprole-based pesticide on transcriptional response and gut microbiota of the crucian carp, Carassius carassius, *Ecotoxicol. Environ. Saf.*, 2023, **263**, 115292.
112. J. Tang, W. Wang, Y. Jiang and W. Chu, Diazinon exposure produces histological damage, oxidative stress, immune disorders and gut microbiota dysbiosis in crucian carp (*Carassius auratus gibelio*), *Environ. Pollut.*, 2021, **269**, 116129.
113. X. Wang, M. Shen, J. Zhou and Y. Jin, Chlorpyrifos disturbs hepatic metabolism associated with oxidative stress and gut microbiota dysbiosis in adult zebrafish, *Comp. Biochem. Physiol. C Toxicol. Pharmacol.*, 2019, **216**, 19–28.
114. A. Kakade, M. Sharma, E.-S. Salama, P. Zhang, L. Zhang, X. Xing, J. Yue, Z. Song, L. Nan and S. Yujun, Heavy metals (HMs) pollution in the aquatic environment: Role of probiotics and gut microbiota in HMs remediation, *Environ. Res.*, 2023, **223**, 115186.
115. W. Yan, N. Hamid, S. Deng, P.-P. Jia and D.-S. Pei, Individual and combined toxicogenetic effects of microplastics and heavy metals (Cd, Pb, and Zn) perturb gut microbiota homeostasis and gonadal development in marine medaka (*Oryzias melastigma*), *J. Hazard. Mater.*, 2020, **397**, 122795.
116. A. L. Suzzi, M. Stat, G. R. MacFarlane, J. R. Seymour, N. L. Williams, T. F. Gaston, M. R. Alam and M. J. Huggett, Legacy metal contamination is reflected in the fish gut microbiome in an urbanised estuary, *Environ. Pollut.*, 2022, **314**, 120222.
117. P. Bist and S. Choudhary, Impact of heavy metal toxicity on the gut microbiota and its relationship with metabolites and future probiotics strategy: A review, *Biol. Trace Elem. Res.*, 2022, **200**, 5328–5350.
118. Y. Zhang, P. Zhang, X. Shang, Y. Lu and Y. Li, Exposure of lead on intestinal structural integrity and the diversity of gut microbiota of common carp, *Comp. Biochem. Physiol. C Toxicol. Pharmacol.*, 2021, **239**, 108877.
119. X. L. Meng, S. Li, C. B. Qin, Z. X. Zhu, W. P. Hu, L. P. Yang, R. H. Lu, W. J. Li and G. X. Nie, Intestinal microbiota and lipid metabolism responses in the common carp (*Cyprinus carpio* L.) following copper exposure, *Ecotoxicol. Environ. Saf.*, 2018, **160**, 257–264.
120. F. Zhang, D. Li, Y. Yang, H. Zhang, J. Zhu, J. Liu, X. Bu, E. Li, J. Qin, N. Yu, L. Chen and X. Wang, Combined effects of polystyrene microplastics

and copper on antioxidant capacity, immune response and intestinal microbiota of Nile tilapia (*Oreochromis niloticus*), *Sci. Total Environ.*, 2022, **808**, 152099.

121. D. Hu and K. C. Hornbuckle, Inadvertent polychlorinated biphenyls in commercial paint pigments, *Environ. Sci. Technol.*, 2010, **44**, 2822–2827.
122. J. Xie, Z. Bian, T. Lin, L. Tao, Q. Wu and M. Chu, Global occurrence, bioaccumulation factors and toxic effects of polychlorinated biphenyls in tuna: A review, *Emerging Contam.*, 2020, **6**, 388–395.
123. D.-L. Li, Y.-J. Huang, S. Gao, L.-Q. Chen, M.-L. Zhang and Z.-Y. Du, Sex-specific alterations of lipid metabolism in zebrafish exposed to polychlorinated biphenyls, *Chemosphere*, 2019, **221**, 768–777.
124. A. Qamar, J. Waheed, Q. Zhang, Z. Namula, Z. Chen and J.-J. Chen, Immunotoxicological effects of dioxin-like polychlorinated biphenyls extracted from Zhanjiang Bay sediments in zebrafish, *Environ. Monit. Assess.*, 2020, **192**, 479.
125. E. R. Lauriano, M. Calò, G. Silvestri, D. Zaccone, S. Pergolizzi and P. Lo Cascio, Mast cells in the intestine and gills of the sea bream, *Sparus aurata*, exposed to a polychlorinated biphenyl, PCB 126, *Acta Histochem.*, 2012, **114**, 166–171.
126. M. Zhang, D. H. Zhu, Q. L. Song, M. M. Chen, Y. Hu, D. J. Kang, Z. B. Chen, H. Y. Lin, F. H. Nie and J. J. Chen, The effect of polychlorinated biphenyls extracted from offshore sediments on the intestinal epithelium and microflora of zebrafish, *Int. J. Environ. Sci. Technol.*, 2023, **20**, 12731–12740.
127. F. Spilsbury, M. J. Foysal, A. Tay and M. M. Gagnon, Gut microbiome as a potential biomarker in fish: Dietary exposure to petroleum hydrocarbons and metals, metabolic functions and cytokine expression in juvenile *Lates calcarifer*, *Front. Microbiol.*, 2022, **13**, 827371.
128. J. M. Walter, A. Bagi and D. M. Pampanin, Insights into the potential of the atlantic cod gut microbiome as biomarker of oil contamination in the marine environment, *Microorganisms*, 2019, **7**, 209.
129. J. B. Xiong, L. Nie and J. Chen, Current understanding on the roles of gut microbiota in fish disease and immunity, *Zool. Res.*, 2019, **40**, 70–76.
130. H. Duan, L. Yu, F. Tian, Q. Zhai, L. Fan and W. Chen, Antibiotic-induced gut dysbiosis and barrier disruption and the potential protective strategies, *Crit. Rev. Food Sci. Nutr.*, 2022, **62**, 1427–1452.
131. J. B. Yoon, S. Hwang, J. H. Yang, S. Lee, W. Y. Bang and K. H. Moon, Dynamics of the gut microbiome and transcriptome in Korea native ricefish (*Oryzias latipes*) during chronic antibiotic exposure, *Genes*, 2022, **13**, 1243.
132. M. Wang, Y. Qin, Y. Liu, H. Yang, J. Wang, S. Ru and P. Cui, Short-term exposure to enrofloxacin causes hepatic metabolism disorder associated with intestinal flora dysbiosis in adult marine medaka (*Oryzias melastigma*), *Mar. Pollut. Bull.*, 2023, **192**, 114966.
133. S. Wei, M. I. Bahl, S. M. D. Baunwall, C. L. Hvas and T. R. Licht, Determining gut microbial dysbiosis: A review of applied indexes for assessment

of intestinal microbiota imbalances, *Appl. Environ. Microbiol.*, 2021, **87**, e00395-21.

134. J. Chong, P. Liu, G. Zhou and J. Xia, Using MicrobiomeAnalyst for comprehensive statistical, functional, and meta-analysis of microbiome data, *Nat. Protoc.*, 2020, **15**, 799–821.
135. D. Gevers, S. Kugathasan, L. A. Denson, Y. Vázquez-Baeza, W. Van Treuren, B. Ren, E. Schwager, D. Knights, S. J. Song and M. Yassour, The treatment-naive microbiome in new-onset Crohn's disease, *Cell Host Microbe*, 2014, **15**, 382–392.
136. D. Medina-Felix, F. Vargas-Albores, E. Garibay-Valdez, L. R. Martínez-Córdova and M. Martínez-Porchas, Short communication: Gastrointestinal dysbiosis induced by *Nocardia* sp. infection in tilapia, *Comp. Biochem. Physiol. Part D Genomics Proteomics*, 2023, 101154.
137. A. Kakade, E.-S. Salama, M. Usman, M. Arif, P. Feng and X. Li, Dietary application of *Lactococcus lactis* alleviates toxicity and regulates gut microbiota in *Cyprinus carpio* on exposure to heavy metals mixture, *Fish Shellfish Immunol.*, 2022, **120**, 190–201.
138. A. Bertucci, C. Hoede, E. Dassié, P.-Y. Gourves, A. Suin, K. Le Menach, H. Budzinski and F. Daverat, Impact of environmental micropollutants and diet composition on the gut microbiota of wild european eels (*Anguilla anguilla*), *Environ. Pollut.*, 2022, **314**, 120207.

CHAPTER 5

Fish as a Bioindicator of Microplastic Contamination in the Mediterranean Sea: State of Knowledge

NEZHA MEJJAD[a*], OUSSAMA BOUOUAROUR[b], ABDELMOURHIT LAISSAOUI[a], IKRAM LOUKILI[c] AND MONCEF BENMANSOUR[a]

[a]National Centre for Nuclear Energy, Science and Technology (CNESTEN), Rabat, Morocco; [b]Faculty of Sciences, Mohammed V University, Rabat, Morocco; [c]Mohammed VI Polytechnic University, Hay Moulay Rachid, Ben Guerir, Morocco
*E-mail: mejjad@cnesten.org.ma

5.1 Introduction

Since the 1970s, the protection and conservation of aquatic environments and resources has become one of the most fundamental issues the world is facing. Population growth all around the world has increased human demand and needs, which imply more exploitation of marine ecosystem resources for different purposes. This has led to several conflicts between marine resources' sustainability and social and economic growth mainly because of human intervention and behaviour towards these natural resources causing resources' depletion, habitat destruction and marine water pollution.[1,2]

Issues in Toxicology No. 50
Fish Species in Environmental Risk Assessment Strategies
Edited by Guillermo Eli Liwszyc and Marcelo L. Larramendy

Published by the Royal Society of Chemistry, www.rsc.org

Marine water pollution is becoming increasingly severe, mainly by microplastics as they are polluting aquatic systems at a high rate while their quantification, sources and long-term effects on biota are still uncertain and unclear.[3] Microplastics may have different sources and origin into aquatic systems.[4] However, the main origin of microplastics into oceans are land-based waste drained by rainfall and rivers, which represent the main source of plastic in the sea.[5] In addition, some marine economic activities such as coastal tourism and maritime transportation contribute to marine plastic pollution as well.[6,7]

The Mediterranean Sea (MS) represents only 1% of the world's water but it concentrates 7% of the planet's microplastic debris.[8] According to the United Nations Environmental Programme (UNEP), the MS is among the most polluted zones by plastic with 730 tons of plastic waste generated per day.[9] Studies carried out in this sea have identified microplastics in surface water, sediment and seafloor.[10,11] The coastal countries hotspots of microplastic pollution in the Mediterranean basin are Spain, Turkey, Italy, France and Egypt.[12] Those countries are the most visited destinations and they receive millions of tourists per year from all around the world, which represents the principal factor behind the growing amount of plastics in the MS.[6] It should be noted that tourism activities in the Mediterranean are linked to activities such as fishing, which also heavily contributes to plastic pollution mainly by fishing nets, usually called "ghost nets" as they injure and kill marine animals.[4]

According to Cantasano, the MS is known as an invisible garbage patch accumulating microplastics of a diameter ranging from 5.0 mm to 0.1 μm.[13] This is because, in the MS, there is no trace of permanent accumulation areas as its basin has a semi-enclosed shape and is characterized by current and cyclonic circulation.[14,15] The invisible pollution by microplastics may have severe impacts mainly on fishes and whales, as they confuse microplastics with plankton because of their small size and fish food smell acquired by their long residence in the ocean.[16]

Despite the increasing interest in quantifying and analysing microplastics and nanoplastic debris in the Mediterranean,[17] still little is known about the consequences of these pollutants' accumulation in the region. The shape of the MS basin represents the main factor challenging the region, which renders the quantification of the amount of plastic entering the sea and their possible impacts on biota not fully understood.[18,19]

Given the fact that fish may confuse microplastics for food, analysing and quantifying the microplastic in the fish matrices would be helpful to overcome the existing gaps and delays regarding the management of plastic waste in the Mediterranean. In addition, using fish as a bioindicator of microplastic pollution is crucial for understanding its impact on biota and human health as well.

In the present chapter, we analyse the state of knowledge about plastic pollution in the Mediterranean Sea using bibliometric analysis software and scientific research evolution related to the quantification of microplastics among the Mediterranean basin countries. We also analyse the techniques used for biomonitoring plastic pollution in the MS and we investigate the impacts of microplastics on biota and human health based on the latest research activities published so far.

5.2 Methods and Materials

5.2.1 Data Collection and Analysis

The analysis was based on examining the available literature related to microplastics in the MS. Approximately 100 scientific papers and 20 technical reports were collected and analysed to investigate the state of knowledge regarding plastics' distribution and amount in the Mediterranean countries. The analysis included the exploration of studies that used fish as bioindicators of microplastics in the MS. In this sense, we also carried out a bibliometric analysis of the studies related to microplastic contamination and accumulation in fish. We explored the following databases, Dimensions.ai and Scopus (see Table 5.1). Two search queries were used "Microplastic in Fish" and "Microplastic in fish in the Mediterranean Sea". It is worth noting that the search was done using the title and abstract as keywords. As shown in Table 5.1, the publication year includes the last ten years, meaning that research activities on microplastics in fish was in its early stages at that period in time. In addition, the number of publications is fewer compared to other types of pollutants (*e.g.*, heavy metals[20]). However, the analysis of the vertical distribution of publications carried out on the investigated topic *versus* years (see Figure 5.1) indicates a steady increase in publication rates over the years reflecting the growing interest in assessing microplastics' pollution accumulation, contamination levels and effect on fish. Indeed, fish are used as a bioindicator of pollution in the aquatic environment,[21,22] which explains this increasing interest in investigating microplastics in aquatic systems' species.

It should be noted that out of 1645 documents published between 2011 and 2023, 193 were published in the *Science of the Total Environment* (11.73%), 187 in *Marine Pollution Bulletin Journal* (11.36%), 141 *Environmental Pollution* (8.57) and 65 in *Chemosphere Journal* (3.4%).

Table 5.1 Number of publications related to our search queries based on Dimension.ai and Scopus databases.

		Search query (number of articles)		
Database	Publication year range	Microplastic in fish	Microplastic in fish in the Mediterranean	Publication year range
Dimensions.ai	2011–2023	1645	92	2013–2023
Scopus	2011–2024	1552	85	2012–2024

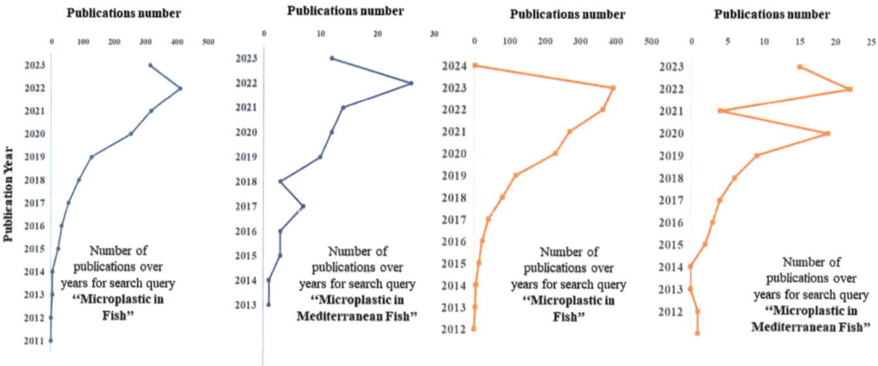

Figure 5.1 Vertical distribution of the number of publications over their publication date for data extracted from Dimension.ai and Scopus databases.

5.2.2 Data Mapping

This study analyses the currently investigated aspects related to the occurrence of microplastics in aquatic environments, their effects on fish, and identifies whether the studies carried out during the last two decades are focusing on fish as bioindicators or as a threat to human health. Such analysis would better define the aspects that need to be developed and those that remain unexplored. In this sense, we used a bibliometric software named VOSviewer Version 1.6.20 (Leiden University, Netherlands); a helpful tool that will allow the creation of network links between the main keywords related to the search query, which allows pointing out the main categories of investigations carried out in relation to our search query: "Microplastics" AND "Fish", "Microplastic" OR "micro-plastic" AND "Fish" and "Mediterranean Sea". The analysis covered the year ranges shown in Table 5.1. As both databases show almost similar statistics and range dates, we have chosen to map the data extracted from https://www.dimensions.ai/. Figure 5.2 shows a flow chart displaying the procedure followed to carry out the mapping of extracted data using VOSviewer software.

According to the number of documents that occurred together related to the search query, it is possible to establish a mapped visualization of keywords, researchers and authors based on co-occurrence analysis that takes into consideration these items' relatedness. The co-occurrence in the present chapter is only assessing keywords co-occurrence to comprehend and examine the current state of knowledge and research tendency in microplastics in the MS using VOSviewer visualization software. "Link" signifies a connection or relationship between two items (keywords), "link strength" means an attribute of each link presented as a positive numerical value, "network" designates a collection of items connected together, and "cluster" signifies groups of items displayed in a network map, each of which has a colour and can only be a part of one cluster with a defined colour.

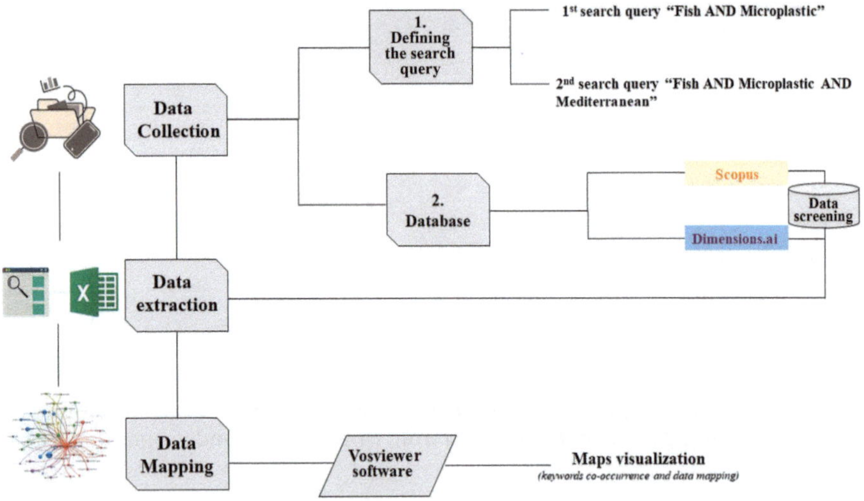

Figure 5.2 Flowchart displaying the procedures followed to carry out the mapping of extracted data using VOSviewer software.

5.3 Results

5.3.1 Plastic Contamination in the Mediterranean Sea: A Bibliometric Analysis

Plastic contamination has become a major concern in the last three decades, where the increased rate of production and consumption has resulted in plastic accumulation in nature because of the lack of awareness regarding the possible effects on the environment and biodiversity.[4] Like all pollutants, plastics reach the ocean as a final receptor that threatens life inside and outside the sea. As shown in Figure 5.3, the accumulation of plastics in the marine environment has environmental and socio-economic consequences. The plastics break down into microplastics and nanoplastics under natural environmental conditions *via* a non-biodegradation or biodegradation process.[23] This debris can be ingested by marine organisms as they acquire the smell of fish if they stay for a long time in the sea. This would then threaten biodiversity and human health. According to Enfrin *et al.* and Tourinho *et al.*, plastics can cause a chain of adverse reactions to marine organisms such as suffocation, intestinal injury, gastric obstruction and malnutrition.[24,25]

At the Mediterranean level, the amount of plastic debris in the sea has increased to such an extent that it has reached environmental proportion crises faced by its riparian countries. The consumption rate, human behaviour, societal awareness, management processes, plastic waste disposal and collection facilities, rules, laws and strategies in these countries vary from one country to another, while they share the same sea. This puts the economy of all Mediterranean countries at risk, as once the life underwater is affected, the related ecosystem services will be affected as well. These would, above

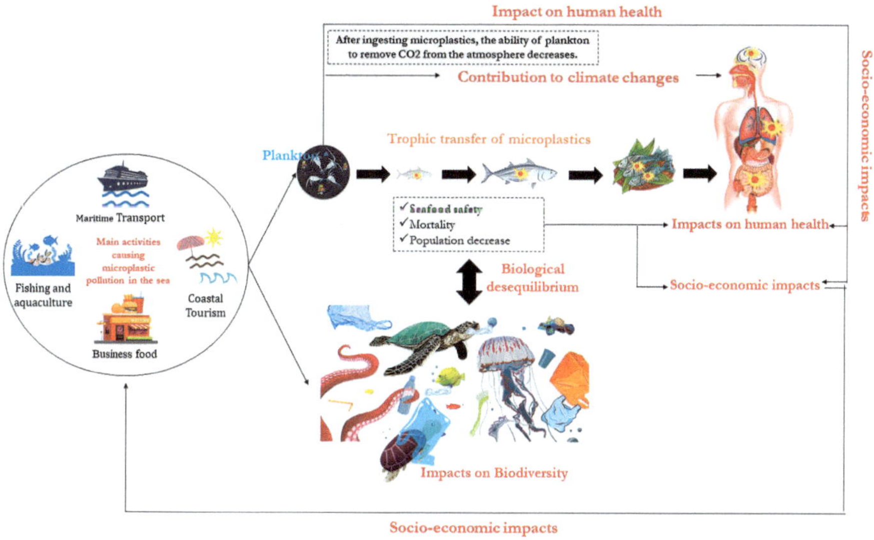

Figure 5.3 Schematic presentation of microplastic contamination effects on environment and socio-economic sectors.

all, encompass fishing and coastal tourism sectors, which represent the economic engine of several Mediterranean countries. The need for the establishment of common management strategies based on a multi-approach monitoring framework of the MS coastal areas in which biota, particularly fish, used as bioindicators of microplastic pollution in aquatic environments, is paramount. The evaluation of plastic debris in fish samples would be of great interest and provide valuable information regarding plastic contamination levels and toxicity.

In order to analyse the current state of knowledge regarding fish contamination by microplastics in the MS, we carried out a mapping of available data related to the study of fish contamination by microplastics in the sea. By 7th June 2023, according to the Scopus database, 80 scientific publications were carried out on marine species in relation to microplastics. The screening of the extracted data shows that 72 documents were scientific articles, 5 were reviews and 2 were book chapters. Figure 5.4 shows maps of network visualization and density visualization of data extracted from the database. In total, 65 keywords or terms were grouped into four clusters and mapped as shown in Figure 5.4 in the form of coloured nodes or circles linked by 1986 links and a total link strength of 11 576. The map (see Figure 5.4A) shows four coloured clusters (red, green, blue and yellow).

The first cluster (red cluster) contains 19 items (keywords) related to fish species and microplastic ingestion. The main keywords co-occurrence for this cluster are study (occurrence: 54), species (occurrence: 49), Mediterranean Sea (occurrence: 44), occurrence (occurrence: 36), fish species (occurrence: 27) and microplastic ingestion (occurrence: 24). The incidence of the

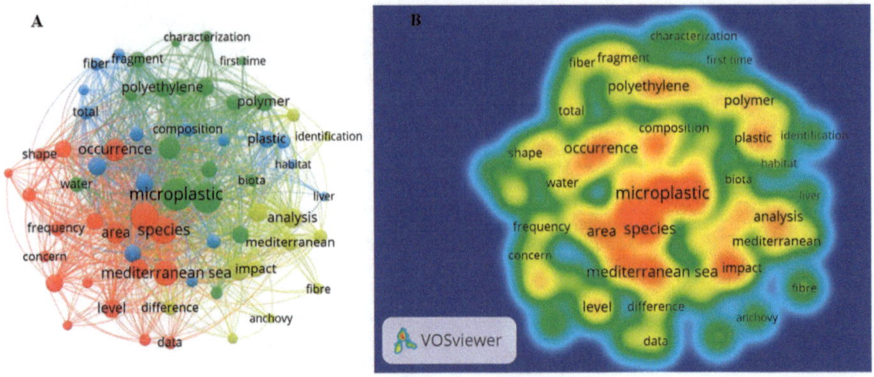

Figure 5.4 Visualization network map (A) and density map (B) of keywords related to the search query.

keyword "occurrence" is related to the detection of microplastics which is crucial for analysing and quantifying their possible effects on marine species. The second cluster (green cluster) consists of 18 keywords concerning microplastic characterization, distribution and occurrence in marine environments including biota. The main keywords include microplastic (occurrence: 65), fish (occurrence: 60), sample (occurrence: 36), concentration (occurrence: 20) and distribution (occurrence: 20). The blue cluster involves plastics' pollution sizes and composition analysis and identification. The keywords co-occurrences show that "size" occurred 26 times, while "plastics" displayed 23 occurrences, and "number" and "marine organisms" occurred 17 times. The yellow cluster includes the analysis and identification of microplastic in marine organisms. The main keywords are as follows: ingestion (occurrence: 36), analysis (occurrence: 29), presence (occurrence: 26), impact (occurrence: 20) and stomach (occurrence: 15). Accordingly, it is crucial to analyse the distribution and assess the concentration of microplastics in fish and their potential impacts based on the assessment of the biological risks prior to establishing remediation approaches and strategies.

The interpretation of the data shows that the analysed publications are mainly focused on the study and identification of microplastics (size, characterization, distribution, composition and types of plastics) and their ingestion by marine organisms (see Figure 5.4B). The analysis points out also that the microplastic ingestion and its effects remain weakly investigated compared to quantification and characterization studies. Fisheries are of great interest to the Mediterranean countries where their economic growth depends highly on this activity.[6] Thus, the sustainability of marine resources implies conducting a continuous assessment of the environmental quality of the marine environment through assessing and quantifying microplastics in marine species, mainly fish. According to the World Wildlife Foundation (WWF) report 2017,[26] the Mediterranean countries are among the highest consumers of fish in the world, whereas poor assessment and monitoring, illegal and irrational fishing and the growth of industrial methods of fishing have heavily affected the state of stocks.[27] Indeed, several studies highlighted the critical situation of fish stocks in the MS where it was reported that the European fisheries' management system has witnessed a failure in the management of the MS stocks' status.[28] In this sense, several studies were carried out to analyse the factors contributing to the deterioration of marine Mediterranean resources including the spreading of new invasive species[29–31] and marine litter, which received strong attention during the latest decades and has direct and indirect impact on fisheries and stocks.

In the MS, several studies reported the occurrence of microplastics in various species including mussels and fish.[32] In a study carried out in Alexandria (Egypt), microplastics were detected in 92% of fish species[33] while more than 75% of fish collected from the Nile River indicated the occurrence of microplastics in their digestive tract.[34] In Tunisia, the analysis of microplastic occurrence in two commercial fish species showed their presence in fish gastrointestinal tracts.[35] Maaghloud *et al.* have indicated the presence of microplastics in three pelagic fish species' stomachs collected between Cape Cantin and Sidi Ifni in Morocco.[36] Similarly, a study carried out on Italian freshwater fishes has shown the occurrence of microplastics and microfibres in the gastrointestinal tracts.[37] In Turkey, Kılıç reported a high rate of microplastic ingestion observed in farmed fish species including the gilt-head seabream *Sparus aurata* (Spariformes, Sparidae), the European seabass *Dicentrarchus labrax* (Moroniformes, Moronidae) and the rainbow trout *Oncorhynchus mykiss* (Salmoniformes, Salmoniae).[38] This suggests increasing concern about the potential effects of microplastic ingestion by fish species on consumers' health and calls for the inclusion of fish species with a high rate of microplastic ingestion on the list of marine microplastic ingestion bioindicators.

5.3.2 Fish Species as a Bioindicator

Several marine organisms have been suggested for monitoring plastic pollution in the literature and they have been studied using biological and ecological criteria.[39] Because of their commercial value, ecological importance, regional distribution and feeding habits, fish are becoming increasingly popular.[40] Few pelagic and demersal species have been identified for their distinct properties. However, the distinction between these two categories may preclude the identification of particular polymer types due to the varied habitats studied.[41]

A fish's suitability as a bioindicator was determined by its occurrence, mobility, commercial value, habitat and dispersion.[42] Some species have been proposed as bioindicators due to their prevalence in harbours and eating patterns. Benthic foragers consume a substantial amount of sediment along with food, but they also often feed throughout the water column.[43]

Many target species for the MS have been proposed.[41] However, only a few researches have indicated that fish biological factors influence the micro-litter ingestion rate.[44] The anchovy *Engraulis encrasicolus*, also known as boquerón (Clupeiformes, Engraulidae), the European sprat, also known as the Russian sardine *Sprattus sprattus* (Clupeiformes, Clupeidae), the Atlantic horse mackerel or common scad *Trachurus trachurus* (Carangiformes, Carangidae) and Atlantic mackerel *Scombrer scombrus* (Scombriformes, Scombridae) are all extensively spread and thus suited as bioindicator species. Unfortunately, all of these species are pelagic. As a result, other benthic/demersal target species must be selected at the regional level.

5.3.2.1 Swordfish Xiphias gladius *and Bluefin Tuna* Thunnus thynnus *Species*

Microplastic ingestion has been detected in many species of fish envisaged for human consumption from the Atlantic, Indian and Pacific Oceans as well as from the Mediterranean Sea. The Atlantic bluefin tuna *Thunnus thynnus* (Scombriformes, Scombridae) is a fast-growing, long-living and highly migratory species that is extremely tolerant of a wide range of environmental conditions and is found in the Atlantic Ocean, Mediterranean basin and Black Sea. *Thunnus thynnus* has also been designated as a top predator that plays an essential part in the pelagic trophic web. The first investigation of plastic occurrence in fish species including the swordfish *Xiphias gladius* (Istiophoriformes, Xiphiidae) and bluefin tuna reported the occurrence of micro-, meso- and macro-plastic with proportions fluctuating between 20% and 44% approximatively.[45] Bray *et al.* found that *T. thynnus* and *X. gladius* ingest the highest amount of microplastics among 49 MS fish species.[41] A recent study carried out on bluefin tuna and swordfish caught from the MS, showed the occurrence of polymers, microplastics and additives (BPA and PTA) in the muscles of *T. thynnus* and *X. gladius* in the same contamination level while those two species are among the most consumed fish.[46]

Accordingly, studies on MS specimens are critical for assessing the state of these taxa as this basin receives a high pollutant input from urban and industrial areas along its shores, as well as cultivated land *via* rivers. In this sense, analysis and assessment of the amount of plastics in the MS fish species are necessary. Studies carried out on two of the most consumed species in the Mediterranean underlined the occurrence of plastics of different types, sizes, shapes and colours in their muscles, stomach and gastrointestinal tract. Further investigations of long-term effects of microplastic on the functioning of the ecosystem on biota and humans are required to confirm whether this assumption is true.

5.3.3 Analysis of Impacts of Plastics on Fish

Several studies have investigated the damaging effects of microplastics on marine organisms which can range from biological functioning trouble to death.[47–51] The mechanisms through which these impacts will be manifested have not been fully identified, yet. Microplastic intoxication is classified based on the nature of the microplastic after ingestion, (1) accumulation in the gastrointestinal tract, resulting in physical injury such as obstruction

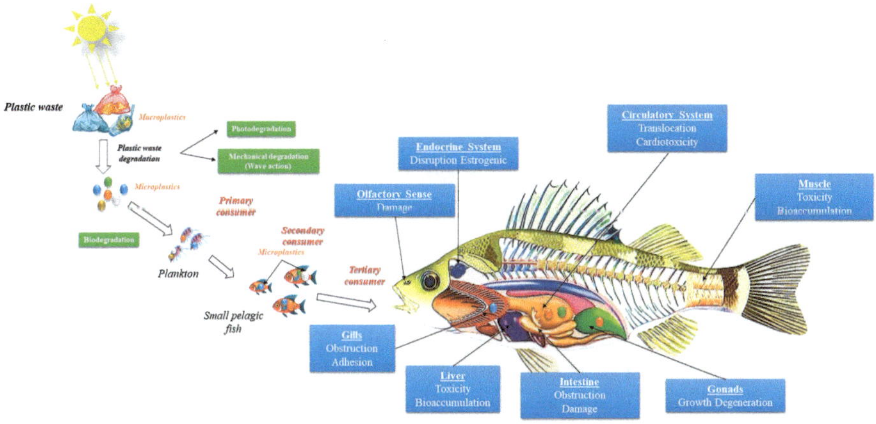

Figure 5.5 Principal effects of microplastics on fish.

and damage, (2) discharge as pseudofaeces, interrupting energy transmission processes in organisms and (3) internal transmission, exposing internal organs and tissues to microplastics (see Figure 5.5).

Microplastics may also cause morphological and functional abnormalities in the digestive tracts of fish, resulting in nutritional and developmental difficulties.[52,53] Many studies have been conducted to demonstrate that microplastics are a threat to fish, with death frequently occurring before they reach maturity due to their intake. The most prevalent impacts of microplastics are oxidative stress, decreased mobility, gene expression disruption and reproductive organ damage.[54–56]

Fish can be also affected by growth suppression or retardation, dysbiosis of their gut, weight loss, disruption of the liver's anti-oxidative condition and damage to reproductive organs.[57] They can be also subjected to stress, behavioural alterations and immune system activity disruption.[58]

Despite the huge number of research related to plastic pollution distribution and potential effects, all of the studies revealed the need for further in-depth investigations related to the effects of plastics on biodiversity and marine ecosystem functioning, which makes it clear that there are gaps of knowledge still remaining, mainly, *e.g.*, the sampling and analytical techniques in plastic pollution studies are inhomogeneous and need to be standardized for providing accurate and precise results. In addition, there is a recommendation to focus on human risk assessment analysis and establish international microplastic daily intake levels.

5.4 Measures and Mitigations

Analysis of the studies carried out on fish in the MS allows the conclusion that the presence of plastic in the aquatic environment is closely related to blue economy activities, mainly coastal tourism, maritime transport and fishing. In this order, measures to tackle the problems must include these sectors' released waste regulations and management. As these activities are inter-related, it is indispensable to establish common regulations and management strategies for the management of plastic waste for reliable and effective results. Mediterranean countries receive millions of tourists each year mainly during the summer period. According to a newly released report by the International Union for Conservation of Nature (IUCN), economic losses are due, in part, to plastic accumulation.[59] In this order, the Mediterranean countries' stakeholders must take into consideration the need to develop frameworks for managing plastic waste entering the Mediterranean area's water while taking into consideration each country's followed strategies and regulations for waste management.

While analysing the origins of plastic, we figured out that all source generators of these pollutants are connected, while the management systems of generated waste are often different (see Figure 5.6). It is true that the use rate and types of plastics used in each activity can be different but waste sorting must be adopted by all activities consuming plastic items while encouraging the owners of hotels, restaurants, coffee-shops, *etc.* to use eco-friendly alternative items to plastics. Plastic waste is spreading in our aquatic systems like an out-of-control virus worsened by their long-life span. Plastics can take hundreds of years to break down, which in turn affects underwater life, and risks human health. Accordingly, preventive measures, management of the existing plastic waste and the implementation of a continuous monitoring and assessment programme of microplastics in the most consumed fish in the Mediterranean would help to reduce plastic waste. In general, Mediterranean consumers prefer small pelagic fish such as anchovies and sardines, with tuna and hake also being very popular.[60] These most consumed marine products must be included as bioindicators in the list of microplastics' quantification and occurrence assessment in the MS.

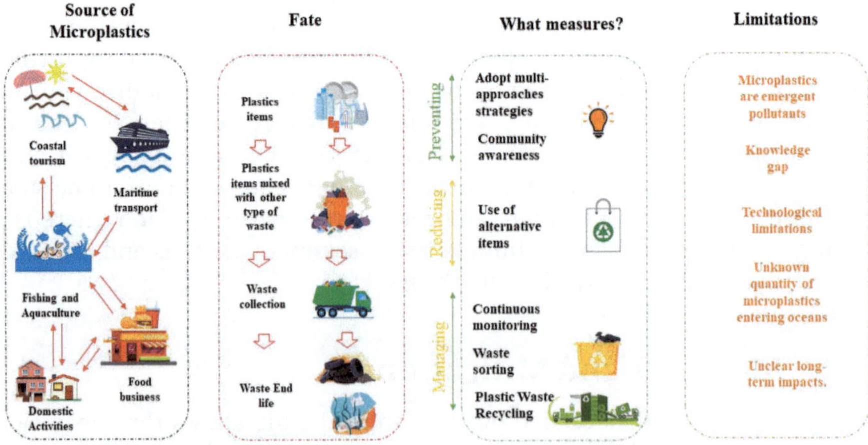

Figure 5.6 Microplastic source, fate, preventive measures and limitations.

It is worth noting that because microplastics are emerging pollutants, various limitations need to be highlighted (see Figure 5.6). Several studies were carried out to quantify the microplastics in oceans in different matrices including sediment and fish. However, almost all studies focus on the detection, assessment and monitoring of microplastics and their potential effects on the environment and human health. Limitations related to the microplastic distribution, temporal trends and their behaviour in the marine environments and biological system were reported in numerous studies,[61–63] while some research studies stated the lack of consistent methodology for microplastics' detection and characterization.[62,64] An international accepted standard measurement system should be implemented and used in order to be able to compare results. Various sampling strategies, sample preparation techniques as well as detection and analysis procedures are employed resulting in unreliable and inconsistent results making the ecological and biological effects of marine microplastic poorly understood.[62,65] In this order, the standardization and harmonization of the analytical procedures also including quality control is crucial for establishing efficient monitoring strategies and risk assessment of microplastics in the seas. In almost every case, particles of plastic are lost during sample preparation or analyses resulting in underestimation of microplastics' contamination. Therefore, it is better to further develop standardized methodological and analytical techniques.

5.5 Discussion and Further Considerations

The present chapter analyses the state of knowledge of microplastic occurrence in Mediterranean fish. The analysis pointed out that most of the studies carried out focused on microplastic detection and characterization in some small pelagic fish such as the anchovy. However, little is known about microplastic quantities, distribution, temporal trends, and impact on the environment and biota. Thus, further investigation is needed to overcome the existing gaps through study of the most consumed fish species by humans to avoid potential adverse effects on human health. This would help establish and ensure strategies for safe foodstuffs from the sea for current and future generations.

It is well known that consumption rate, human behaviour, societal awareness, management processes, plastic waste disposal and collection facilities, rules, laws as well as management strategies in Mediterranean countries vary from one country to another, even if sharing the same sea. Accordingly, it is crucial to implement a multi-approach strategy that takes into consideration all the above-mentioned considerations among Mediterranean countries and stakeholders to tackle the microplastic accumulation rate increases in a holistic manner.

The analysis has also underlined the challenges of analytical procedures that make the estimation and quantification of microplastic concentrations even more complex. Various techniques are used to quantify and detect microplastics in different types of samples. This is one of the limiting factors towards an adequate and efficient determination of microplastic occurrences. Adoption of the same techniques with consistent application would enable concrete and efficient management strategies to be established. The establishment of preventive measures and actions to limit and reduce the quantity of microplastic reaching the MS is essential, as the presently used techniques are not capable of providing accurate results, so far.

It should be noted that an increase in microplastic contamination in the MS would worsen the state of its fisheries' stocks, which are already suffering from a lack of proper management, resulting in the deterioration of marine resources. Thus, increasing the concern about the alarming situation of plastic pollution in the MS by means of raising awareness among Mediterranean communities, providing waste sorting and collection facilities, creating and producing eco-friendly alternatives of plastic items and calling for the adoption of circular economy concepts are some of the main steps to be followed towards a reduction of plastics reaching the sea.

References

1. L. S. L. Teh and U. R. Sumaila, Trends in global shared fisheries, *Mar. Ecol.: Prog. Ser.*, 2015, **530**, 243–254.
2. N. Mejjad and M. Rovere, Understanding the impacts of blue economy growth on deep-sea ecosystem services, *Sustainability*, 2021, **13**, 12478.
3. Z. L. R. Botterell, N. Beaumont, T. Dorrington, M. Steinke, R. C. Thompson and P. K. Lindeque, Bioavailability and effects of microplastics on marine zooplankton: A review, *Environ. Pollut.*, 2019, **245**, 98–110.
4. N. Mejjad, A. Laissaoui, A. Fekri and O. El Hammoumi, Marine plastic pollution in Morocco: state of the knowledge on origin, occurrence, fate, and management, *Environ. Sci. Pollut. Res.*, 2023, **30**, 107371–107389.
5. L. J. Meijer, T. Van Emmerik, R. van der Ent, C. Schmidt and L. Lebreton, More than 1000 rivers account for 80% of global riverine plastic emissions into the ocean, *Sci. Adv.*, 2021, **7**, eaaz5803.
6. N. Mejjad, A. Rossi and A. Pavel, The coastal tourism industry in the Mediterranean: A critical review of the socio-economic and environmental pressures and impacts, *Tourism Manage. Perspect.*, 2022, **44**, 101007.
7. G. G. N. Thushari and J. D. M. Senevirathna, Plastic pollution in the marine environment, *Heliyon*, 2020, **6**, e04709.
8. E. Alessi, G. Di Carlo, S. Campogianni, E. B. Di Carlo and B. Jeffries, *Out of the plastic trap: Saving the Mediterranean from plastic pollution,* World Wide Fund for Nature, Mediterranean Marine Initiative, 2018.
9. United Nations Environmental Programme (UNEP), Mediterranean Action Plan: Pollution in the Mediterranean, 2019, https://www.unep.org/unepmap/resources/factsheets/pollution#:%7E:text=The%20Mediterranean%20is%20polluted%20by,recorded%20marine%20litter%20on%20beaches, accessed on 09 September 2023.
10. S. Dunphy, *Ocean Currents Create Microplastic "hotspots" in the Mediterranean Sea*, Europeanscientist, Paris, 2020.

11. I. A. Kane, M. A. Clare, E. Miramontes, R. Wogelius, J. J. Rothwell, P. Garreau and F. Pohl, Seafloor microplastic hotspots controlled by deep-sea circulation, *Science*, 2020, **368**, 1140–1145.
12. UNEP/MAP, *Marine Litter Assessment in the Mediterranean*, UNEP/MAP, Athens, 2015.
13. N. Cantasano, Marine pollution by microplastics in the Mediterranean Sea, *J. Mar. Sci. Eng.*, 2022, **7**, 858.
14. S. Liubartseva, G. Coppini, R. Lecci and E. Clementi, Tracking plastics in the Mediterranean: 2D Lagrangian model, *Mar. Pollut. Bull.*, 2018, **1**, 151–162.
15. J. Mansui, G. Darmon, T. Ballerini, O. van Canneyt, Y. Ourmières and C. Miaud, Predicting marine litter accumulation patterns in the Mediterranean basin: Spatio-temporal variability and comparison with empirical data, *Prog. Oceanogr.*, 2020, **182**, 102268.
16. M. S. Savoca, C. W. Tyson, M. McGill and C. J. Slager, Odours from marine plastic debris induce food search behaviours in a forage fish, *Proc. R. Soc. B*, 2017, **284**, 20171000.
17. F. Colloca, G. Scarcella and S. Libralato, Recent trends and impacts of fisheries exploitation on Mediterranean stocks and ecosystems, *Front. Mar. Sci.*, 2017, **4**, 267643.
18. M. Llorca, D. Álvarez-Muñoz, M. Ábalos, S. Rodríguez-Mozaz, L. H. M. L. M. Santos, V. M. León, J. A. Campillo, C. Martínez-Gómez, E. Abad and M. Farré, Microplastics in Mediterranean coastal area: toxicity and impact for the environment and human health, *Trends Environ. Anal. Chem.*, 2020, **27**, e00090.
19. A. Baudena, E. Ser-Giacomi, I. Jalón-Rojas, F. Galgani and M. L. Pedrotti, The streaming of plastic in the Mediterranean Sea, *Nat. Commun.*, 2022, **13**, 2981.
20. N. Mejjad, A. Laissaoui, B. El Mansouri, A. Fekri, A. Moumen, K. El Khalidi and O. El Hammoumi, in *Coasts, Estuaries and Lakes*, ed. N. Jayaraju, G. Sreenivasulu, M. Madakka, and M. Manjulatha, Springer, Cham, Switzerland, 2023, pp. 3–24.
21. R. Z. Habib and T. Thiemann, Microplastic in commercial fish in the Mediterranean Sea, the Red Sea and the Arabian Gulf. Part 1: The Mediterranean Sea, *J. Water Resour. Prot.*, 2021, **13**, 563–587.
22. S. Sharma, V. Sharma and S. Chatterjee, Microplastics in the Mediterranean Sea: Sources, pollution intensity, sea health, and regulatory policies, *Front. Mar. Sci.*, 2021, **8**, 634934.
23. M. S.-L. Yee, L.-W. Hii, C. K. Looi, W.-M. Lim, S.-F. Wong, Y.-Y. Kok, B.-K. Tan, C.-Y. Wong and C.-O. Leong, Impact of microplastics and nanoplastics on human health, *Nanomaterials*, 2021, **11**, 496.
24. M. Enfrin, L. F. Dumée and J. Lee, Nano/microplastics in water and wastewater treatment processes - Origin, impact and potential solutions, *Water Res.*, 2019, **161**, 621–638.
25. P. S. Tourinho, V. Kočí, S. Loureiro and C. A. van Gestel, Partitioning of chemical contaminants to microplastics: Sorption mechanisms,

environmental distribution and effects on toxicity and bioaccumulation, *Environ. Pollut.*, 2019, **252**, 1246–1256.

26. World Wildlife Foundation (WWF), Seafood and the Mediterranean: local tastes, global markets, 2017, Available at https://wwfeu.awsassets.panda.org/downloads/wwf_mediterranean_fishing_report_2017_singles.pdf, accessed on 15 September 2023.
27. J. Lleonart and L. Recasens, in *Resource and Environmental Issues Relevant to Mediterranean. General Fisheries Council for the Mediterranean Fisheries and the Environment in the Mediterranean Sea, 142, Etud. Rev. CGPM*, No. 66, ed. J. F. Caddy, Food and Agriculture Organization of the United Nations (FAO), Rome, 1966, pp. 5–18.
28. M. Cardinale, G. C. Osio and G. Scarcella, Mediterranean Sea: A Failure of the European Fisheries Management System, *Front. Mar. Sci.*, 2017, **4**, 72.
29. B. S. Galil, A. Marchini, A. Occhipinti-Ambrogi, D. Minchin, A. Narščius, H. Ojaveer and S. Olenin, International arrivals: Widespread bioinvasions in European Seas, *Ethol. Ecol. Evol.*, 2014, **26**, 152–171.
30. V. Parravicini, E. Azzurro, M. Kulbicki and J. Belmaker, Niche shift can impair the ability to predict invasion risk in the marine realm: an illustration using Mediterranean fish invaders, *Ecol. Lett.*, 2015, **18**, 246–253.
31. Z. Wang, M. Saebi, E. K. Grey, J. J. Corbett, D. Chen, D. Yang and Z. Wan, Ballast water-mediated species spread risk dynamics and policy implications to reduce the invasion risk to the Mediterranean Sea, *Mar. Pollut. Bull.*, 2021, **174**, 113285.
32. A. Mancia, T. Chenet, G. Bono, M. L. Geraci, C. Vaccaro, C. Munari, M. Mistri, A. Cavazzini and L. Pasti, Adverse effects of plastic ingestion on the Mediterranean small-spotted catshark (*Scyliorhinus canicula*), *Mar. Environ. Res.*, 2020, **155**, 104876.
33. A. A. M. El-Sayed, M. I. A. Ibrahim, S. Shabaka, M. M. Ghobashy, M. A. Shreadah and S. A. Abdel Ghani, Microplastics contamination in commercial fish from Alexandria City, the Mediterranean Coast of Egypt, *Environ. Pollut.*, 2022, **313**, 120044.
34. F. R. Khan, Y. Shashoua, A. Crawford, A. Drury, K. Sheppard, K. Stewart and T. Sculthorp, The plastic Nile': First evidence of microplastic contamination in fish from the Nile River (Cairo, Egypt), *Toxics*, 2020, **8**, 22.
35. S. Abidli, N. Akkari, Y. Lahbib and N. Trigui El Menif, The first evaluation of microplastics in sediments from the complex Lagoon-Channel of Bizerte (Northern Tunisia), *Water, Air, Soil Pollut.*, 2017, **7**, 262.
36. H. Maaghloud, R. Houssa, S. Ouansafi, F. Bellali, K. El Bouqdaoui, N. Charouki and A. Fahde, Ingestion of microplastics by pelagic fish from the Moroccan Central Atlantic coast, *Environ. Pollut.*, 2020, **261**, 114194.
37. G. Forgione, F. Izzo, M. Mercurio, D. Cicchella, L. Dini, G. Giancane and M. Paolucci, Microplastics pollution in freshwater fishes in the South of Italy: Characterization, distribution, and correlation with environmental pollutants, *Sci. Total Environ.*, 2023, **864**, 161032.

38. E. Kılıç, N. Yücel and S. Mübarek Şahutoğlu, First record of microplastic occurrence at the commercial fish from Orontes River, *Environ. Pollut.*, 2022, **307**, 119576.
39. A. Macali and E. Bergami, Jellyfish as innovative bioindicator for plastic pollution, *Ecol. Indic.*, 2020, **115**, 106375.
40. M. C. Fossi, C. Pedà, M. Compa, C. Tsangaris, C. Alomar, F. Claro, C. Ioakeimidis, F. Galgani, T. Hema, S. Deudero, T. Romeo, P. Battaglia, F. Andaloro, I. Caliani, S. Casini, C. Panti and M. Baini, Bioindicators for monitoring marine litter ingestion and its impacts on Mediterranean biodiversity, *Environ. Pollut.*, 2018, **237**, 1023–1040.
41. L. Bray, N. Digka, C. Tsangaris, A. Camedda, D. Gambaiani, G. A. de Lucia, M. Matiddi, C. Miaud, L. Palazzo, A. Pérez-del-Olmo, J. A. Raga, C. Silvestri and H. Kaberi, Determining suitable fish to monitor plastic ingestion trends in the Mediterranean Sea, *Environ. Pollut.*, 2019, **247**, 1071–1077.
42. C. Zhang, S. Wang, Z. Pan, D. Sun, S. Xie, A. Zhou, J. Wang and J. Zou, Occurrence and distribution of microplastics in commercial fishes from estuarine areas of Guangdong, South China, *Chemosphere*, 2020, **260**, 127656.
43. A. K. Whitfield, J. Panfili and J. D. Durand, A global review of the cosmopolitan flathead mullet *Mugil cephalus* Linnaeus 1758 (Teleostei: Mugilidae), with emphasis on the biology, genetics, ecology and fisheries aspects of this apparent species complex, *Rev. Fish Biol. Fish.*, 2012, **22**, 641–681.
44. A. Sbrana, T. Valente, U. Scacco, J. Bianchi, C. Silvestri, L. Palazzo, G. A. de Lucia, C. Valerani, G. Ardizzone and M. Matiddi, Spatial variability and influence of biological parameters on microplastic ingestion by *Boops boops* (L.) along the Italian coasts (Western Mediterranean Sea), *Environ. Pollut.*, 2020, **263**, 114429.
45. T. Romeo, B. Pietro, C. Pedà, P. Consoli, F. Andaloro and M. C. Fossi, First evidence of presence of plastic debris in stomach of large pelagic fish in the Mediterranean Sea, *Mar. Pollut. Bull.*, 2015, **1**, 358–361.
46. F. Di Giacinto, L. Di Renzo, G. Mascilongo, V. Notarstefano, G. Gioacchini, E. Giorgini, T. Bogdanović, S. Petričević, E. Listeš, M. Brkljača, F. Conti, C. Profico, B. Zambuchini, G. Di Francesco, C. Giansante, G. Diletti, N. Ferri and M. Berti, Detection of microplastics, polymers and additives in edible muscle of swordfish (*Xiphias gladius*) and bluefin tuna (*Thunnus thynnus*) caught in the Mediterranean Sea, *J. Sea Res.*, 2023, **192**, 102359.
47. U. Subaramaniyam, R. S. Allimuthu, S. Vappu, D. Ramalingam, R. Balan, B. Paital, N. Panda, P. K. Rath, N. Ramalingam and D. K. Sahoo, Effects of microplastics, pesticides and nano-materials on fish health, oxidative stress and antioxidant defense mechanism, *Front. Physiol.*, 2023, **14**, 1217666.
48. M. S. Bhuyan, Effects of microplastics on fish and in human health, *Front. Environ. Sci.*, 2022, **10**, 827289.
49. L. Alberghini, A. Truant, S. Santonicola, G. Colavita and V. Giaccone, Microplastics in fish and fishery products and risks for human health: A review, *Int. J. Environ. Res. Public Health*, 2022, **20**, 789.

50. E. M. Cunningham, S. M. Ehlers, K. Kiriakoulakis, P. Schuchert, N. H. Jones, L. Kregting, L. C. Woodall and J. T. A. Dick, The accumulation of microplastic pollution in a commercially important fishing ground, *Sci. Rep.*, 2022, **12**, 4217.
51. C. J. Thiele, M. D. Hudson, A. E. Russell, M. Saluveer and G. Sidaoui-Haddad, Microplastics in fish and fishmeal: an emerging environmental challenge?, *Sci. Rep.*, 2021, **11**, 2045.
52. K. Jabeen, B. Li, Q. Chen, L. Su, C. Wu, H. Hollert and H. Shi, Effects of virgin microplastics on goldfish (*Carassius auratus*), *Chemosphere*, 2018, **213**, 323–332.
53. J.-N. Huang, B. Wen, L. Xu, H.-C. Ma, X.-X. Li, J.-Z. Gao and Z.-Z. Chen, Micro/nano-plastics cause neurobehavioral toxicity in discus fish (*Symphysodon aequifasciatus*): insight from brain-gut-microbiota axis, *J. Hazard. Mater.*, 2022, **421**, 126830.
54. X. Mu, S. Qi, J. Liu, L. Yuan, Y. Huang, J. Xue, L. Qian, C. Wang and Y. Li, Toxicity and behavioral response of zebrafish exposed to combined microplastic and bisphenol analogues, *Environ. Chem. Lett.*, 2021, **20**, 41–48.
55. Y. Zhao, Z. Qin, Z. Huang, Z. Bao, T. Luo and Y. Jin, Effects of polyethylene microplastics on the microbiome and metabolism in larval zebrafish, *Environ. Pollut.*, 2021, **282**, 117039.
56. Y.-K. Zhang, B.-K. Yang, C.-N. Zhang, S.-X. Xu and P. Sun, Effects of polystyrene microplastics acute exposure in the liver of swordtail fish (*Xiphophorus helleri*) revealed by LC-MS metabolomics, *Sci. Total Environ.*, 2022, **850**, 157772.
57. Z. Wang, M. Saebi, E. K. Grey, J. J. Corbett, D. Chen, D. Yang and Z. Wan, Ballast water-mediated species spread risk dynamics and policy implications to reduce the invasion risk to the Mediterranean Sea, *Mar. Pollut. Bull.*, 2022, **174**, 113285.
58. C. Espinosa, M. Á. Esteban and A. Cuesta, Dietary administration of PVC and PE microplastics produces histological damage, oxidative stress and immunoregulation in European sea bass (*Dicentrarchus labrax* L.), *Fish Shellfish Immunol.*, 2019, **68**, 574–583.
59. D. Mittempergher, L. Raes and A. Jain, *The Economic Impact of Marine Plastic Pollution in Antigua and Barbuda: Impacts on the Fisheries and Tourism Sectors, and the Benefits of Reducing Mismanaged Waste*, International Union for Conservation of Nature (IUCN), Switzerland, 2023, pp. 1–96.
60. A. Saidi, G. Sacchi, C. Cavallo, G. Cicia, R. Di Monaco, S. Puleo and T. D. Giudice, Drivers of fish choice: an exploratory analysis in Mediterranean countries, *Agric. Food Econ.*, 2022, **10**, 29.
61. S. Zhao, L. Zhu, L. Gao and D. Li, Limitations for microplastic quantification in the ocean and recommendations for improvement and standardization, in *Microplastic Contamination in Aquatic Environments: An Emerging Matter of Environmental Urgency*, ed. E. Y. Zeng, Elsevier, 1st edn, 2018, pp. 27–49.
62. F. Barbosa, J. A. Adeyemi, M. Z. Bocato, A. Comas and A. Campiglia, A critical viewpoint on current issues, limitations, and future research

needs on micro- and nanoplastic studies: From the detection to the toxicological assessment, *Environ. Res.*, 2020, **182**, 109089.
63. H. Joyce, R. Nash, J. Frias, J. White, A. Cau, E. Carreras-Colom and F. Kavanagh, Monitoring microplastic pollution: The potential and limitations of *Nephrops norvegicus*, *Ecol. Indic.*, 2023, **154**, 110441.
64. F. Stock, G. Reifferscheid, N. Brennholt and E. Kostanaia, *Plastics in the Aquatic Environment - Part I*, Springer Nature, Switzerland, 2021.
65. F. Stock, C. Kochleus, B. Bänsch-Baltruschat, N. Brennholt and G. Reifferscheid, Sampling techniques and preparation methods for microplastic analyses in the aquatic environment – A review, *TrAC, Trends Anal. Chem.*, 2019, **113**, 84–92.

CHAPTER 6

Colossoma macropomum, *the "Tambaqui", (Characiformes, Serrasalmidae): A Natural Heritage of the Amazon Threatened by Contamination With Organophosphorus Compounds*

HALLANA CRISTINA MENEZES DA SILVA[a*], DANIELE APARECIDA MATOSO[a,b], ANA LÚCIA SILVA-GOMES[c] AND ROBERTO FERREIRA ARTONI[d]

[a]Programa de Pós-Graduação em Biologia Evolutiva, Laboratório de Genética e Evolução. Departamento de Genética e Biologia Estrutural, Universidade Estadual de Ponta Grossa, Paraná, Brazil; [b]Laboratório de Biotecnologia e Citogenômica Animal. Departamento de Genética, Universidade Federal do Amazonas, Amazonas, Brazil; [c]Laboratório de Parasitologia de Animais Aquáticos. Departamento de Parasitologia, Universidade Federal do Amazonas, Amazonas, Brazil; [d]Laboratório de Genética e Evolução. Departamento de Biologia Estrutural, Molecular e Genética, Universidade Estadual de Ponta Grossa, Paraná, Brazil
*E-mail: hallana19@gmail.com

Issues in Toxicology No. 50
Fish Species in Environmental Risk Assessment Strategies
Edited by Guillermo Eli Liwszyc and Marcelo L. Larramendy

Published by the Royal Society of Chemistry, www.rsc.org

6.1 Organophosphorus Compounds and Their Use and Contamination in Fish Farms

In Brazil, from all the types of animal protein production, fish farming has grown the most in the last eight years, at an average rate of 5.6% over this period, which generates a commercial value of approximately 1.6 billion USD per year. The ranking in the country is separated by two types of production: the production of farmed fish and the production of native fish. In national production, the most produced fish is the Nile tilapia *Oreochromis niloticus* (Cichiliformes, Cichilidae), representing 63.5% of farmed fish, which is equivalent to about 534 005 tons per year. The states with the greatest production of tilapia are Paraná, São Paulo, Minas Gerais, Santa Catarina and Mato Grosso do Sul.[1] Leading the production of native species in Brazil and especially that of the state of Amazonas, is the tambaqui *Colossoma macropomum* (Characiformes, Serrasalmidae) (see Figure 6.1). In fish farming, the production of tambaqui is commonly carried out in semi-excavated tanks, which have a clay bottom. The advantage of this production system, compared to the damming of rivers and streams, is the higher biomass production per unit of volume, better monitoring of fish, faster removal of the fish and the lower probability of infection by parasites and other pathologies.[2] To ensure the success of the production system, one of the main goals of producers is to avoid infection with parasites, which can greatly lower the quality as well as the quantity of fish produced.[3]

Regardless of the species, protozoa are the most common parasites in farmed fish, and are responsible for significant losses in production.[3] Infections by protozoans such as *Ichthyophthirius multifilis* (Ciliophora) and *Piscinoodinium pillulare* (Dinoflagellida), monogenoides such as *Mymarothecium boegeri* and *Anacanthorus spathul* (Dactylogyridae), freshwater proboscis bearing leeches such as Glossiphonidae ssp. (Hirudinea),[4] acanthocephalans such as *Neochinorhynchus buttnerae*[5] and digeneans as *Austrodiplostomum compactum*[6] are common parasites found in tambaqui production.

Figure 6.1 The tambaqui *Colossoma macropomum* (Characiformes, Serrasalmidae).

In order to avoid production losses caused by parasitic infections, it is important for the producer to know about good sanitary management practices, especially in the initial life cycles of tambaqui (culture of larvae and fry).[7] In addition, antiparasitic drugs are quite commonly used to prevent the proliferation of parasites, especially when it comes to animals introduced from other fish farms. To control fish mortality due to infection by these parasites and avoid loss of production, some fish farmers use drugs based on organochlorine and organophosphorus compounds.[8]

Among the most commonly used organophosphates in Brazil, we can highlight trichlorfon (IUPAC: 2,2,2-trichloro-1-dimethoxyphosphorylethanol; CID: 87195248), which is the most commonly used compound in fish farming, and malathion (IUPAC: diethyl 2-dimethoxyphosphinothioylsulfanylbutanedioate, CID: 22435987), which is the most commonly used compound in agriculture; both are employed for eliminating parasites in fish and plants, respectively. Currently, both agrochemicals have been widely studied regarding their effects on aquatic animals *via* direct or indirect contamination.[8–15]

Trichlorfon, shown in Figure 6.2, is marketed in Brazil under the names Dipterex 500®, Methyl Paration®, Neguvon® and Masoten® (Ordinance No. 10 08.03.85 of the Ministry of the Environment – D. O. U. 14.03.85, Ordinance No. 318 23.06.87 of the Ministry of Health – D. O. U. 26.06.87). Dipterex 500® is indicated for the control of pests in plants such as avocado, pineapple, lettuce, cotton, carrots, coconut, cabbage, beans, watermelon, corn, soybeans, among others. Neguvon® is indicated for the treatment of internal and external parasitosis in cattle, horses, sheep and pigs and is composed of 97% trichlorfon. Masoten® is best known for treating fish in fish farming activities and each 100 g of commercial formulation contains 80% trichlorfon.

Masoten® is indicated for the treatment of parasitosis of freshwater fish such as the common fish louse, the fish lice *Argulus* sp. (Ichthyostraca, Argulidae), the gill lice copepod *Ergasilus* sp. (Maxillopoda, Ergasilidae), anchor worms copedod *Lernaea* sp. (Cyclopoida, Lernaeidae), protozoans including *Trichodinas* sp. (Ciliophora, Trichodinidae), the monogeneans gill fluke *Dactylogyrus* sp. (Monogenea, Dactylogyridae) and the salmon fluke *Gyrodactylus* sp. (Monogenea, Gyrodactylidae) as well as some species of trematodes (commercial leaflet Bayer S.A.). The dosage used in fish farming depends on

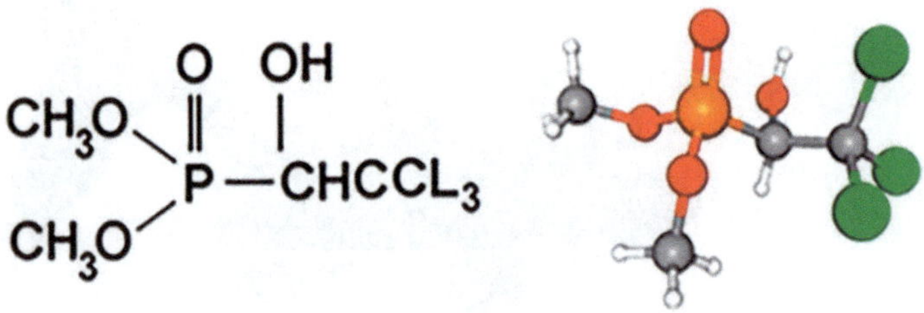

Figure 6.2 Chemical structure of trichlorfon.[8] Courtesy of the National Library of Medicine.

the organism to be eliminated and the size of the tanks in which the product is to be administered, as well as whether the treatment will be before or after the parasite infestation. Usually, aquatic animals treated with trichlorfon are bathed in the substance, a practice commonly known as immersion baths, whereby trichlorfon is previously prepared and added to the water of the tank.[16]

Some xenobiotic compounds, when metabolized in phase I, end up generating metabolites that are considered more toxic than the primary compound. As is the case with trichlorfon which, after its metabolism at the end of phase I, ends up generating secondary metabolites, such as dimethyl trichlorfon, dimethyl dichlorvos, dimethyl hydrogen phosphate, methyl hydrogen phosphate, phosphoric acid and trichloroethanol, which are more toxic. With a long lifetime, the best form of degradation of trichlorfon is demethylation, breakdown of the P–C bond of the chemical structure and steric hydrolysis of dichlorvos.[17]

Resolution No. 20 of the National Environment Council (CONAMA), established on July 18th, 1986, states the permitted dosage of 1.0 mg L^{-1} for organophosphorus compounds in fresh water. However, what is observed in practice is that the levels used by fish farmers vary greatly, ranging from 0.13 mg L^{-1} to 25 mg L^{-1}.[8] According to toxicological studies carried out on trichlorfon, the Brazilian Health Regulatory Agency (ANVISA), linked to the Ministry of Health, part of the Brazilian National Health System (SUS) as the coordinator of the Brazilian Health Regulatory System (SNVS), characterized the toxicity of the compound as Class II, highly toxic. However, in 2008, under the Resolution No. 10, dated February 22nd, 2008, the Ministry of Health *via* ANVISA requested the reassessment of trichlorfon, since several studies showed that the compound could cause neurotoxicity, with carcinogenic potential and reproductive toxicity. After this toxicological re-evaluation, the Resolution RDC No. 37, August 16th, 2010 prohibited the use of the compound in Brazil due to its genotoxicity, neurotoxicity, immunotoxicity, endocrine disruption and general toxicity that compromises reproduction. In addition, neurological damage in humans was more pronounced than that demonstrated in animals. Despite the toxicological reassessment, Resolution No. 583 (May 10th, 2018) again allowed the use of the compound in agricultural practices.

Malathion [diethyl(dimethoxythiophosphorylthio)succinate] is the best-known organophosphorus compound and is marketed as Malathion® (see Figure 6.3). In Brazil, this compound is classified as an insecticide and is used mainly for pest control in lettuce, cotton, rice, coffee, beans, corn, soybeans, among others (Information Note No. 215, February 2nd, 2023 [DOU of 08/03/2023]). In 2019, in the Resolution No. 2.080 (July 31st, 2019), the insecticide was reclassified as Category 5, thus being considered unlikely to cause acute harm, *i.e.*, a blue band product by the ANVISA. However, according to the International Agency for Research on Cancer (IARC) (UN, WHO), the compound was classified as 2a, probably carcinogenic to humans.[18]

Currently, Malathion®, in spite of the above-mentioned and widely known negative effects, is one of the most widely used insecticides in the Amazon

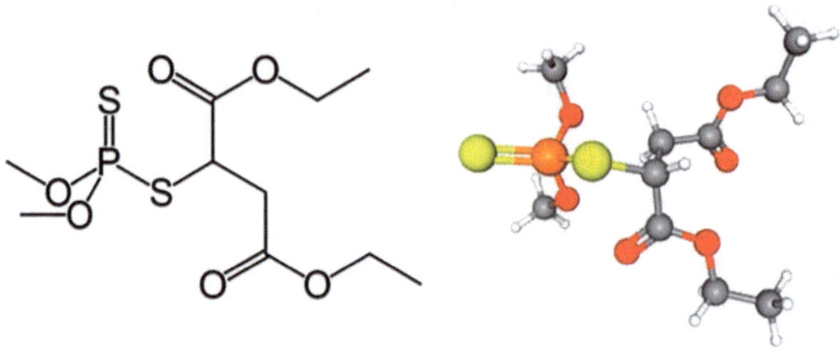

Figure 6.3 Chemical structure of Malathion®.[8] Courtesy of the National Library of Medicine.

region. Several studies have detected the presence of this insecticide not only in sprayed fruits and vegetables from plantations, but also its accumulation in soil and water from streams and rivers near to crops, where the compound had been used.[19] Recent studies have shown that there is contamination by pesticides in urban and peri-urban areas of the Brazilian Amazon basin, and rivers such as the Amazon and the Negro are contaminated with the compound.[15] Just like trichlorfon and other organophosphates, the main mechanism of action of this compound is the blocking of the enzyme acetylcholinesterase (AChE).

It is known that organophosphates can cause some harmful effects in fish such as impaired enzymatic activity of AChE, impaired activity of some of the enzymes responsible for detoxification, increased enzymatic activity in cytochrome P450 pathways, damage to organ structure, oxidative stress, imbalance of energy metabolism, immune response, alteration in tumour response and apoptosis pathways. One of the main organs affected by organophosphates is the liver, which is the most important organ when it comes to metabolism and detoxification of organisms.[20]

6.2 Absorption and Metabolism of Organophosphates

Organophosphorus compounds have their main application in agriculture, in order to control pests, but most of this class of pesticides is considered toxic since a range of studies have pointed out that this type of compound can be bioaccumulated in the environment.[21] These compounds were initially manufactured during World War II and Malathion® was one of the first organophosphates to be considered "safe" for commercialization. After the authorization of this compound, at least more than 200 other chemicals have been authorized worldwide.[22]

Organophosphorus compounds are recognized mainly by the presence of phosphorus (P) in their structure, with three atoms (H, O and C) linked individually and a covalent bond that supports the structure. They derive from

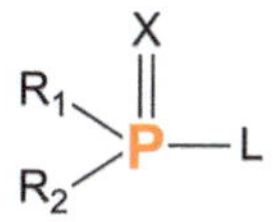

Figure 6.4 Chemical structure of organophosphorus compounds.[21]

two main groups: phosphoric acid (H_3PO_4) and phosphonic acid (H_3PO_3). The nomenclature of these compounds is based on the atoms surrounding the central P. In the structure of the organophosphate, the leaving group (L) is the one considered most reactive and most variable when organophosphorus phosphorylates AChE. The least reactive groups within the structure are the R1 and R2 groups (see Figure 6.4), which are most often composed of alkoxy groups.[22]

The main effect of organophosphorus compounds is the immediate inhibition of the action of the enzyme acetylcholinesterase (AChE). This enzyme is characterized as a cholinesterase, terminating nerve impulse transmission by modifying the neurotransmitter acetylcholine (ACh) into acetic acid (Ac) and choline (Ch) by a process of hydrolysis. When the organophosphorus-AChE bond occurs, an irreversible complex is formed, causing the action of the enzyme to be blocked.[8] In tambaqui, this action has been demonstrated in the work of Duncan *et al.*[11] These authors demonstrated that the loss of AChE enzymatic activity was achieved in fish exposed to trichlorfon. This immediate blockage of AChE may be more noticeable in fish, in tissues such as muscle and gills, where muscle stiffness is noticeable and the fish have swimming difficulty.[23] In tissues such as the brain, its action is immediately noticed by the loss of natatory balance.[23]

Xenobiotics, using the definition of the mid-1960s, include any chemical to which an organism is exposed to and is foreign to that organism's normal metabolism. Thus, the following are considered as xenobiotics: pesticides, occupational chemicals, clinical drugs, drugs of abuse, chemicals with general authorized use, among others.[24] For their metabolization, organophosphates are subjected to several metabolic pathways, which make up phase I and phase II of the detoxification process, which are mediated by several groups of enzymes. This is due to the variation in the atoms bonded to P and C in the structure of the compound. The "leaving group" is the first group to be "turned off", which phosphorylates its esterase targets. After this phosphorylation, secondary metabolites are formed, which also need to be metabolized and then excreted from the cell. Phase I of the metabolism of these compounds consists of the oxidation, reduction and hydrolysis phase of the compound, whereas phase II consists of conjugation reactions.[25]

In the oxidation step, which occurs in phase I of metabolism, the two main enzymes to be mentioned are cytochromes P450 (CYP450) and flavin-containing monooxygenases (FMOs). Overall, cytochrome P450 enzymes comprise one of the most important classes of enzymes responsible for the metabolism of xenobiotics. This is because, in the family, they have different isoforms, and this ensures a wide substrate specificity, catalyzing a variety of oxidations in phosphorus, sulfur, nitrogen and carbon. CYP450 pathways generally occur in the endoplasmic reticulum and have a higher metabolization capacity in the liver of vertebrates.[25] Flavin-containing monooxygenases are an additional group of enzymes that are responsible for the oxidation step and are most often found in the liver of mammals. Unlike CYP, they have fewer isoforms in the family and, therefore, a more limited substrate specificity.[25]

The next step in phase I is the step of reduction of the compounds, which are the results of CYP-mediated reactions and are considered rare in mammals. The importance of reduction reactions is precisely to provide more reducing environments, facilitating the metabolism of the compound. The last step of phase I is the catalytic hydrolysis reaction of phosphates/oxons, which result in the elimination of the leaving group. This step is catalyzed by enzymes known as phosphotriesterases (A-esterases), which are hydrolases directed to hydrolyze organophosphates.[25]

Phase II of xenobiotic metabolism consists of conjugation reactions, whose main objective is to make the metabolites that resulted from phase I more soluble in water for their subsequent excretion.[25] These conjugations occur more frequently in the hydrolyzed leaving groups in phase I. These hydrolytic metabolites are already detoxified products, especially with regard to AChE activity, so they do not need another phase of metabolism. For this reason, phase II reactions are considered much less toxic than phase I reactions. One of the groups of main enzymes that can be mentioned in this step is that of glutathione-S-transferases (GSTs).[25]

Some studies have been developed by Brazilian research groups from the Federal University of Amazonas (UFAM), the National Institute of Amazonian Research (INPA) and the State University of Ponta Grossa (UEPG) that seek to more fully understand the correlation of organophosphate contaminants and their influence on the genome of native Amazonian species, such as the tambaqui *C. macropomum* and the likely effects for their cultivation in captivity. In this sense, the next section of this chapter will carry out a brief review on the development of such studies and the main conclusions reached so far.

6.3 Genomic Studies on the Native Amazonian Species *Colossoma macropomum*, the Tambaqui, and Organophosphorus Contaminants

Malathion was one of the first organophosphates to be released for use and its application is mainly as an insecticide in agriculture. Several studies seek to evaluate the indirect effects of this compound on aquatic organisms, since crops may be located near rivers and lakes. Describing the effects of this compound on the tambaqui *C. macropomum*, which is a non-target organism native to freshwater habitats in the Amazon and Orinoco basins of tropical South America, Souza *et al.*, first defined the $LC50_{96\text{ h}}$ of malathion for the species. The value of the lethal concentration that was capable of killing 50% of the animals in 96 hours of exposure was 15.77 ± 3.30 mg L^{-1}, with a confidence interval of 95%.[13]

After the definition of the $LC50_{96\ h}$, Souza *et al.* evaluated parameters such as mitochondrial physiology, production of reactive oxygen species (ROS), biotransformation and antioxidant response and oxidative, genotoxic and neurotoxic damage in malathion-exposed tambaqui specimens.[13] Their results showed that, in response to exposure, there was an increase in the activity of biotransformation enzymes and antioxidants, which reduced the production of ROS. Mitochondrial metabolism and AChE enzyme activity remained constant during the period of exposure and the DNA was not damaged, which is somewhat surprising, since the immediate inhibition of AChE is described as the main mechanism of action of all organophosphates.[22] As already seen in other species of aquatic animals, there were severe lesions in the liver and gills as well as increased expression of the proto-oncogene *ras*. Furthermore, the *tp53* gene showed no significant difference in expression between the groups studied.[13]

Although exposure to malathion did not cause inhibition of AChE enzyme activity, Souza *et al.* observed that phase II biotransformation mechanisms were activated in the first hours of exposure, mainly in the liver, indicating, accordingly, that this organophosphate, as expected, enters phase II for its metabolization. Oxidative damage was also observed in the first hours of exposure and mitochondrial bioenergetics was also affected. These studies prove that indirect contamination of tambaqui specimens can cause serious damage to fish.[14]

Trichlorfon is another organophosphorus compound that has been widely studied, mainly for its use in fish farming. To determine the $LC50_{96\ h}$ value, as well as to evaluate the effects of exposure to trichlorfon, Silva *et al.* exposed juvenile tambaqui specimens to five different concentrations of the compound.[23] The obtained results evidenced a $LC50_{96\ h}$ value equivalent to 0.870 mg L^{-1}. The fish analyzed showed impaired swimming ability at the highest concentration as well as damage to some organs. Duncan *et al.* carried out an analysis also in tambaqui specimens searching the enzymatic activity of AChE and GST in the brain, muscle, intestine and liver of trichlorfon-exposed fish.[11] AChE showed a 90% inhibition of activity in the brain, muscle and intestine. On the other hand, GST showed no variation in its activity in any of the tissues studied.[11] The blocking of the enzymatic activity was noticeable in the first hours of exposure of the exposed fish to a higher concentration, with muscle stiffness, as a result of the AChE blockage in the muscle and loss of swimming ability, indicating as a result the blocking activity of AChE in the brain.[11]

Several studies have demonstrated that trichlorfon has a damaging effect on various tissues that were evaluated, as well as gene expression and metabolic activation of various degradation pathways and it also alters the expression of genes that concern energy metabolism and proto-oncogenes.[13,14,20,26–29] Furthermore, other studies have shown that trichlorfon has the ability to activate mobile genetic retroelements, thus indicating the high rate of cellular damage that the compound may be exerting on the cells of tambaqui.[30]

Retroelements of the *Rex* family, as indicated by their names, are mobile genetic elements that retrotranspose *via* an RNA-intermediate that is reverse-transcribed to DNA by the encoded reverse transcriptase and integrated into a new location within the host genome by an integrase enzyme. Furthermore, retrotransposable elements are those whose dispersion in the genome can be increased as a result of environmental stresses.[31,32] Costa *et al.*[30] studied the behaviour of the *Rex1*, *Rex3* and *Rex6* retroelements in the tambaqui genome.[29] The specimens were exposed to 30% (0.261 mg L^{-1}) and 50% (0.435 mg L^{-1}) for the trichlorfon $LC50_{96\ h}$.[23] The observed results showed that the *Rex1* and *Rex6* retroelements had a dispersed distribution within the chromosomes. *Rex3* was the retroelement that presented the greatest dispersion and increased duplications in the chromosomes of the analyzed specimens, indicating that this retroelement is more sensitive to trichlorfon exposure. It is believed that this result was observed because the organism needs to have genetic flexibility in order to adapt to environmental changes such as stress from exposure to trichlorfon.[30]

In a similar study, Nonato *et al.* performed an enzymatic analysis of the methylation profile of the tambaqui 18S rDNA gene exposed to 30% and 50% of trichlorfon $LC50_{96\ h}$.[12] For this, restriction enzymes and a methylation-sensitive PCR were used. The results showed an increase in methylated fragments of the 18S rDNA gene at both studied concentrations. This result suggests that trichlorfon may have a significant effect on the genome of the tambaqui, which deactivates genes when necessary due to stress caused by the exposure to trichlorfon.

In order to analyze the effects of exposure on the tambaqui liver transcriptome, Silva *et al.* (unpublished results) exposed tambaqui specimens to 50% of the $LC50_{96\ h}$ for trichlorfon, which was mixed in water. As a result, several metabolic pathways were analyzed that presented differentially expressed genes when compared to the control group, with emphasis on cellular transport pathways, immune system and apoptosis genes. Finally, the authors suggest the hypothesis presented in Figure 6.5. According to them, the molecule enters the cells through genes of solute-carrying enzymes (different isoforms), thereby activating the immune response through genes of enzymes such as *aerolysin*, *progranulin*, *Q9 alpha* and *F10 alpha*. After activation of the immune system, proto-oncogenes such as *tp53*, *aldh1l1*, *padi2* and *pim2* are activated trying to repair the damage primarily caused into DNA. If this damage cannot be repaired, apoptosis genes, such as *cidec* and *bada*, are activated.[33]

Regarding the evaluation of gene expression levels of those genes related to the response of the liver cells from trichlorfon-exposed tambaqui specimens, Carvalho *et al.* (unpublished data) evaluated the expression of *AChE*, *GST*, *CYP2J6* and *CYP2C8*. The specimens were exposed during 96 h to 30% and 50% of the $LC50_{96\ h}$ determined for the species. The results showed a decrease in gene expression levels of *GST*, gene encoding for the enzyme responsible for phase II biodegradation. The expression levels of *AChE*, gene encoding for the main active enzyme of the compound, were not altered by

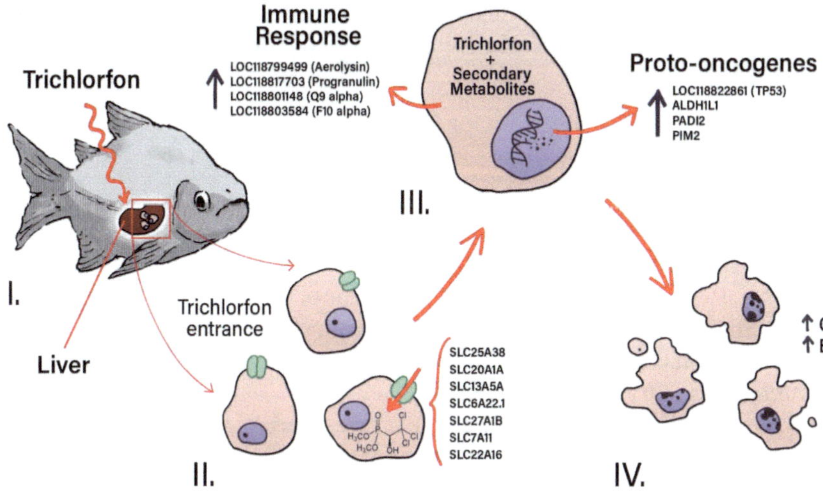

Figure 6.5 Summary of the liver response in tambaqui to exposure to trichlorfon.[34]

trichlorfon exposure. Similar results were also observed for the expression of *CYP2J6* and *CYP2C8* genes encoding for the enzymes responsible for the biotransformation of xenobiotics.

6.4 Conclusions and Final Remarks

All these differentially expressed genes, as well as the activated metabolic pathways, demonstrate the harmful effects that trichlorfon molecules cause in tambaqui, focusing on the pathways that concern proto-oncogenes, which are important for the maintenance of the healthy cellular environment. All these studies clearly and objectively show that the organophosphates trichlorfon and malathion, as well as other xenobiotics, have profound effects on the genomic homeostasis of tambaqui. This fish, as already mentioned, is the most farmed and frequently sold fish in the northern region of Brazil. It is also one of the main sources of animal protein of the riverine and native population. Providing healthy, high-quality food must therefore be an urgent goal for food security and the sustainable future of local communities as well as the preservation of the environment. In this sense, it is recommended that these organophosphate compounds should be used sparingly in fish farms since their uncontrolled use can produce undesirable phenotypes that can affect the quality of the final fish product.

Acknowledgements

The present work was supported in part by INCT ADAPTA II funded by CNPq – Brazilian National Research Council (465540/2014-7), FAPEAM – Amazonas State Research Foundation (062.1187/2017), and CAPES – Coordination for the Improvement of Higher Education Personnel Post Graduate Program in Genetics, Conservation and Evolutive Biology – National Institute of Amazonian Research.

References

1. F. Souza and I. Dias, Peixe BR da Piscicultura, in *Anuário 2024*, Associação Brasileira de Piscicultura, 2024, pp. 1–63.
2. J. Costa, R. Freitas, A. L. Gomes, G. Bernadino, D. Carneiro and M. I. Martins, Effect of stocking density on economic performance for *Colossoma macropomum* (Cuvier, 1816), juvenile in earthen Earthen ponds, *Lat. Am. J. Aquat. Res.*, 2016, **44**, 165–170.
3. J. R. Garcez, V. S. L. da Nóbrega, T. P. Torres and A. A. Signor, Cultivo de tambaqui (*Colossoma macropomum*) em tanques-rede: Aspectos técnicos, *Res. Soc. Dev.*, 2021, **10**, e45810817560.
4. E. F. Santos, M. Tavares-Dias, D. A. Pinheiro, D. Anavias, L. R. Neves, L. Rigôr, R. Marinho das Graças Barbosa and M. Kelly Días Reis, Fauna parasitária de tambaqui Colossoma macropomum (Characidae) cultivado em tanque-rede no estado do Amapá, Amazônia oriental, *Acta Amazonica*, 2013, **43**, 105–111.
5. A. L. Silva-Gomes, J. Gomes Coelho Filho, W. Viana-Silva, M. I. Braga-Oliveira, G. Bernardino and J. I. Costa, The impact of *Neoechinorhynchus buttnerae* (Golvan, 1956) (Eoacanthocephala: Neochinorhynchidae) outbreaks on productive and economic performance of the tambaqui *Colossoma macropomum* (Cuvier, 1818), reared in ponds, *Lat. Am. J. Aquat. Res.*, 2017, **45**, 496–500.
6. E. F. de Farias, H. C. M. da Silva, A. P. C. de Carvalho, R. de M. Martins, A. Belem-Costa, W. P. Duncan, A. L. S. Gomes and D. A. Matoso, Molecular identification of *Austrodiplostomum* sp., an eye parasite among farmed tambaqui in Amazonia, *Genet. Mol. Biol.*, 2023, **46**, e20210345.
7. E. Affonso, F. Pinheiro de Barros, E. Martins Brasil, M. Tavares-Dias and E. Akifuni Ono, Macapá, in *Manejo e Sanidade de Peixes em Cultivo*, ed. M. Tavares-Dias, and E. Amapá, Brazil, 2009, pp. 346–360.
8. H. C. M. da Silva, A. P. C. de Carvalho, A. L. S. Gomes, R. F. Artoni and D. A. Matoso, Impact of trichlorfon organophosphate use in pisciculture: A review, *Stud. Environ. Anim. Sci.*, 2022, **3**, 1399–1423.
9. M. D. Baldissera, C. F. Souza, B. Parmeggiani, R. G. Vendrusculo, L. C. Ribeiro, D. K. Muenchen, C. C. Zeppenfeld, A. D. Meinhart, R. Wagner, R. Zanella, O. D. Prestes, A. S. da Silva, G. Leipnitz and B. Baldisserotto, Protective effects of diet containing rutin against trichlorfon-induced muscle bioenergetics disruption and impairment on fatty acid profile of silver catfish, *Ecotoxicol. Environ. Saf.*, 2020, **205**, 111127.
10. M. D. Baldissera, C. F. Souza, R. Zanella, O. D. Prestes, A. D. Meinhart, A. S. Da Silva and B. Baldisserotto, Behavioral impairment and neurotoxic responses of silver catfish *Rhamdia quelen* exposed to organophosphate

pesticide trichlorfon: Protective effects of diet containing rutin, *Comp. Biochem. Physiol., Part C: Toxicol. Pharmacol.*, 2021, **239**, 108871.

11. W. P. Duncan, J. J. S. Idalino, A. G. da Silva, R. F. Moda, H. C. M. da Silva, D. A. Matoso and A. L. S. Gomes, Acute toxicity of the pesticide trichlorfon and inhibition of acetylcholinesterase in *Colossoma macropomum* (Characiformes: Serrasalmidae), *Aquacult. Int.*, 2020, **28**, 815–830.
12. L. da S. Nonato, H. C. M. Silva, A. L. Silva-Gomes, J. B. Traldi, R. F. Artoni and D. A. Matoso, Methylation profile of 18S rDNA gene in brain and muscle of tambaqui exposed to parasiticide trichlorfon, *Int. J. Zool. Anim. Biol.*, 2023, **6**, 1–7.
13. S. Silva de Souza, R. N. Machado, J. C. da Costa, D. F. Campos, G. Sebrenski da Silva and V. M. Fonseca de Almeida-Val, Severe damages caused by Malathion exposure in *Colossoma macropomum*, *Ecotoxicol. Environ. Saf.*, 2020, **205**, 111340.
14. S. S. de Souza, J. da S. Castro, D. F. Campos, R. S. Pereira, G. A. Bataglion, G. S. da Silva and V. M. F. de Almeida-Val, Temporal exposure to malathion: Biochemical changes in the Amazonian fish tambaqui, *Colossoma macropomum*, *Aquat. Toxicol.*, 2021, **241**, 105997.
15. A. Rico, R. de Oliveira, G. Silva de Souza Nunes, C. Rizzi, S. Villa, B. De Caroli Vizioli, C. C. Montagner and A. V. Waichman, Ecological risk assessment of pesticides in urban streams of the Brazilian Amazon, *Chemosphere*, 2022, **291**, 132821.
16. L. R. R. Rauco, *Diss. Mestrado*, Univ. Est. Paulista, Brazil, 2002, pp. 1–86.
17. Internacional Programme On Chemical Safety, *Environmental Health Criteria 132 Trichlorfon*, World Heal Organization, 1992.
18. Internation Agency for Reseach of Cancer, *IARC Monographs on the Evaluation of Carcinogenic Risks to Humans, Some Organophosphate Insecticides and Herbicides*, IARC, Wold Health Organization, Lyon, France, 2017, vol. 112.
19. P. S. Campos, *Diss. Mestrado*, Univ. Fed. do Amaz., Brazil, 2009, pp. 1–78.
20. R. Jia, Y. Hou, W. Feng, B. Li and J. Zhu, Alterations at biochemical, proteomic and transcriptomic levels in liver of tilapia (*Oreochromis niloticus*) under chronic exposure to environmentally relevant level of glyphosate, *Chemosphere*, 2022, **294**, 133818.
21. H. Dzudzevic Cancar, S. Soylemez, Y. Akpinar, M. Kesik, S. Göker, G. Gunbas, M. Volkan and L. Toppare, A novel acetylcholinesterase biosensor: Core-shell magnetic nanoparticles incorporating a conjugated polymer for the detection of organophosphorus pesticides, *ACS Appl. Mater. Interfaces*, 2016, **8**, 8058–8067.
22. H. W. Chambers, E. C. Meek and J. E. Chambers, *Chemistry of Organophosphorus Insecticides*, Elsevier Inc., 3rd edn, 2010, vol. 2.
23. H. C. M. da Silva, A. G. da Silva, J. J. S. Idalino, F. B. de Sousa, A. L. S. Gomes, W. P. Duncan and D. A. Matoso, Trichlorfon acute lethal toxicity to juvenile tambaqui (*Colossoma macropomum*), *Aquacult. Res.*, 2020, **51**, 863–866.

24. E. Hodgson, *Introduction to Biotransformation (Metabolism)*, Elsevier Inc., 3rd edn, 2010, vol. 1.
25. J. E. Chambers, E. C. Meek and H. W. Chambers, *The Metabolism of Organophosphorus Insecticides*, Elsevier Inc., 3rd edn, 2010, vol. 2.
26. G. S. da Silva, L. V. de Matos, J. O. da S. Freitas, D. F. de Campos and V. M. F. de Almeida e Val, Gene expression, genotoxicity, and physiological responses in an Amazonian fish, *Colossoma macropomum* (Cuvier 1818), exposed to Roundup® and subsequent acute hypoxia, *Comp. Biochem. Physiol. C Toxicol. Pharmacol.*, 2019, **222**, 49–58.
27. R. Chen, R. Hou, X. Hong, S. Yan and J. Zha, Organophosphate flame retardants (OPFRs) induce genotoxicity *in vivo*: A survey on apoptosis, DNA methylation, DNA oxidative damage, liver metabolites, and transcriptomics, *Environ. Int.*, 2019, **130**, 104914.
28. M. Zhu, J. Gong, M. Zhan, C. Xi, G. Shen and H. Shen, Transcriptome analysis reveals the molecular mechanism of long-term exposure of *Eriocheir sinensis* to low concentration of trichlorfon, *Comp. Biochem. Physiol., Part D: Genomics Proteomics*, 2021, **40**, 100916.
29. X. Wang, X. Chang, L. Zhao, J. Feng, H. Li and J. Liang, Trichlorfon exposure in common carp (*Cyprinus carpio* L.) leads to oxidative stress, neurotoxicity, and immune responses, *Aquaculture*, 2022, **548**, 737681.
30. M. dos Santos Costa, H. C. Menezes da Silva, S. Cardoso Soares, R. M. Favarato, E. Feldberg, A. L. Silva Gomes, R. Ferreira Artoni and D. A. Matoso, A perspective of molecular cytogenomics, toxicology, and epigenetics for the increase of heterochromatic regions and retrotransposable elements in tambaqui (*Colossoma macropomum*) exposed to the parasiticide trichlorfon, *Animals*, 2022, **12**, 1945.
31. F. Araújo da Silva, C. H. Schneider, E. Feldberg, F. Beggiato Baccaro, N. D. Mourna Carvalho and M. C. Gross, *Zebrafish*, 2016, **13**, 197–208.
32. H. C. Menezes da Silva, L. Braga Ribeiro, A. J. da Mota, E. Feldberg and D. A. Matoso, Impacts of stress caused by copper sulfate ($CuSO_4$) on the genome of the tambaqui (*Colossoma macropomum*): Quantification of *Rex1* and heterochromatic profile, *Braz. Arch. Biol. Technol.*, 2023, **66**, e23220170.
33. Silva *et al.*, 2024, unpublished data.

CHAPTER 7

Auxinic Herbicides and Ecotoxicological Risk for Fish Species Inhabiting the Pampas Region of South America

CELESTE RUIZ DE ARCAUTE[a,b*]

[a]Cátedra de Citología, Facultad de Ciencias Naturales y Museo, Universidad Nacional de La Plata, Calle 64 No. 3, B1904AMA La Plata, Argentina;
[b]Consejo Nacional de Investigaciones Científicas y Técnicas (CONICET), Argentina
*E-mail: cruizdearcaute@fcnym.unlp.edu.ar

7.1 Introduction

Herbicides are chemical compounds used to control undesirable plants or weeds.[1] They are applied to crops of economic importance and in the forest management in urban, peri-urban and rural areas. Weeds exert numerous detrimental effects on crops; they decrease the nutritional, water and light resources of crops and harbour potentially harmful organisms, which in turn contaminate crops, with an estimated 20–40% reduction in the world's agricultural production.[2,3] Within the pesticide market, the herbicide segment leads the global sales. Glyphosate (GLY) is the world's top selling herbicide, followed by atrazine, *S*-metolachlor, 2,4-dichlorophenoxyacetic acid (2,4-D), acetochlor, pendimethalin, metolachlor, propanil, dicamba (DIC) and trifluralin.[4] In Argentina, in particular, the use of these products has increased

Issues in Toxicology No. 50
Fish Species in Environmental Risk Assessment Strategies
Edited by Guillermo Eli Liwszyc and Marcelo L. Larramendy

Published by the Royal Society of Chemistry, www.rsc.org

more than 100% in the last 15 years due to the number and variety of herbicide-tolerant crops.[5]

Most pesticides must be prepared in formulations suitable for marketing which contain active and inert ingredients. The active ingredient (a.i.) prevents, destroys, repels or mitigates pests. In the plant, it acts as a regulator, defoliant, desiccant or nitrogen stabilizer. The rest of the components are excipients without biocidal action *per se*, acting as solvents, adjuvants, emulsifiers, surfactants and/or preservatives that improve the penetration, extend the useful life or minimize the degradation of the pesticide in both target and non-target plants.[6] Several studies have shown that inert compounds may have greater toxicity than the a.i. itself.[7–11] Therefore, the toxic properties of a given commercial formulation not only depends on the a.i., but also on its possible interactions with the inert compounds. Furthermore, numerous studies have also demonstrated toxicity exerted by the excipients present within commercial formulations.[8–10,12,13] Nevertheless, in Argentina, their true identity is still considered to be a trade secret, *i.e.*, confidential information that manufacturers protect under federal or state laws.[14]

Auxinic (or hormonal) herbicides, including 2-methyl-4-chlorophenoxyacetic acid (MCPA), 2,4-D, DIC, picloram and quinclorac, among others (see Table 7.1) were artificially synthesized for the first time in 1940.[15,16] They act by imitating the action of the natural auxinic hormone group, indoleacetic acid (IAA) being the predominant form. At low concentrations, auxinic herbicides stimulate cell division, promoting the rapid growth of stems, petiolus and leaves.[17] On the other hand, at high concentrations, they function as herbicides by inhibiting leaf growth and selectively eliminate Dicotyledons without affecting Monocotyledons through a mechanism that has not yet been totally fully elucidated.[15,16] Plants exposed to auxinic herbicides show abnormal growth and destruction of the parenchyma, cambium and phloem, which normally trigger the death of the organism.[18–20]

In the last decades, fish populations around the world have decreased as a result of pesticide contamination of natural and agricultural areas.[21,22] In this context, fish fauna has become a suitable model to evaluate the toxicity exerted by many pesticides.[23] Since these vertebrates are distributed practically in all areas of the aquatic environment, they are excellent indicators of pesticide-induced toxic, physiological, genotoxic, mutagenic and carcinogenic effects.[24–28] Therefore, they have long been used as experimental models to detect and predict environmental toxicity through acute and chronic

Table 7.1 Families of auxinic herbicides.

Family	Herbicides
Phenoxy-carboxylic acids	2,4-D, MCPA
Benzoic acids	DIC, chloramben
Pyridino-carboxylic acids	Picloram, clopyralid
Pyrimidine-carboxylic acids	Aminocyclopyrachlor
Quinoline-carboxylic acids	Quinclorac

bioassays. Species like the zebrafish *Danio rerio* (Cypriniformes, Cyprinidae), the common medaka *Oryzias latipes* (Beloniformes, Adrianichthyidae), the rainbow trout *Oncorhynchus mykiss* (Salmoniformes, Salmonidae), the guppy *Poecilia reticulata* (Cyprinodontiformes, Poeciliidae), the goldfish *Carassius auratus* (Cypriniformes, Cyprinidae), the common carp *Cyprinus carpio* (Cypriniformes, Cyprinidae) and the spotted snakehead fish *Channa punctata* (Anabantiformes, Channidae), among several others, have been used as vertebrate model organisms in different toxicity tests worldwide.[29–33]

The family Poeciliidae, from the order Cyprinodontiformes, is mostly composed of small fresh or brackish water fishes. It includes several well-known live-bearing aquarium species, *e.g.*, the guppy *P. reticulata*, the molly *P. sphenops*, the platy or platyfish *Xiphophorus maculatus*, the swordtail *X. helleri* and ten spotted live-bearer fish *Cnesterodon decemmaculatus*, whose distribution ranges from Southern USA to Northern areas of the Río de La Plata (River Plate) and Africa, including Madagascar. Moreover, two species from this family, *Poecilia* sp. and *Gambusia* sp., are widely used for mosquito larvae control as a strategy for malaria prevention, currently covering areas in tropical and subtropical regions around the world.[34,35]

The South American Pampas Region is a great grass plain located in territories of three countries, Argentina, Brazil and Uruguay,[36–38] covering more than 50 000 000 ha of arable lands mainly sown with genetically modified crops.[39] A large variety of water bodies in the area of the Rio de La Plata Basin (shared by Argentina and Uruguay) as well as other South American river basins are inhabited by high-density populations of the poeciliid *C. decemmaculatus* (see Figure 7.1), including the Lagoa dos Patos lagoon in Brazil, the basins of the Salado and Negro Rivers in Argentina and coastal streams of both countries.[28,40] *Cnesterodon decemmaculatus*, as a member of the Poeciliidae family, is characterized for being livebearers, indicating that females retain their eggs inside the body and give birth to live, free-swimming young. The species is commonly used as a model organism to measure the lethal and sublethal effects caused by exposure to agrochemicals, both *in situ* and under laboratory experimental conditions. This species is easy to breed and maintain in the laboratory. Furthermore, its high fertility rate,

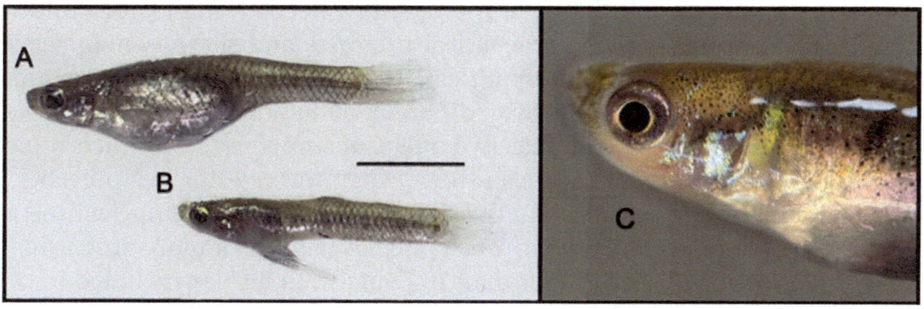

Figure 7.1 Adult specimens of *Cnesterodon decemmaculatus*, female (A) and male (B). Details of the axial part of the body (C). Bar represents 10 mm.

rapid development and broad range of tolerance to temperature, salinity and pH makes it suitable for laboratory bioassays.[40–42] In this sense, it has been widely used to assess the lethal and sublethal effects induced by several pesticides, including paraquat,[43,44] pirimicarb,[45,46] endosulfan,[47] chlorpyrifos,[45,47–53] cypermethrin[48,54–56] and GLY.[37,45,52,54,56–60] Other Pampasic species commonly employed for the evaluation of pesticide-induced toxic effects are the Argentinian silverside, also known as pejerrey, *Odontesthes bonariensis* (Atheriniformes, Atherinopsidae),[61–63] the one sided livebearer *Jenynsia lineate* (Cyprinodontiformes, Anablepidae),[64,65] the boga *Leporinus obtusidens* (Cyprinodontiformes, Anablepidae),[66,67] the guppy *P. vivipara* (Cyprinodontiformes, Poeciliidae)[68,69] as well as the silver catfish, also known as bagre negro *Rhamdia quelen* (Siluriformes, Heptapteridae).[70,71]

7.2 Risk Assessment Endpoints in Fish Species

Acute lethality assays provide an index of the toxicity induced by a given pollutant, without considering its sublethal effects.[72,73] Therefore, they only offer partial information about the toxic effects of pollutants on the biota. Sublethal alterations may have serious ecological impacts on behaviour, reproduction and fitness. These effects often occur at lower concentrations than those causing mortality.[74,75] Thus, these alterations are used as sublethal toxicity endpoints and serve as a tool for the environmental risk assessment and analyses of the toxicological impact of certain contaminants.[76,77] Fish may also exhibit other alterations as a result of xenobiotic exposure, such as those observed during social behaviour, which affect their ability to search for food and avoid predators, putting their survival at risk.[78,79] Accordingly, the effects of environmental contaminants by employing these endpoints have been studied by several research groups.[80–82]

Among genotoxic evaluation assays, the use of the micronucleus (MN) assay is well documented.[46,83–87] This method allows one to assess the clastogenic/aneugenic damage induced by xenobiotics at the chromosome level in different types of proliferating cells.[23,88] Particularly, the analysis of MNs has become an accepted and widely used methodology in genotoxicological studies of fish and other aquatic organisms exposed to a wide variety of different xenobiotics.[9,89,90] Furthermore, it has become mandatory during the approval process of pharmaceutical products and several chemicals.[91] Besides, various nuclear abnormalities (NAs) are usually analysed as indicators of DNA damage induced by different pollutants,[92] such as binucleated cells (BNs), when the cell has two nuclei of similar size with their intact nuclear membrane; blebbed nuclei (BLs), when the cell nucleus presents a small evagination of genetic material that protrudes from the nuclear membrane; lobed nuclei (LBs), when the cell nucleus has one or more protrusions of the nuclear membrane, larger than BL. Finally, to be also included in the list are notched nuclei (NTs), when nuclei present vacuoles and a pronounced invagination of their nuclear membrane.[8,89,93] These NAs are usually designated as "other nuclear alterations", following established criteria.[8,89,93]

The single cell gel electrophoresis (SCGE) or comet assay is widely used to monitor polluted areas (*in situ* assays),[94,95] as well as to evaluate the genotoxic effects induced by xenobiotics after direct or indirect exposure under laboratory conditions (*in vivo* assays).[45,83,96] As such, the bioassay is a highly sensitive, versatile and rapid method for the detection of several DNA lesions such as single- and double-DNA strand breaks and alkaline-labile sites induced by different environmental pollutants.[97] The technique can be modified by the addition of lesion-specific endonucleases that induce DNA breaks at the identified damaged bases. These endonuclease-induced additional breaks provide information about the type of lesion induced at DNA base levels and can be used to estimate oxidative DNA damage in genotoxicological evaluations.[97–99] Two commonly used endonucleases are the formamidopyrimidine-DNA glycosylase (Fpg), which identifies and eliminates damaged purines from DNA and the endonuclease III (Endo III), which transforms oxidized pyrimidines into strand breaks.[100,101] Furthermore, the use of this worldwide employed methodology in a wide variety of cell types and species used as biotic matrices and types of DNA damage, fully demonstrates its versatility, as recently summarized by Collins and co-workers.[102]

7.3 Studies Evaluating Auxinic Herbicides in Pampasic Fish Species

A list of studies evaluating this group of herbicides employing species that inhabit the Pampas region, as biological matrices, is presented in Table 7.2. Two auxin-based herbicide formulations from Argentina, the benzoic acid DIC (Banvel®, 57.7% DIC, Syngenta S.A.)[103] and the 2,4-D amine salt (DMA®, 54.8% 2,4-D, Dow Agrosciences S.A.)[103] were investigated by our research group to determine their acute lethal and sublethal effects on *C. decemmaculatus* specimens exposed under laboratory conditions. Whereas mortality was the endpoint selected for analysing lethal effects, induction of MNs and SCGE assays were employed as genotoxicity endpoints for sublethal effects. Behavioural alterations were also investigated in both studies.[103,104]

Mortality results revealed that $DIC_{96\ h}$ median lethal concentration, *i.e.*, $LC50_{96\ h}$, was 1639 mg L^{-1} (range, 1471–1808 mg DIC per L). Several behavioural changes were observed in DIC-exposed *C. decemmaculatus* specimens starting at the lowest concentration tested, namely, gathering at the bottom of the aquarium (GBA), slowness in motion (SM), abnormal swimming (AS) and slow reaction (SR).[103] The genotoxic evaluation with the MN assay showed an increased frequency of MNs in fish exposed for 48 h to the highest DIC concentration assayed (1229 mg DIC per L) whereas no MN induction was observed after 96 h of exposure in any tested concentration. Moreover, analysed NAs demonstrated that NTs frequency increased with all the three DIC concentrations evaluated. However, only the highest DIC concentration (1229 mg DIC per L) induced BNs and LBs. On the other hand, none of the other NAs were induced after 96 h of exposure. The SCGE assay

Table 7.2 Studies evaluating auxinic herbicides in fish species of the Pampas Region[a]

Herbicide	Species	a.i. concentration (%)	Type of exposure	Time of exposure	Biomarker evaluated	LOEC (mg L^{-1})	References
DIC	*Cnesterodon decemmaculatus*	57.7	Acute	96 h	Mortality ($LC50_{96\ h}$)	1639	100
					Behaviour	1152	
					MN assay	1229	
					Comet assay	410	
DIC	*Cnesterodon decemmaculatus*	57.7	Acute	96 h	Enzyme-modified Comet assay	410	103
					Enzymatic biomarkers		
DIC	*Cnesterodon decemmaculatus*	57.7	Chronic	28 d	Comet assay	410	102
DIC	*Jenynsia lineata*	48	Acute	48 h	Enzymatic biomarkers	3×10^{-5}	62
2,4-D	*Rhamdia quelen*	86.8	Acute	96 h	Mortality ($LC50_{96\ h}$)	745	68
					Behaviour	600	
					Enzymatic biomarkers	600	
					Biochemical biomarkers	400	
2,4-D	*Leporinus obtudiensis*	86.8	Acute	96 h	Enzymatic biomarkers	10	65
					Biochemical biomarkers	1	
2,4-D	*Poecilia vivipara*	86.8	Acute	48 h	Mortality ($LC50_{48\ h}$)	34.64	66
					Histological changes	0.01	
2,4-D	*Cnesterodon decemmaculatus*	58.4	Acute	96 h	Mortality ($LC50_{96\ h}$)	1008	101
					Behaviour	800	
					MN assay	252	
					Comet assay	252	
2,4-D	*Cnesterodon decemmaculatus*	58.4	Acute	96 h	Enzyme-modified Comet assay	252	103
					Enzymatic biomarkers		
2,4-D	*Cnesterodon decemmaculatus*	58.4	Chronic	28 d	Comet assay	252	102
DIC + 2,4-D	*Cnesterodon decemmaculatus*	57.7 + 58.4	Acute	96 h	Mortality ($LC50_{96\ h}$)	819.5 + 504	106

[a]DIC: dicamba; 2,4-D: 2,4-dichlorophenoxiacetic acid; LOEC: lowest observed effect concentration; h: hours; d: days.

results indicated an increased genetic damage index (GDI) with all the DIC concentrations assayed after exposure for both 48 and 96 h ($P < 0.001$).[103]

Mortality results after 2,4-D amine (DMA®) exposure showed an $LC50_{96\ h}$ of 1008 mg L^{-1} (range, 929–1070 mg DMA® per L). The behavioural changes observed in 2,4-D-exposed *C. decemmaculatus* (starting at a concentration equivalent to 800 mg DMA® per L) were GBA, SM, AS and SR.[104] The MN assay revealed an increased frequency of MNs in fish exposed to all three 2,4-D concentrations for 48 and 96 h. The analysis of other NAs evidenced an increase in NTs frequency only in fish exposed to 504 mg DMA® per L for 96 h. In addition, an enhancement of BLs was observed at 756 and 504 mg DMA® per L after 48 and 96 h of exposure, respectively. Results from the SCGE assay showed an increased GDI with all the 2,4-D amine concentrations assayed after exposure lasting both 48 and 96 h ($P < 0.001$).[104]

We further evaluated the chronic genotoxic effects induced on *C. decemmaculatus* by a continuous exposure to Banvel® (41 and 82 mg L^{-1} DIC) and DMA® (25.2 and 50.4 mg L^{-1} 2,4-D) for 28 days.[105] The SCGE assay was performed in samples obtained on days 0, 7, 14, 21 and 28 after exposure. While 2,4-D induced DNA damage after 7–28 days of exposure regardless of the concentration used, DIC induced a genotoxic effect after exposure during 7–14 days (2.5% $LC50_{96\ h}$) and 7 days (5% $LC50_{96\ h}$), demonstrating the genotoxic capability of the long-term exposure of these auxinic herbicides when applied at low concentrations, at least on the species employed as biotic matrix.[105]

Finally, we evaluated the oxidative stress induced after exposure of *C. decemmaculatus* specimens to the commercial formulations Banvel® (410 mg 2,4-D per L) and DMA® (252 mg DIC per L) during 96 h.[106] The oxidative stress-induced DNA damage was detected with the enzyme-modified SCGE assay, and the antioxidant response of both auxinic herbicides was assessed by measuring catalase (CAT), glutathione-S-transferase (GST), acetylcholinesterase (AChE) and glutathione content (GSH) activities. The results obtained demonstrated that DIC and 2,4-D induced oxidative damage at a purine level but not at pyrimidine DNA bases. Results demonstrated that while GST and CAT activities increased, GSH and AChE levels decreased after exposure to the herbicides. These results demonstrated that both auxinic herbicides could jeopardize the survival or fitness of *C. decemmaculatus* due to reactive oxygen species (ROS) generation.[106]

When multiple crops are cultivated simultaneously, different types of agrochemicals are required. This practice has generated a new group of contaminants that have called the attention of toxicologists during the last decades.[56,95,107–109] Consequently, we evaluated the toxicity exerted by the mixture of DIC and 2,4-D in the commercial formulations Banvel® and DMA® on *C. decemmaculatus* during 96 h.[109] Results revealed that an equitoxic mixture of DIC and 2,4-D was nearly an additive concentration. Furthermore, a synergistic interaction was seen after assaying non-equitoxic combinations of DIC or 2,4-D. Thus, regardless of the concentrations used, DIC- and 2,4-D-induced toxicity was synergized by the presence of the counterpart herbicide within mixtures.[109]

The toxic effects induced by the 2,4-D-based herbicide U46D-Fluid (86.8%, Dow AgroSciences S.A.) were evaluated in another species of Pampasic fish, the silver catfish *R. quelen*, after acute exposure to different 2,4-D concentrations (range, 1–780 mg 2,4-D per L).[70] The authors evaluated mortality, activity, feeding behaviour and metabolic parameters such as AChE, hepatic glycogen, glucose and lactate levels. Results showed an increase in AChE activity in the brain and, on the other hand, a decrease of such activity in muscular tissue.[70] Moreover, hepatic glycogen was reduced after 400–700 mg 2,4-D per L treatment. Glucose levels were also reduced in liver and muscle samples, whereas plasma glucose levels increased. In addition, muscle lactate levels increased at all 2,4-D concentrations tested. Behavioural changes such as lethargy and erratic swimming were also observed. These findings suggested metabolic and histological responses to compensate for the stress caused by 2,4-D exposure.[70]

Vigário and Sabóia-Morais evaluated the acute effects induced by U46D-Fluid (86.8% 2,4-D, Dow AgroSciences S.A.) on the guppy *P. vivipara*.[68] Fish exposed to 10–40 µl U46D-Fluid per L for 48 h revealed a value equivalent to 34.64 µl L^{-1} as the $LC50_{48\ h}$ for the pesticide on the species. Gills and liver epithelia showed an increased mucous viscosity and the formation of plaques that could inhibit gas exchange and osmoregulation, vacuolized and swelled hepatocytes, among other modifications, demonstrating the induction of histopathological effects by this widely employed herbicide.[68]

Similarly, Braga da Fonseca *et al.* evaluated the acute effects of a 2,4-D-commercial formulation (86.8% 2,4-D a.i., BASF S.A.) on piava, *L. obtusidens,* also known as boga.[67] The endpoints evaluated were AChE activity and several metabolic parameters. After 96 h of exposure to 2,4-D concentrations ranging from 1 to 10 µg L^{-1}, alterations in brain and muscle AChE activity as well as in blood and tissue parameters such as glucose, lactate, glycogen and protein levels were observed.[67]

Finally, the acute toxic effects exerted by the DIC-based herbicide Dombell® (48% DIC, Cibeles S.A.) on adult female specimens of the one-sided livebearer *J. lineate* were evaluated.[64] Changes in CAT, GST, superoxide dismutase (SOD) activities, malondialdehyde (MDA) and hydrogen peroxide (H_2O_2) levels were analysed to detect oxidative stress after exposure to 0.03–30 µg DIC per L concentrations.[64] The results obtained with 0.03 µg DIC per L evidenced an increase of MDA levels and the inhibition of SOD and GST activities in brain and liver, respectively, demonstrating the sublethal effects induced by DIC after short-term exposure to environmentally relevant concentrations (see Table 7.2).

7.4 Final Remarks

Auxinic herbicides are widely used in agricultural regions worldwide, reaching aquatic environments and affecting the biota. Studies evaluating the lethal and sublethal effects of these compounds showed that even environmentally relevant low concentrations are able to induce alterations in metabolic parameters.[64,68] Another aspect to be considered is the fact that agrochemicals are usually applied in the field as formulated products. This practice mixes the a.i. with wetting agents, organic solvents and emulsifiers to improve penetration, adsorption and performance. These additives may exert toxicity by themselves and synergize or antagonize the toxicity of the a.i. In mixtures, chemicals interact with each other and with biological matrices, modifying the toxicity of the individual substances and causing lethality,[56,105,107–109] oxidative stress[110] and genotoxicity,[58,104,107,111] among other related effects. Since the full identity of all these ingredients is considered to be a confidential trade secret kept by the Argentinian regulatory agencies, they are not usually disclosed and discussed when evaluating the effects of commercial formulations on living organisms. Nevertheless, an overall risk assessment evaluation should also determine the additional effects induced by the excipient(s) in a commercial formulation.

As previously stated, the treatments applied in the above-mentioned studies covered a wide range of herbicide concentrations, some of which were above the values detected in environmental studies. In Argentina, there is scarce information available on the environmental concentrations of these auxinic herbicides. For instance, the range of DIC concentrations was 15.8–156.7 ng L^{-1} in a lagoon from Buenos Aires Province and 7.5–215.9 ng L^{-1} in rivers of Cordoba (Argentina) Province,[64,112] whereas 2,4-D concentrations ranging from 1 to 4 μg L^{-1} were detected in rivers of the Province of Buenos Aires.[113,114]

Taking into consideration the fact that in the field, it would be unlikely to find the concentrations used in several of these studies, these values would only be probably observed after specific events such as direct application or accidental discharge. However, we cannot rule out that fish populations, workers and populations of rural areas as well as the biota in general may be accidentally exposed to these concentration ranges of agrochemicals. Bearing in mind the different components of commercial formulations, further studies would be necessary to reveal whether the observed effects are specific to DIC and 2,4-D, or the result of some (if not all) of the adjuvants included in the commercial formulations.

Fish interact closely with the physical, biological and chemical aquatic ecosystem and are sensitive to different types of disturbances and they are an important part of the aquatic ecosystem. The studies previously mentioned show that the Neotropical fish species employed are valid indicators of xenobiotic exposure and an appropriate tool to be applied for diagnosing the anthropogenic impact on the environment.

Acknowledgments

Funding was provided by the National University of La Plata (Grants 11/N926 and 11/N1014) and the National Council for Scientific and Technological Research (CONICET, PIP No. 3086), both from Argentina.

References

1. Organisation for Economic Co-operation and Development (OECD), *Glossary of Statistical Terms*, 2008, http://ls-fts.unog.ch/sites/default/files/Traduction/Stat_Documents/OECD-Statistics%20Glossary-ENG.pdf, accessed on November 2023.
2. FAO, FIDA, OMS, PMA and UNICEF, El estado de la seguridad alimentaria y la nutrición en el mundo, 2017, http://www.fao.org/3/a-I7695s.pdf, accessed on November 2023.
3. A. Kubiak, A. Wolna-Maruwka, A. Niewiadomska and A. A. Pilarska, The problem of weed infestation of agricultural plantations vs. the assumptions of the european diversity strategy, *Agronomy*, 2022, **22**, 1–30.
4. United States Environmental Protection Agency (U.S. EPA), *Pesticides Industry Sales and Usage. 2008-2012 Market Estimates*, W. U.S. Government Printing Office, DC, Editor, U.S. EPA, Washington, DC, 2017, p. 24.
5. Instituto Nacional de Tecnología Agropecuaria (INTA), Los productos fitosanitarios en los sistemas productivos de la Argentina, in *Una mirada desde el INTA*, 2022, pp. 1–62.
6. S. Beggel, I. Werner, R. E. Connon and J. P. Geist, Sublethal toxicity of commercial insecticide formulations and their active ingredients to larval fathead minnow (*Pimephales promelas*), *Sci. Total Environ.*, 2009, **408**, 3169–3175.
7. C. A. Brühl, S. Pieper and B. Weber, Amphibians at risk? Susceptibility of terrestrial amphibian life stages to pesticides, *Environ. Toxicol. Chem.*, 2011, **30**, 2465–2472.
8. N. Nikoloff, G. S. Natale, D. Marino, S. Soloneski and M. L. Larramendy, Flurochloridone-based herbicides induced genotoxicity effects on *Rhinella arenarum* tadpoles (Anura: Bufonidae), *Ecotoxicol. Environ. Saf.*, 2014, **100**, 275–281.
9. J. M. Pérez-Iglesias, C. Ruiz de Arcaute, N. Nikoloff, L. Dury, S. Soloneski, G. S. Natale and M. L. Larramendy, The genotoxic effects of the imidacloprid-based insecticide formulation Glacoxan Imida on Montevideo tree frog *Hypsiboas pulchellus* tadpoles (Anura, Hylidae), *Ecotoxicol. Environ. Saf.*, 2014, **104**, 120–126.
10. C. Ruiz de Arcaute, J. M. Pérez-Iglesias, N. Nikoloff, G. S. Natale, S. Soloneski and M. L. Larramendy, Genotoxicity evaluation of the insecticide imidacloprid on circulating blood cells of Montevideo tree frog *Hypsiboas pulchellus* tadpoles (Anura, Hylidae) by comet and micronucleus bioassays, *Ecol. Indic.*, 2014, **45**, 632–639.
11. United States Environmental Protection Agency (U.S. EPA), United States Code, Title 7 - Agriculture, Chapter 6 - Insecticides and Environmental Pesticide Control, Subchapter II - Environmental Pesticide Control, 2012, http://uscode.gov/view.xhtml?path=/prelim@title7/chapter6&edition=prelim, accessed on November 2023.
12. L. de Brito Rodrigues, G. Gonçalves Costa, E. Lundgren Thá, L. R. da Silva, R. de Oliveira, D. Morais Leme, M. M. Cestari, C. Koppe Grisolia,

M. Campos Valadares and G. A. R. de Oliveira, Impact of the glyphosate-based commercial herbicide, its components and its metabolite AMPA on non-target aquatic organisms, *Mutat. Res.*, 2019, **842**, 94–101.

13. World Health Organization (WHO), Public health impacts of pesticides used in agriculture (WHO in collaboration with the United Nations Environment Programme, Geneva, 1990), Genova, Switzerland, 1990, https://iris.who.int/bitstream/handle/10665/39772/9241561394.pdf?sequence=1&isAllowed=y, accessed on November 2023.
14. Servicio Nacional de Sanidad y Calidad Agroalimentaria (SENASA), Manual de procedimientos, criterios y alcances para el registro de productos fitosanitarios en la República Argentina, Resolución 350/99, 1999, http://www.argentina.gob.ar/sites/default/files/manual_proced_res_350-99.pdf, accessed on February 2024.
15. K. Grossmann, Auxinic herbicides: current status of mechanism and mode of action, *Pest Manage. Sci.*, 2010, **66**, 113–120.
16. X. Tan, L. I. Calderon-Villalobos, M. Sharon, C. Zheng, C. V. Robinson, M. Estelle and N. Zheng, Mechanism of auxin perception by the TIR1 ubiquitin ligase, *Nature*, 2007, **446**, 640–645.
17. T. R. Bunch, J. A. Gervais, K. Buhl and D. Stone, Dicamba technical fact sheet, National Pesticide Information Center, Oregon State University Extension Services, 2012, http://npic.orst.edu/factsheets/archive/dicamba_tech.html, accessed on November 2023.
18. Cámara de Sanidad Agropecuaria y Fertilizantes (CASAFE), *La Argentina 2050. La revolución tecnológica del agro. Hacia el desarrollo integral de nuestra sociedad*, CASAFE, 2009.
19. Cámara de Sanidad Agropecuaria y Fertilizantes (CASAFE), Mercado de productos fitosanitarios/Año 2016 vs 2015, 2018, http://www.casafe.org/pdf/2018/ESTADISTICAS/Informe-Mercado-Fitosanitarios-2016.pdf, accessed on November 2023.
20. K. Grossmann, Mediation of herbicide effects by hormone interactions, *J. Plant Growth Regul.*, 2003, **22**, 109–122.
21. D. L. Ogada, The power of poison: Pesticide poisoning of Africa's wildlife, *Ann. N. Y. Acad. Sci.*, 2014, **1322**, 1–20.
22. K. B. Tierney, D. H. Baldwin, T. J. Hara, P. S. Ross, N. L. Scholz and C. J. Kennedy, Olfactory toxicity in fishes, *Aquat. Toxicol.*, 2010, **96**, 2–26.
23. T. Çavaş, The Use of Fish as Model Aquatic Organisms in Genotoxicity Studies, in *Ecotoxicology and Genotoxicology: Non-traditional Aquatic Models*, ed. M. L. Larramendy, The Royal Society of Chemistry, Cambridge, 2017, pp. 245–277.
24. A. Annabi, K. Said and I. Messaoudi, Monitoring and assessment of environmental disturbance on natural *Gambusia affinis* populations - Histopathological analysis, *Environ. Monit. Assess.*, 2015, **187**, 318–326.
25. J. A. Benitez, R. M. Cerón-Bretón, J. G. Cerón-Bretón and J. Rendón-Von-Osten, The environmental impact of human activities on the Mexican coast of the Gulf of Mexico: Review of status and trends, *WIT Trans. Ecol. Environ.*, 2014, **181**, 37–50.

26. C. Greene, L. Kuehne, C. Rice, K. Fresh and D. Penttila, Forty years of change in forage fish and jellyfish abundance across greater Puget Sound, Washington (USA): Anthropogenic and climate associations, *Mar. Ecol.: Prog. Ser.*, 2015, **525**, 153–170.
27. D. Schlenk, R. Lavado, J. E. Loyo-Rosales, W. Jones, L. Maryoung, N. Riar, I. Werner and D. Sedlak, Reconstitution studies of pesticides and surfactants exploring the cause of estrogenic activity observed in surface waters of the San Francisco Bay Delta, *Environ. Sci. Technol.*, 2012, **46**, 9106–9111.
28. S. Soloneski and M. L. Larramendy, in *Ecotoxicology and Genotoxicology - Non-traditional Aquatic Models, The Use of the Ten Spotted Live-bearer Fish Cnesterodon decemmaculatus (Jenyns, 1842) (Pisces, Poeciliidae) in the Genotoxic Evaluation of Environmental Pollutants*, ed. M. L. Larramendy, The Royal Society of Chemistry, Cambridge, 2017, pp. 327–346.
29. A. A. S. de Oliveira, L. C. Vieira, S. C. Dreossi, D. J. Dorta, C. Gravato, M. E. da Silva Ferreira and D. P. de Oliveira, Integrating morphological, biochemical, behavioural, and molecular approaches to investigate developmental toxicity triggered by tebuthiuron in zebrafish (*Danio rerio*), *Chemosphere*, 2023, **340**, 139894.
30. J. Jin, T. Kurobe, B. G. Hammock, C. H. Lam, L. Lin and S. J. Teh, Toxic effects of fluridone on early developmental stages of Japanese Medaka (*Oryzias latipes*), *Sci. Total Environ.*, 2020, **700**, 134495.
31. B. Lacy, M. Rivera, L. Flores and M. S. Rahman, Combined effects of high temperature and pesticide mixture exposure on free-swimming behaviors and hepatic cytochrome P450 1A expression in goldfish, *Carassius auratus*, *J. Toxicol. Environ. Health, Part A*, 2023, **86**, 144–165.
32. D. D. Zhang, W. W. Liu, L. Li, G. Zhu and J. Ma, Single and combined effects of chlorpyrifos and glyphosate on the brain of common carp: based on biochemical and molecular perspective, *Int. J. Mol. Sci.*, 2023, **24**, 12934.
33. I. A. Badroo, H. P. Nandurkar and A. H. Khanday, Toxicological impacts of herbicide paraquat dichloride on histological profile (gills, liver, and kidney) of freshwater fish *Channa punctatus* (Bloch), *Environ. Sci. Pollut. Res.*, 2020, **27**, 39054–39067.
34. M. Roni and C. C. Harilal, Biotoxicity of *Acrostichum aureum* L. synthesized zinc oxide nanoparticles against *Aedesalbopictus*, and impact on predation efficiency of mosquito fish *Gambusia affinis*, *Inorg. Chem. Commun.*, 2023, **156**, 111224.
35. T. Moungthipmalai, C. Puwanard, J. Aungtikun, S. Sittichok and M. Soonwera, Ovicidal toxicity of plant essential oils and their major constituents against two mosquito vectors and their non-target aquatic predators, *Sci. Rep.*, 2023, **13**, 2119.
36. A. L. Cabrera and A. Willink, Biogeografía de América Latina, in *Programa Regional de Desarrollo Científico y Tecnológico*, Departamento de Asuntos Científicos, Secretaría General de la Organización de los Estados Americanos Washington, District of Columbia, 1973, p. 120.

37. A. F. Bonifacio and A. C. Hued, Single and joint effects of chronic exposure to chlorpyrifos and glyphosate based pesticides on structural biomarkers in *Cnesterodon decemmaculatus*, *Chemosphere*, 2019, **236**, 124311.
38. W. Pengue, Transgenic crops in Argentina: the ecological and social debt, *Bull. Sci. Technol. Soc.*, 2005, **25**, 314–322.
39. International Service for the Acquisition of Agri-biotech Applications (ISAAA), Global Status of Commercialized Biotech/GM Crops: 2016, *ISAAA Brief*, Ithaca, New York, 2016, vol. 52, pp. 1–125.
40. L. Ferrari, *Manual de procedimiento básico para la cría de Cnesterodon decemmaculatus en laboratorio*, EdUnLu, Luján, 2017.
41. R. C. Menni, S. E. Gómez and F. López Armengol, Subtle relationships: freshwater fishes and the chemistry of water in southern South America, *Hydrobiologia*, 1996, **328**, 173–197.
42. R. C. Menni, Peces y ambientes en la Argentina continental, in *Monografías del Museo Argentino de Ciencias Naturales*, Buenos Aires, 2004.
43. W. D. Di Marzio and M. C. Tortorelli, Effects of paraquat on survival and total cholinesterase activity in fry of *Cnesterodon decemmaculatus* (Pisces, Poeciliidae), *Bull. Environ. Contam. Toxicol.*, 1994, **52**, 274–278.
44. W. D. Di Marzio, J. L. Alberdi, M. E. Sáenz and M. D. C. Tortorelli, Effects of paraquat (Osaquat® formulation) on survival and total cholinesterase activity in male and female adults of *Cnesterodon decemmaculatus* (Pisces, Poeciliidae), *Environ. Toxicol. Water Qual.*, 1998, **13**, 55–59.
45. J. Vera-Candioti, S. Soloneski and M. L. Larramendy, Single-cell gel electrophoresis assay in the ten spotted live-bearer fish, *Cnesterodon decemmaculatus* (Jenyns, 1842), as bioassay for agrochemical-induced genotoxicity, *Ecotoxicol. Environ. Saf.*, 2013, **98**, 368–373.
46. J. Vera-Candioti, S. Soloneski and M. L. Larramendy, Pirimicarb-based formulation-induced genotoxicity and cytotoxicity on the fresh water fish *Cnesterodon decemmaculatus* (Jenyns, 1842) (Pisces, Poeciliidae), *Toxicol. Ind. Health*, 2013, **31**, 1–11.
47. H. A. Mugni, A. Paracampo, N. Marrochi and C. Bonetto, Cypermethrin, chlorpyrifos and endosulfan toxicity to two non-target freshwater organisms, *Fresenius Environ. Bull.*, 2012, **21**, 2085–2089.
48. H. Mugni, P. Demetrio, A. Paracampo, M. Pardi, G. Bulus and C. Bonetto, Toxicity persistence in runoff water and soil in experimental soybean plots following chlorpyrifos application, *Bull. Environ. Contam. Toxicol.*, 2012, **89**, 208–212.
49. J. Vera-Candioti, S. Soloneski and M. L. Larramendy, Chlorpyrifos-based insecticides induced genotoxic and cytotoxic effects in the ten spotted live-bearer fish, *Cnesterodon decemmaculatus* (Jenyns, 1842), *Environ. Toxicol.*, 2014, **29**, 1390–1398.
50. A. Paracampo, M. Solis, C. Bonneto and H. Mugni, Acute toxicity of chlorpyrifos to the non-target organism *Cnesterodon decemmaculatus*, *Int. J. Environ. Health Res.*, 2015, **25**, 96–103.

51. A. F. Bonifacio, M. L. Ballesteros, R. I. Bonansea, I. Filippi, M. V. Amé and A. C. Hued, Environmental relevant concentrations of a chlorpyrifos commercial formulation affect two neotropical fish species, *Cheirodon interruptus* and *Cnesterodon decemmaculatus*, *Chemosphere*, 2017, **188**, 486–493.
52. A. F. Bonifacio, J. Cazenave, C. Bacchetta, M. L. Ballesteros, M. De Los Ángeles Bistoni, M. V. Amé, L. Bertrand and A. C. Hued, Alterations in the general condition, biochemical parameters and locomotor activity in *Cnesterodon decemmaculatus* exposed to commercial formulations of chlorpyrifos, glyphosate and their mixtures, *Ecol. Indic.*, 2016, **67**, 88–97.
53. D. L. Bernal-Rey, M. Dos Santos Afonso and R. J. Menendez-Helman, Effects of chlorpyrifos on acetylcholinesterase activity in two freshwater fish species (*Cnesterodon decemmaculatus* and *Gambusia affinis*), *Int. J. Environ. Health*, 2017, **8**, 316–330.
54. W. D. Di Marzio, Estudio de los efectos de sustancias contaminantes sobre peces de agua dulce, Tesis Doctoral, Facultad de Ciencias Naturales y Museo, Universidad Nacional de La Plata, 1999.
55. P. Carriquiriborde, J. Díaz, H. Mugni, C. Bonneto and A. E. Ronco, Impact of cypermethrin on stream fish populations under field-use in biotech-soybean production, *Chemosphere*, 2007, **68**, 613–621.
56. J. C. Brodeur, S. Melpel, A. B. Anglesio, D. Cristos, M. A. D'Andrea and M. B. Poliserpi, Toxicities of glyphosate- and cypermethrin-based pesticides are antagonic in the tenspotted livebearer fish (*Cnesterodon decemmaculatus*), *Chemosphere*, 2016, **155**, 429–435.
57. A. F. Bonifacio, M. J. Zambrano and A. C. Hued, Integrated ecotoxicological assessment of the complex interactions between chlorpyrifos and glyphosate on a non-target species *Cnesterodon decemmaculatus* (Jenyns, 1842), *Chemosphere*, 2020, **261**, 127782.
58. W. F. Carvalho, C. Ruiz de Arcaute, L. Torres, D. de Melo e Silva, S. Soloneski and M. L. Larramendy, Genotoxicity of mixtures of glyphosate with 2,4-dichlorophenoxyacetic acid chemical forms towars *Cnesterodon decemmaculatus* (Pisces, Poeciliidae), *Environ. Sci. Pollut. Res.*, 2020, **27**, 6515–6525.
59. J. Vera-Candioti, S. Soloneski and M. L. Larramendy, Evaluation of the genotoxic and cytotoxic effects of glyphosate-based herbicides in the ten spotted live-bearer fish *Cnesterodon decemmaculatus* (Jenyns, 1842), *Ecotoxicol. Environ. Saf.*, 2013, **89**, 166–173.
60. R. J. Menéndez-Helman, G. V. Ferreyroa, M. dos Santos Afonso and A. Salibián, Glyphosate as an acetylcholinesterase inhibitor in *Cnesterodon decemmaculatus*, *Bull. Environ. Contam. Toxicol.*, 2012, **88**, 6–9.
61. A. Gárriz and L. A. Miranda, Effects of metals on sperm quality, fertilization and hatching rates, and embryo and larval survival of pejerrey fish (*Odontesthes bonariensis*), *Ecotoxicology*, 2020, **29**, 1072–1082.
62. A. Gárriz, P. S. del Fresno, P. Carriquiriborde and L. A. Miranda, Effects of heavy metals identified in Chascomús shallow lake on the

endocrine-reproductive axis of pejerrey fish (*Odontesthes bonariensis*), *Gen. Comp. Endocrinol.*, 2019, **273**, 152–162.

63. P. Carriquiriborde, J. I. Fernandino, C. G. López, E. de San Benito, G.-V. J. M. D. Cristos and V. L. Trudeau, Atrazine alters early sexual development of the South American silverside, *Odontesthes bonariensis*, *Aquat. Toxicol.*, 2023, **254**, 106366.
64. D. M. Villagran, L. R. Lombardero, A. C. Crupkin, J. R. Mendieta, S. R. Medici, F. Lavarello, M. F. Simoniello and M. L. Menone, Can the herbicide dicamba produce oxidative stress in the native south american fish *Jenynsia lineata* at environmentally relevant concentrations?, *Bull. Environ. Contam. Toxicol.*, 2023, **111**, 41.
65. L. Bertrand and F. G. Iturburu, Pesticides bioassays using neotropical aquatic species: Trends during the last twenty years and future challenges in Argentina, *Chemosphere*, 2023, **326**, 138369.
66. Y. Moreira Ribeiro, D. Peruci Moreira, A. A. Weber, C. Ferreira Sales, R. Magno Costa Melo, N. Bazzoli, E. Rizzo and A. L. Paschoalini, Adverse effects of herbicides in freshwater Neotropical fish: A review, *Aquat. Toxicol.*, 2022, **252**, 106293.
67. M. Braga da Fonseca, L. Glusczak, B. Silveira Moraes, C. Cavalheiro de Menezes, A. Pretto, M. A. Tierno, R. Zanella, F. Ferreira Gonçalves and V. L. Loro, The 2,4-D herbicide effects on acetylcholinesterase activity and metabolic parameters of piava freshwater fish (*Leporinus obtusidens*), *Ecotoxicol. Environ. Saf.*, 2008, **69**, 416–420.
68. A. F. Vigário and S. M. T. Sabóia-Morais, Effects of the 2,4-D herbicide on gills epithelia and liver of the fish *Poecilia vivipara* xenobiotics, genotoxic potential, *Pesqui. Vet. Bras.*, 2014, **34**, 523–528.
69. D. V. de Paulo, C. F. Mariz Jr., M. K. de Melo Alves, R. N. Alves, R. M. Batista, G. Fillman and P. S. M. Carvalho, Histological and behavioral toxicity of tributyltin in the tropical guppy *Poecilia vivipara*, *Environ. Toxicol. Chem.*, 2020, **39**, 1953–1963.
70. R. Cattaneo, V. L. Loro, R. Spanevello, F. A. Silveira, L. Luz, D. S. Miron, M. B. Fonseca, B. S. Moraes and B. Clasen, Metabolic and histological parameters of silver catfish (*Rhamdia quelen*) exposed to commercial formulation of 2,4-dichlorophenoxiacetic acid (2,4-D) herbicide, *Pestic. Biochem. Physiol.*, 2008, **92**, 133–137.
71. M. F. Gomes, P. V. de Carvalho Soares, L. R. Rocha Martins, J. R. Esquivel Garcia, F. Y. Yamamoto and A. Martins de Freitas, Sublethal effects of triclosan and triclocarban at environmental concentrations in silver catfish (*Rhamdia quelen*) embryos, *Chemosphere*, 2021, **263**, 127985.
72. United States Environmental Protection Agency (U.S. EPA), *Methods for Acute Toxicity Tests with Fish, Macroinvertebrates, and Amphibians, 660/3-75-009*, U.S. Government Printing Office, Washington, DC, 1975, p. 62.
73. United States Environmental Protection Agency (U.S. EPA), *Methods for Measuring the Acute Toxicity of Effluents and Receiving Waters to Freshwater and Marine Organisms, 821-R-02-012*, U.S. Government Printing Office, Washington, DC, 5th edn, 2002, pp. 1–266.

74. C. Cabrera, E. Ortega, M. L. Lorenzo and M. C. Lopez, Cadmium contamination of vegetable crops, farmlands, and irrigation waters, *Rev. Environ. Contam. Toxicol.*, 1998, **154**, 55–81.
75. K. M. Gaworecki and S. J. Klaine, Behavioral and biochemical responses of hybrid striped bass during and after fluoxetine exposure, *Aquat. Toxicol.*, 2008, **88**, 207–213.
76. A. S. Kane, J. D. Salierno, G. T. Gipson, T. C. Molteno and C. Hunter, A video-based movement analysis system to quantify behavioral stress responses of fish, *Water Res.*, 2004, **38**, 3993–4001.
77. M. Faimili, F. Garaventa, V. Piazza, G. Greco, C. Corrà, F. Magillo, M. Pittore, E. Giacco, L. Gallus, C. Falugi and G. Tagliafierro, Swimming speed alteration of larvae of *Balanus amphitrite* (Darwin) as a behavioral endpoint toxicological bioassays, *Mar. Biol.*, 2006, **149**, 87–96.
78. R. Sarikaya and M. Yilmaz, Investigation of acute toxicity and the effect of 2,4-D (2,4-dichlorophenoxyacetic acid) herbicide on the behavior of the common carp (*Cyprinus carpio* L., 1758; Pisces, Cyprinidae), *Chemosphere*, 2003, **52**, 195–201.
79. M. Yilmaz, A. Gül and E. Karaköse, Investigation of acute toxicity and the effect of cadmium chloride ($CdCl_2.H_2O$) metal salt on behavior of the guppy (*Poecilia reticulata*), *Chemosphere*, 2004, **56**, 375–380.
80. S. Bretaud, P. Saglio and J. P. Toutant, Behavioral dysfunctions correlate to altered physiology in rainbow trout (*Oncorhynchus mykiss*) exposed to cholinesterase-inhibiting chemicals, *Arch. Environ. Contam. Toxicol.*, 2001, **40**, 70–76.
81. D. Dogan and C. Can, Hematological, biochemical, and behavioral responses of *Oncorhynchus mykiss* to dimethoate, *Fish Physiol. Biochem.*, 2011, **37**, 951–995.
82. V. M. Pereira, J. W. Bortolotto, L. W. Kist, M. B. de Azevedo, R. S. Fritsch, R. da Luz Oliveira, T. C. B. Pereira, C. D. Bonan, M. R. Vianna and M. R. Bogo, Endosulfan exposure inhibits bran AChE activity and impairs swimming performance in adult zebrafish (*Danio rerio*), *Neurotoxicology*, 2012, **33**, 469–475.
83. D. Ali, N. S. Nagpure, S. Kumar, R. Kumar, B. Kushwaha and W. S. Lakra, Assessment of genotoxic and mutagenic effects of chlorpyrifos in freshwater fish *Channa punctatus* (Bloch) using micronucleus assay and alkaline single-cell gel electrophoresis, *Food Chem. Toxicol.*, 2009, **47**, 650–656.
84. D. G. Cavalcante, C. B. Martinez and S. H. Sofia, Genotoxic effects of Roundup on the fish *Prochilodus lineatus*, *Mutat. Res.*, 2008, **655**, 41–46.
85. T. Cavaş and S. Ergene-Gözükara, Induction of micronuclei and nuclear abnormalities in *Oreochromis niloticus* following exposure to petroleum refinery and chromium processing plant effluents, *Aquat. Toxicol.*, 2005, **74**, 264–271.
86. J. Vera Candioti, G. Natale, S. Soloneski, A. E. Ronco and M. L. Larramendy, Sublethal and lethal effects on *Rhinella arenarum* (Anura, Bufonidae) tadpoles exerted by the pirimicarb-containing technical formulation insecticide Aficida®, *Chemosphere*, 2010, **78**, 249–255.

87. T. Cavaş, *In vivo* genotoxicity evaluation of atrazine and atrazine-based herbicide on fish *Carassius auratus* using the micronucleus test and the comet assay, *Food Chem. Toxicol.*, 2011, **49**, 1431–1435.
88. M. Fenech, M. Kirsch-Volders, A. T. Natarajan, J. Surralles, J. W. Crott, J. Parry, H. Norppa, D. A. Eastmond, J. D. Tucker and P. Thomas, Molecular mechanisms of micronucleus, nucleoplasmic bridge and nuclear bud formation in mammalian and human cells, *Mutagenesis*, 2011, **26**, 125–132.
89. T. Cavaş and S. Ergene-Gözükara, Micronuclei, nuclear lesions and interphase silver-stained nucleolar organizer regions (AgNORs) as cyto-genotoxicity indicators in *Oreochromis niloticus* exposed to textile mill effluent, *Mutat. Res.*, 2003, **538**, 81–91.
90. F. D. Gökalp Muranli and U. Güner, Induction of micronuclei and nuclear abnormalities in erythrocytes of mosquito fish (*Gambusia affinis*) following exposure to the pyrethroid insecticide lambda-cyhalothrin, *Mutat. Res.*, 2011, **726**, 104–108.
91. M. D. Mudry and M. A. Carballo, *Genética Toxicológica*, Editorial de los Cuatro Vientos, Buenos Aires, 2006.
92. K. R. Carrasco, K. L. Tilbury and M. S. Mayers, Assessment of the piscine micronuclei test as an *in situ* biological indicator of chemical contaminants effects, *Can. J. Fish. Aquat. Sci.*, 1990, **47**, 2123–2136.
93. I. Strunjak-Perovic, R. Coz-Rakovac, N. Topic Popovic and M. Jadan, Seasonality of nuclear abnormalities in gilthead sea bream *Sparus aurata* (L.) erythrocytes, *Fish Physiol. Biochem.*, 2009, **35**, 287–291.
94. R. R. Otter, J. Meier, K. M. Kubach, J. M. Lazorchak and S. J. Klaine, The effects of urbanization on *Lepomis macrochirus* using the comet assay, *Ecotoxicol. Environ. Saf.*, 2012, **84**, 299–303.
95. J. S. Barbosa, T. M. Cabral, D. N. Ferreira, L. F. Agnez-Lima and S. R. Batistuzzo de Medeiros, Genotoxicity assessment in aquatic environment impacted by the presence of heavy metals, *Ecotoxicol. Environ. Saf.*, 2010, **73**, 320–325.
96. D. Calcavante, C. Bueno dos Reis Martinez and S. H. Sofia, Genotoxic effects of Roundup® on the fish *Prochilodus lineatus*, *Mutat. Res.*, 2008, **655**, 41–46.
97. A. Azqueta, I. C. Costa-Amaral and A. R. Collins, in *The Comet Assay in Toxicology, High-throughput Measurement of DNA Breaks and Oxidised Bases with the Comet Assay*, ed. D. Anderson, and A. Dhawan, The Royal Society of Chemistry, Cambridge, 2016, pp. 67–92.
98. A. R. Collins and A. Azqueta, DNA repair as a biomarker in human biomonitoring studies; further applications of the comet assay, *Mutat. Res.*, 2012, **736**, 122–129.
99. J. M. Pérez-Iglesias, C. Ruiz de Arcaute, G. S. Natale, S. Soloneski and M. L. Larramendy, Evaluation of imazethapyr-induced DNA oxidative damage by alkaline Endo III- and Fpg-modified single-cell gel electrophoresis assay in *Hypsiboas pulchellus* tadpoles (Anura, Hylidae), *Ecotoxicol. Environ. Saf.*, 2017, **142**, 503–508.

100. A. R. Collins, The comet assay for DNA damage and repair: principles, applications, and limitations, *Mol. Biotechnol.*, 2004, **26**, 249–261.
101. A. R. Collins, M. Dusinská, C. M. Gedik and R. Stetina, Oxidative damage to DNA: do we have a reliable biomarker?, *Environ. Health Perspect.*, 1996, **104**, 465–469.
102. A. Collins, P. Møller, G. Gajski, S. Vodenková, A. Abdulwahed, D. Anderson, E. E. Bankoglu, S. Bonassi, E. Boutet-Robinet, G. Brunborg, C. Chao, M. S. Cooke, C. Costa, S. Costa, A. Dhawan, J. de Lapuente, C. Del Bo', J. Dubus, M. Dusinska, S. J. Duthie, N. El Yamani, B. Engelward, I. Gaivão, L. Giovannelli, R. Godschalk, S. Guilherme, K. B. Gutzkow, K. Habas, A. Hernández, O. Herrero, M. Isidori, A. N. Jha, S. Knasmüller, I. M. Kooter, G. Koppen, M. Kruszewski, C. Ladeira, B. Laffon, M. L. Larramendy, L. Le Hégarat, A. Lewies, A. Lewinska, G. E. Liwszyc, A. López de Cerain, M. Manjanatha, R. Marcos, M. Milić, V. Moraes de Andrade, M. Moretti, D. Muruzabal, M. Novak, R. Oliveira, A. K. Olsen, N. Owiti, M. Pacheco, A. K. Padey, S. Pfuhler, B. Pourrut, K. Reisinger, E. Rojas, E. Rundén-Pran, J. Sanz-Serrano, S. Shaposhnikov, V. Sipinen, K. Smeets, H. Stopper, J. P. Teixeira, V. Valdiglesias, M. Valverde, F. van Acker, F. J. van Schooten, M. Vasquez, J. F. Wentzel, M. Wnuk, A. Wouters, B. Žegura, T. Zikmund, S. A. S. Langie and A. Azqueta, Measuring DNA modifications with the comet assay: a compendium of protocols, *Nat. Protoc.*, 2023, **18**, 929–989.
103. C. Ruiz de Arcaute, S. Soloneski and M. L. Larramendy, Evaluation of the genotoxicity of a herbicide formulation containing 3,6-dichloro-2-metoxybenzoic acid (dicamba) in circulating blood cells of the tropical fish *Cnesterodon decemmaculatus*, *Mutat. Res.*, 2014, **773**, 1–8.
104. C. Ruiz de Arcaute, S. Soloneski and M. Larramendy, Toxic and genotoxic effects of the 2,4-dichlorophenoxyacetic acid (2,4-D)-based herbicide on the Neotropical fish *Cnesterodon decemmaculatus*, *Ecotoxicol. Environ. Saf.*, 2016, **128**, 222–229.
105. C. Ruiz de Arcaute, M. L. Larramendy and S. Soloneski, Genotoxicity by long-term exposure to the auxinic herbicides 2,4-dichlorophenoxyacetic acid and dicamba on *Cnesterodon decemmaculatus* (Pisces: Poeciliidae), *Environ. Pollut.*, 2018, **243**, 670–678.
106. C. Ruiz de Arcaute, N. A. Ossana, J. M. Pérez-Iglesias, S. Soloneski and M. L. Larramendy, Auxinic herbicides induce oxidative stress on *Cnesterodon decemmaculatus* (Pisces: Poeciliidae), *Environ. Sci. Pollut. Res.*, 2019, **26**, 20485–20498.
107. C. M. J. Brodeur, M. B. Poliserpi, M. F. D´andrea and M. Sanchez, Synergy between glyphosate- and cypermethrin-based pesticides during acute exposures in tadpoles of the common South American toad *Rhinella arenarum*, *Chemosphere*, 2014, **112**, 70–76.
108. C. Ruiz de Arcaute, J. C. Brodeur, S. Soloneski and M. L. Larramendy, Toxicity to *Rhinella arenarum* tadpoles (Anura, Bufonidae) of herbicides mixtures commonly used to treat fallow containing resistant weeds: glyphosate-dicamba and glyphosate-flurochloridone, *Chemosphere*, 2020, 125623.

109. C. Ruiz de Arcaute, S. Soloneski and M. L. Larramendy, Synergism of mixtures of dicamba and 2,4-dichlorophenoxyacetic acid herbicide formulations on the Neotropical fish *Cnesterodon decemmaculatus* (Pisces, Poeciliidae), *Environ. Pollut.*, 2018, **236**, 33–39.
110. A. P. Cuzziol Boccioni, G. Lener, J. Peluso, P. M. Peltzer, A. M. Attademo, C. M. Aronzon, M. F. Simoniello, L. D. Demonte, M. R. Repetti and R. C. Lajmanovich, Comparative assessment of individual and mixture chronic toxicity of glyphosate and glufosinate ammonium on amphibian tadpoles: A multibiomarker approach, *Chemosphere*, 2022, **307**, 136554.
111. S. Soloneski, C. Ruiz de Arcaute and M. L. Larramendy, Genotoxic effect of a binary mixture of dicamba and glyphosate-based commercial herbicide formulations on *Rhinella arenarum* (Hensel, 1867) (Anura, Bufonidae) late-stage larvae, *Environ. Sci. Pollut. Res.*, 2016, **23**, 17811–17821.
112. S. Corcoran, C. D. Metcalfe, T. Sultana, M. V. Amé and M. L. Menone, Pesticides in surface waters in Argentina monitored using polar organic chemcal integrative samplers, *Bull. Environ. Contam. Toxicol.*, 2020, **104**, 21–26.
113. J. Peluso, C. M. Aronzon, M. Acquaroni and C. S. Pérez-Coll, Biomarkers of genotoxicity and health status of *Rhinella fernandezae* populations from the lower Paraná River Basin, Argentina, *Ecol. Indic.*, 2020, **117**, 106588.
114. D. J. Pérez, E. Okada, E. De Gerónimo and M. L. Menone, Spatial and temporal trends and flow dynamics of glyphosate and other pesticides within an agricultural watershed in Argentina, *Environ. Toxicol. Chem.*, 2017, **36**, 3206–3216.

CHAPTER 8

Native Nigerian Fish Species Used as Models in Ecotoxicological Studies: A Rapid Systematic Review

TEMITOPE OLAWUNMI SOGBANMU[a*] AND
AKEEM BABATUNDE DAUDA[b]

[a]Ecotoxicology and Conservation Unit, Department of Zoology, Faculty of Science, University of Lagos, Akoka Campus, Lagos, Nigeria; [b]Department of Fisheries and Aquaculture, Federal University Dutsin-Ma, Katsina State, Nigeria
*E-mail: tsogbanmu@unilag.edu.ng

8.1 Introduction

Fish species are important aquatic models utilised in ecotoxicology studies. Being the simplest and oldest vertebrate species as well as their relative ease of culture or sampling in the wild, they have been utilised as model organisms in various studies including ecotoxicological ones.[1,2] In order to ascertain the potential effects of substances often introduced by anthropogenic activities on ecosystem structure and function, biological indicators, such as fish species, are often used as biological models.[3] Similarly, laboratory-based studies are conducted to determine the effect of potentially harmful substances on living organisms in which model fish species are used as experimental

Issues in Toxicology No. 50
Fish Species in Environmental Risk Assessment Strategies
Edited by Guillermo Eli Liwszyc and Marcelo L. Larramendy

Published by the Royal Society of Chemistry, www.rsc.org

organisms, among others.[4] International Regulatory Institutions such as the Organisation for Economic Cooperation and Development (OECD) and United States Environmental Protection Agency (USEPA) have ratified fish models for ecotoxicological studies, toxicity testing and biomonitoring.[4,5]

However, most of these model fish species are native to temperate countries and climes that are not representative of native species across the world.[6] While the basic physiology of most model species has been established and hence utilisable for ecotoxicological studies in various contexts and climes, there have been debates on the use of native species in ecotoxicological studies or biomonitoring programmes.[7,8] Due to differences in their adaptation, sensitivities and physiological niches in temperate and tropical climes, responses to various stressors will differ among fish species as in most organisms.[9]

In Africa, few species have been recognised under globally accepted guidelines for their use in ecotoxicological studies. In Nigeria, the most populous African country with diverse ecozones, rich aquatic resources and biodiversity,[10] there are a few fish species native to the country which have been documented in biomonitoring and toxicity testing studies.[11] These may be considered as prime candidates to be globally acceptable as model fish species in ecotoxicological studies.

8.1.1 Ecotoxicological Studies

The quest for socio-economic development by humans and consequent increase in anthropogenic activities since the industrial age has resulted in the introduction of harmful substances into the environment in amounts at which harm is caused to living organisms including humans and their property.[12] This phenomenon is commonly described as pollution. Substances are released or discharged into the ecosystem in a voluntary or accidental fashion from human activities such as industrial emissions, transportation activities, agricultural production, textile manufacturing and natural resource mining resulting in perturbations of these hitherto natural ecosystems. When these perturbations persist in time at a constant rate or sudden incident exceeds the capacity of the ecosystem to recover through natural mechanisms, toxic effects are observed.[7] The effects are measurable as indices in living organisms such as animals and some of these have been utilised as bioindicators of toxic exposure or effects in various ecosystems as well as model species in field and laboratory toxicological studies.[3,7,11]

8.1.1.1 Fish Species Used in Ecotoxicological Studies

Fish species are some of the most common model animals utilised for assessing the potential toxic effects of substances under laboratory conditions or as bioindicators of stressors or adverse impacts in aquatic ecosystems.[1,6] Some common globally ratified fish models that are used in ecotoxicological studies include the zebrafish *Danio rerio* (Cypriniformes, Cyprinidae), the

Japanese rice fish, also known as medaka, *Oryzias latipes* (Beloniformes, Adrianichthidae) and the rainbow trout *Oncorhynchus mykiss* (Salmoniformes, Salmonidae). Also included in this category are, an anadromous fish species indigenous to Pacific coastal streams, the fathead minnow *Pimephales promelas* (Cypriniformes, Cyprinidae), an omnivorous freshwater fish native and ubiquitous to North America, the three-spined stickleback, *Gasterosteus aculeatus* (Scorpaeniformes, Gasterosteidae), the Atlantic killifish *Fundulus heteroclitus* (Cyprinodontiformes, Fundulidae) and the marine medaka *Oryzias melastigma* (Beloniformes, Adrianichthyidae), among others.[4,13–15] However, most of these model fish species are non-native to Africa, including Nigeria, thus limiting their utility for comparative ecotoxicological studies. Except for the African sharptooth catfish *Clarias gariepinus* (Siluriformes, Claridae) (see Figure 8.1) and Nile tilapia *Oreochromis niloticus* (Cichliformes, Cichlidae) (see Figure 8.2), which are gaining utility in ecotoxicity studies across the world,[16,17] native Nigerian fish species are not well known as experimental

Figure 8.1 African sharptooth catfish *Clarias gariepinus* (Siluriformes, Claridae).

Figure 8.2 Nile tilapia *Oreochromis niloticus* (Cichliformes, Cichlidae).

models in ecotoxicological studies outside of their native contexts. This chapter tries to point out their usefulness as appropriate species.

8.1.1.2 Ecotoxicological Studies in Nigeria Using Fish Species

Nigeria is the most populous country in the African continent with a population of nearly 226 million people.[18] Expectedly, this large population with its economy still developing contributes variously to anthropogenic pollution with impacts on different ecosystems. For example, plastic pollution has become a major menace in Nigeria coupled with inadequate solid waste management, thus making it one of the greatest environmental challenges after climate change.[19,20] Similarly, as a major producer of crude oil, Nigeria has suffered the environmental degradation and ecological disasters associated with crude oil spills and mining of oil across its coastal and marine ecosystems, particularly in the Niger Delta.[21] Consequently, environmental regulations within the country require the use of indigenous model animal species in biomonitoring studies for ecotoxicological effects as well as laboratory toxicity studies before approval of new chemicals or evaluation of wastewater.[22] Furthermore, fish species, as other animals and living organisms from various regions or climate, will respond differently under varying environmental or climatic conditions to stressors.[9] Hence, ecotoxicological studies in Nigeria majorly utilise native model fish species among other animals for evaluation of the toxic effects of stressors *in situ* and *ex situ*.

The limited number of native fish species acceptable as models is not due to the inexistence of a variety of fish species but essentially due to limited information on the number, biology and ecology of available fish species, as well as the culturable status of the species. For instance, the freshwater ecosystem, which is the most studied ecosystem compared to marine and brackish waters, is reported to have 316 native fish species out of which information on ecology is only available for 121 species (38.3%).[23] While field studies can utilise all known species in the targeted ecosystem, laboratory studies are limited to the use of culturable fish species only. Globally, over 598 aquatic species have been or are currently being farmed with finfish accounting for 369 (61.7%) of the species.[24,25] The farming of fish globally is done for several purposes including fish for consumption, ornamental and research purposes. In Nigeria, there are discrepancies on both the number of culturable and actually cultured species. Ajani *et al.* noted that there are 157 fish species from 51 families that are cultivable in Nigeria.[26] However, a very limited number of species are actually being cultured and they are primarily cultured for food. According to Dauda *et al.*, only a few fish species are being cultured in the country, among which the native species are from only two families, namely Clariidae (*Clarias* sp. and *Heterobranhcus* sp.) and Cichlidae (*Coptodon* sp., *Oreochromis* sp. and *Sarotherodon* sp.).[27] However, more recently, there has been commercial culture of *Heterotis niloticus*, the African bonytongue or Arowona (Osteoglossiformes) from the family Osteoglossidae in the Southern part of Nigeria. In this chapter, we reviewed native fish species in Nigeria that are utilized for ecotoxicology studies.

8.2 Review Approach

In order to obtain a fair and unbiased representation of the relevant literature on the subject, a review question (what native/indigenous fish species are used as models in ecotoxicity studies in Nigeria?) was crafted. Following the Preferred Reporting Items for Systematic Reviews and Meta-Analyses (PRISMA) guidelines,[28] a Population, Intervention, Comparator and Outcome (PICO) framework was adopted with Population (native/indigenous fish species/pisces/ichthyofauna; Nigeria); Intervention (ecotoxicological studies); Comparator (non-native, non-fish species, outside Nigeria or Non-Nigerian) and Outcome (Use/Application). Using this framework, a search string was developed using keywords combined with Boolean operators [(fish species OR Pisces OR ichthyofauna) AND (native OR indigenous) AND (Nigeria) AND (ecotox* OR toxic*)]. A literature search was conducted using Google Scholar search engine from anytime until November 7, 2023. Only studies that were published in English and conducted in Nigeria were included.[29] Articles with studies conducted outside Nigeria, even if with fish species native to Nigeria, were excluded. Besides, articles which were not published in English were excluded. Only primary studies (original research) were considered or included. Studies that were not comprehensively reported were also excluded.[30]

8.3 Ecotoxicological Studies Using Model Fish Species in Nigeria Search Results

The search in Google Scholar returned 3940 results, in which the included articles were limited to the first 200 results.[31] Of these 200 articles, a total of 155 articles were rejected based on the exclusion criteria (non-English language papers, non-native fish species, studies conducted outside Nigeria, studies not relevant to the review focus and lack of comprehensiveness) leaving a total of 45 articles that were included in this review (see Figure 8.3). The major themes of the review based on the results were native fish species used in biomonitoring studies to evaluate the ecological effects of stressors used as bioindicators and native fish species employed as model species in laboratory toxicological studies to establish the effects of substances (singly or jointly) on biomarkers of the fish species.

8.3.1 Native Nigerian Fish Species Used as Models in Laboratory-based Toxicological Studies

In Nigeria, ecotoxicological studies conducted under laboratory conditions have evaluated various chemicals (singly or jointly) and complex mixtures of substances which are associated with, or emanate from anthropogenic activities. Most of the laboratory-based toxicological studies utilise the native African sharptooth catfish, also known as the African mud or North African catfish *C. gariepinus* (see Table 8.1; Figure 8.1). Other common native species

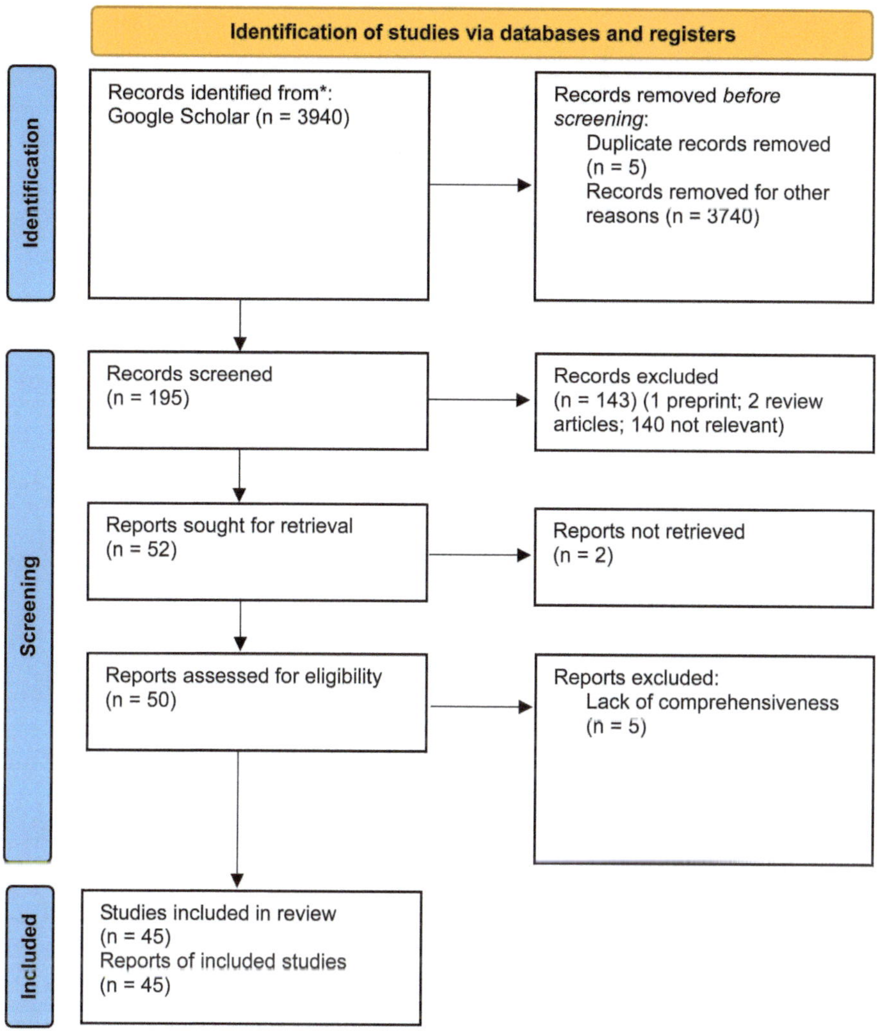

Figure 8.3 PRISMA flow diagram of studies included in the review.

utilised in toxicological studies are the Nile tilapia, *O. niloticus* (see Table 8.1; Figure 8.2) as well as the Guinean tilapia *C. guineensis* (Cichliformes, Cichlidae) (see Table 8.1; Figure 8.3).

In particular, various life stages (embryos, fry, juveniles and adults) of *C. gariepinus* have been and are being utilised to evaluate toxicological effects of substances such as plant-based ichthyotoxins,[32–37] effluent (abattoir and pharmaceutical),[3,38,39] pesticides,[36,40–46] hydrocarbons (crude oil and diesel),[47,48] pharmaceuticals (Ibuprofen),[49] sediment organic and porewater extracts,[50] among others (see Table 8.1). The wide use of *C. gariepinus* in ecotoxicological studies is due to its ability to breed in captivity, high resistance to disease, fast growth and hardy nature, among other attributes.[37]

Table 8.1 Native fish species utilised as models or bioindicators in ecotoxicological studies in Nigeria[a]

Species	Scientific name	Toxicant	Ecotoxicological endpoints evaluated at different levels of biological organisation in fish species			Field (ecosystem name)/Laboratory study		Location in Nigeria	Source
			Cell	Tissue/Organ	Individual/Population				
African sharp-tooth catfish/African mud catfish/North African catfish	*Clarias gariepinus*	Plant-based ichthyotoxins *Carica papaya* (seeds), *Anacardium occidentale* (bark), and *Manihot esculenta* (leaves), *Raphia vinifera* (nuts)		Liver biochemical indices	96hLC$_{50}$	LS	LS	NA	Ozoagudike and Bawa-Allah (2021)[35]
African sharp-tooth catfish/African mud catfish/North African catfish	*Clarias gariepinus*	Pawpaw *Carica papaya* (seed powder)	Haematological indices		Avoidance behaviour	LS	LS	NA	Ayotunde *et al.* (2011)[32]
African sharp-tooth catfish/African mud catfish/North African catfish	*Clarias gariepinus*	Pawpaw *Carica papaya* (seed aqueous extract)		Hepatic biochemical indices		LS	LS	NA	Eyo *et al.* (2013)[33]
African sharp-tooth catfish/African mud catfish/North African catfish	*Clarias gariepinus*	Desert rose *Adenium obesum* (ethanol extract of stem bark)		Liver histopathological changes	Acute toxicity	LS	LS	NA	Abalaka *et al.* (2015)[34]

African sharp-tooth catfish/African mud catfish/North African catfish	*Clarias gariepinus*	Long justicia *Justicia extensa* (leaf ethanolic extract)		Gill and liver histology	LC_{50} and behavioural endpoints	LS	LS	NA	Ekpenyong *et al.* (2022)[37]
African sharp-tooth catfish/African mud catfish/North African catfish	*Clarias gariepinus*	Abattoir effluent	Haematological and genotoxic (nuclear abnormalities) indices	Liver function enzymes	$96hLC_{50}$	LS	LS	NA	Oyeniran *et al.* (2021)[39]
African sharp-tooth catfish/African mud catfish/North African catfish	*Clarias gariepinus*	Abattoir effluent	Biochemical (oxidative stress) indices	Gill and liver histology	$96hLC_{50}$	LS	LS	NA	Olaniran *et al.* (2019)[3]
African sharp-tooth catfish/African mud catfish/North African catfish	*Clarias gariepinus*	Pharmaceutical effluent		Gill and liver histopathology		LS	LS	NA	Agboola and Fawole (2014)[38]
African sharp-tooth catfish/African mud catfish/North African catfish	*Clarias gariepinus*	Atrazine	Haematological and biochemical (AST, ALT, LDH) indices	Gill and liver histology		LS	LS	NA	Michael (2018)[44]
African sharp-tooth catfish/African mud catfish/North African catfish	*Clarias gariepinus*	Emamectin benzoate (EMB)	Haematological indices	Gill and liver histology	$96hLC_{50}$	LS	LS	NA	Ogueji *et al.* (2023)[46]

(continued)

Table 8.1 (*Continued*)

Species	Scientific name	Toxicant	Ecotoxicological endpoints evaluated at different levels of biological organisation in fish species			Field (ecosystem name)/Laboratory study		Location in Nigeria	Source
			Cell	Tissue/Organ	Individual/Population				
African sharp-tooth catfish/African mud catfish/North African catfish	*Clarias gariepinus*	Paraquat dichloride			LC_{50}	LS	LS	NA	Aghoghovwia *et al.* (2019)[45]
African sharp-tooth catfish/African mud catfish/North African catfish	*Clarias gariepinus*	Diuron	Haematological indices			LS	LS	NA	Ajani and Awogbade (2012)[41]
African sharp-tooth catfish/African mud catfish/North African catfish	*Clarias gariepinus*	Gammalin 20		Liver, gill and intestinal histology	96hLC_{50} and behavioural effects	LS	LS	NA	Ezemonye and Ogbomida (2010)[40]
African sharp-tooth catfish/African mud catfish/North African catfish	*Clarias gariepinus*	Cyperdicot (insecticide) and Vitamin E supplement		Gill, liver and kidney histology	96hLC_{50} and behavioural effects	LS	LS	NA	Odo *et al.* (2016)[43]
African sharp-tooth catfish/African mud catfish/North African catfish	*Clarias gariepinus*	Atrazine, *Carica papaya* and *Mangifera indica*	Haematological indices	Gill and liver biochemical and histological indices		LS	LS	NA	Owolabi and Abdulkareem (2021)[36]

African sharp-tooth catfish/African mud catfish/North African catfish	*Clarias gariepinus*	Chloropyrifos (Termifos)			$96hLC_{50}$ and behavioural changes	LS	LS	NA	Nwani and Echi (2013)[42]
African sharp-tooth catfish/African mud catfish/North African catfish	*Clarias gariepinus*	Crude oil and diesel			Acute toxicity	LS	LS	NA	Ezike *et al.* (2019)[48]
African sharp-tooth catfish/African mud catfish/North African catfish	*Clarias gariepinus*	WSF of Nigerian light crude oil			$96hLC_{50}$	LS	LS	NA	Olaifa (2012)[47]
African sharp-tooth catfish/African mud catfish/North African catfish	*Clarias gariepinus*	Sediment organic and porewater extracts			Developmental abnormalities; decreased hatching success	LS	LS	NA	Sogbanmu *et al.* (2020)[50]
African sharp-tooth catfish/African mud catfish/North African catfish	*Clarias gariepinus*	Ibuprofen	Haematological indices		$96hLC_{50}$ and behavioural changes	LS	LS	NA	Ogueji *et al.* (2018)[49]
African sharp-tooth catfish/African mud catfish/North African catfish	*Clarias gariepinus*	Industrial effluent		Liver, kidney and gill biochemical indices		FW	Asejire River	Oyo	Arojojoye and Adeosun (2016)[60]

(*continued*)

Table 8.1 (*Continued*)

Species	Scientific name	Toxicant	Ecotoxicological endpoints evaluated at different levels of biological organisation in fish species			Field (ecosystem name)/Laboratory study	Location in Nigeria	Source
			Cell	Tissue/Organ	Individual/Population			
African sharp-tooth catfish/African mud catfish/North African catfish	*Clarias gariepinus*	PAHs		Bioaccumulation		FW Ogbese River	Ondo	Ololade *et al.* (2021)[63]
African sharp-tooth catfish/African mud catfish/North African catfish	*Clarias gariepinus*	Probably bitumen and other pollutants			Negative allometric growth pattern	FW Oluwa River	Ondo	Ayandiran and Fawole (2014)[61]
African sharp-tooth catfish/African mud catfish/North African catfish	*Clarias gariepinus*	Bitumen pollution	Haematological and biochemical markers in blood	Gill, liver and muscle histology		FW Oluwa River	Ondo	Ayandiran and Dahunsi (2016)[62]
African sharp-tooth catfish/African mud catfish/North African catfish	*Clarias gariepinus*	Heavy metals		Bioaccumulation		FW Ikpoba reservoir	Edo	Egun *et al.* (2023)[59]
African sharp-tooth catfish/African mud catfish/North African catfish	*Clarias gariepinus*	Pesticides-impacted river		Histology of gills and liver		FW River Owan	Edo	Ogbeide *et al.* (2019)[64]

African sharp-tooth catfish/ African mud catfish/North African catfish	*Clarias gariepinus*	Heavy metals and physico-chemical parameters	Micronucleus, comet assay, haematological and bio-chemical (oxidative stress) indices in the blood			FW	Apodu, Asa reservoirs and Asa River	Kwara	Anifowoshe *et al.* (2022)[65]
African sharp-tooth catfish/ African mud catfish/North African catfish	*Clarias gariepinus*	Heavy metals		Gill, stomach, intestine and liver histology		FW	Lake Alau	Borno	Dimari and Hati (2009)[66]
Nile tilapia	*Oreochromis niloticus*	Dichlorvos and Paraquat			LC_{50}	LS	LS	NA	Soriwei *et al.* (2021)[52]
Nile tilapia	*Oreochromis niloticus*	Glyphosate		Gill, liver, kidney and brain histology	$96hLC_{50}$	LS	LS	NA	Ayoola (2008)[51]
Nile tilapia	*Oreochromis niloticus*	*Cactus euphorbia kamerunica* (piscicidal plant)			$96hLC_{50}$	LS	LS	NA	Fai and Fagade (2005)[54]
Nile tilapia	*Oreochromis niloticus*		Micronucleus in gill and kidney erythrocytes			FW	Anambra River	Anambra	Obiakor *et al.* (2014)[70]

(continued)

Table 8.1 (*Continued*)

Species	Scientific name	Toxicant	Ecotoxicological endpoints evaluated at different levels of biological organisation in fish species			Field (ecosystem name)/Laboratory study		Location in Nigeria	Source
			Cell	Tissue/Organ	Individual/Population				
Nile tilapia	*Oreochromis niloticus*	Trace metals		Bioaccumulation in gill and muscle		FW	Okumeshi River	Delta	Ekeanyanwu *et al.* (2010)[55]
Guinean tilapia	*Coptodon guineensis*	Chlorophyrifos	Haematological indices		$96hLC_{50}$; operculum beat frequency; tail beat frequency	LS	LS	NA	Chindah *et al.* (2004)[53]
Guinean tilapia	*Coptodon guineensis*	18 OCPs			Bioaccumulation	BW	Lagos Lagoon	Lagos	Akinsanya *et al.* (2015)[81]
Guinean tilapia	*Coptodon guineensis*	PAH		Bioaccumulation		MW	Landing beaches of Bodo City, B-Dere and Kaa water side	Rivers	Nkpaa *et al.* (2013)[83]
Redbelly tilapia	*Cotpodon zilli*	Heavy metals		Bioaccumulation		FW	Ikpoba reservoir	Edo	Egun *et al.* (2023)[59]

Redbelly tilapia	*Cotpodon zilli*	Pesticides-impacted river		Histology of gills and liver		FW	River Owan	Edo	Ogbeide *et al.* (2019)[64]
Redbelly tilapia	*Coptodon zilli*	Heavy metals and physico-chemical parameters	Micronucleus, comet assay, haematological and biochemical (oxidative stress) indices in the blood			FW	Apodu, Asa reservoirs and Asa River	Kwara	Anifowoshe *et al.* (2022)[65]
Redbelly tilapia	*Coptodon zilli*	Heavy metals		Bioaccumulation in digestive system		FW	Owubu Creek	Rivers	Friday *et al.* (2013)[69]
Mango tilapia	*Sarotherodon galilaeus*	Heavy metals		Gill, stomach, intestine and liver histology		FW	Lake Alau	Borno	Dimari and Hati (2009)[66]
African sicklefish	*Drepane africana*	Bioluminescent bacteria		Biomonitoring in fish gut		MW	Oniru beach	Lagos	Agwu and Aguta (2022)[80]
Madeiran sardinella	*Sardinella maderensis*	Bioluminescent bacteria		Biomonitoring in fish gut		MW	Oniru beach	Lagos	Agwu and Aguta (2022)[80]
Flathead grey mullet	*Mugil cephalus*	Bioluminescent bacteria		Biomonitoring in fish gut		MW	Oniru beach	Lagos	Agwu and Aguta (2022)[80]

(*continued*)

Table 8.1 (*Continued*)

Species	Scientific name	Toxicant	Ecotoxicological endpoints evaluated at different levels of biological organisation in fish species			Field (ecosystem name)/Laboratory study		Location in Nigeria	Source
			Cell	Tissue/Organ	Individual/Population				
Ray finned Nile fish	*Eleotris soaresi*	19 OCPs			Bioaccumulation	BW	Lagos Lagoon	Lagos	Akinsanya *et al.* (2015)[81]
Sompat grunt	*Pomadasys jubelini*	20 OCPs			Bioaccumulation	BW	Lagos Lagoon	Lagos	Akinsanya *et al.* (2015)[81]
African bony tongue	*Heterotis niloticus*	Heavy metals		Oxidative stress indices, histology of liver and intestine		BW	Lekki Lagoon	Lagos	Akinsanya *et al.* (2020)[75]
African bony tongue	*Heterotis niloticus*	Heavy metals		Gill, stomach, intestine and liver histology		FW	Lake Alau	Borno	Dimari and Hati (2009)[66]
African obscure snakehead fish	*Parachanna obscura*	OCPs			Bioaccumulation	BW	Lekki Lagoon	Lagos	Akinsanya *et al.* (2020)[76]
Mandi/red tailed synodontis	*Synodontis clarias*	Pyrethroids (alpha-cypermethrin)		Bioaccumulation in intestine and liver		BW	Lekki Lagoon	Lagos	Alimba *et al.* (2015)[78]

Mandi/red tailed synodontis	*Synodontis clarias*		Gill and blood micronucleus	Gill, liver, kidney and ovary histology	BW FW	Lekki Lagoon; Ogun River	Lagos and Ogun	Akinsanya *et al.* (2020)[79]
Mandi/red tailed synodontis	*Synodontis clarias*		Micronucleus in gill and kidney erythrocytes		FW	Anambra River	Anambra	Obiakor *et al.* (2014)[70]
Black mudfish	*Neochanna diversus*	Pesticides-impacted river		Histology of gills and liver	FW	River Owan	Edo	Ogbeide *et al.* (2019)[64]
Longneck croaker	*Pseudotolithus typus*	Heavy metals and PAHs		Bioaccumulation in gut	BW	Qua Iboe River Estuary	Akwa Ibom	Ekanem *et al.* (2023)[72]
Silver catfish	*Chrysichthys nigrodigitatus*	Heavy metals and PAHs		Bioaccumulation in gut	BW	Qua Iboe River Estuary	Akwa Ibom	Ekanem *et al.* (2023)[72]
Silver catfish	*Chrysichthys nigrodigitatus*	Trace metals		Bioaccumulation in gill and muscle	FW	Okumeshi River	Delta	Ekeanyanwu *et al.* (2020)[55]
Silver catfish	*Chrysichthys nigrodigitatus*	Heavy metals		Bioaccumulation in liver, kidney, heart and gills and biochemical indices	FW	Ibaka (Mbo) and Ifiayong Rivers, Akwa Ibom State	Akwa Ibom	Akpakpan *et al.* (2014)[73]

(*continued*)

Table 8.1 (*Continued*)

Species	Scientific name	Toxicant	Ecotoxicological endpoints evaluated at different levels of biological organisation in fish species			Field (ecosystem name)/Laboratory study	Location in Nigeria	Source
			Cell	Tissue/Organ	Individual/Population			
Mudskipper	*Periophthalmus koelreuteri*	PAHs		Bioaccumulation in tissues		FW BW Kporghor and Iko Rivers; BW – Sime (Tai) coastal water	Akwa Ibom; Rivers	Nwaichi and Ntorgbo (2016)[82]
Sicklefin mullet	*Neochelon falcipinnis*	PAH		Bioaccumulation		MW Landing beaches of Bodo City, B-Dere and Kaa water side	Rivers	Nkpaa *et al.* (2013)[83]

[a]LS: laboratory study; FW: freshwater; BW: brackish water; MW: marine water; NA: not applicable; PAH: polycyclic aromatic hydrocarbons; OCPs: organochlorine pesticides; LC_{50}: median lethal concentration; $96hLC_{50}$ median lethal concentration within 96 h.

The toxicological indices that are evaluated mostly span across the cell, tissue/organ, and individual to population levels of biological organisation. Often, multiple toxicological indices are evaluated under acute and sub-acute exposures of the fish species to the toxicants (see Table 8.1). These indices include haematological, erythrocytic biochemical (oxidative stress, antioxidant enzymes), genotoxic (micronucleus and comet assays) indices (cellular level); tissue/organ biochemical (oxidative stress, antioxidant enzymes, liver function and kidney function enzymes) and histological biomarkers as well as behavioural and median lethal concentration indices (individual/population) (see Table 8.1).

Similarly, *O. niloticus* and *C. guineensis* have been utilised to evaluate the toxicological effects of pesticides[51–53] and piscicidal plants used in fishing.[54] The toxicological effects established include haematological (cellular level), tissue/organ histological alterations as well as median lethal concentration and behavioural effects (individual/population level) (see Table 8.1). Many species in the Cichlid family have been used in biomonitoring because of their ubiquity and ability to tolerate adverse conditions.[55] However, for laboratory-based toxicological studies, the Nile tilapia *O. niloticus* is more commonly used.[56] The Nile tilapia is found in a wide range of habitats including slow and fast-moving freshwater as well as brackish water ecosystems. They are omnivorous and feed majorly on mud, detritus and phytoplankton.[57]

8.3.2 Native Nigerian Fish Species Used as Bioindicators in Biomonitoring Studies (Field-based Ecotoxicological Studies)

Clarias gariepinus is a common species in African tropical waters and a commercial fish found in the wild in Nigerian freshwater ecosystems as well as being easily cultured across Nigeria.[58,59] It is a migratory (potamodromous), benthopelagic, nocturnal freshwater fish with an omnivorous feeding habit. They are finfishes (class Osteichthyes) belonging to the order Siluriformes and family Clariidae.[39] Due to their ubiquity, established and well described biology and ecology, they are often encountered during biomonitoring studies in aquatic ecosystems such as Asejire River in Oyo State,[60] Ogbese and Oluwa Rivers, Ondo State,[61–63] River Owan and Ikpoba reservoir, Edo State,[59,64] Asa River, Apodu and Asa reservoirs, Kwara State[65] as well as Lake Alau, Borno State[66] (see Table 8.1). Ecotoxicological indices ranging from haematological and biochemical indices in the blood (cellular level of analysis), to tissue/organ histological alterations and bioaccumulation of pollutants have been evaluated in relation to industrial effluent, polycyclic aromatic hydrocarbons, pesticides and bitumen pollution in the freshwater ecosystems studied.

The family Cichlidae, from the order Cichliformes, is a very important freshwater fish family and it has been reported to be the dominant family in most of the studies on fish abundance and distribution in the inland water bodies of Nigeria.[67,68] The two most commonly reported species belonging to this family are the Nile tilapia *O. niloticus* (see Figure 8.2) and the redbelly tilapia *Cotpodon zillii* (previously known as *Tilapia zillii*) (see Figure 8.4), both native to Nigeria. *Cotpodon zillii* caught from Ikpoba reservoir[59] and River Owan, Edo State,[64] Asa reservoir and River, Kwara State[65] as well as Owubu creek, Rivers State[69] have been shown to bioaccumulate heavy metals. Also, they show nuclear abnormalities, genotoxic effects, oxidative stress, tissue and organ histological alterations in response to mainly pesticides and heavy metals in these freshwater ecosystems. *Oreochromis niloticus* sampled from the Okumeshi River, Delta State[55] and Anambra River, Anambra State[70] showed erythrocytic nuclear abnormalities and bioaccumulation of heavy metals in the gills and muscles (see Table 8.1). Among other important species, also native to Nigeria, from the Cichlidae family, is the mango tilapia *Sarotherodon galilaeus* (see Figure 8.5) sampled in Lake Alau, Borno State,

Figure 8.4 Redbelly tilapia *Coptodon zillii* (Cichliformes, Cichlidae).

Figure 8.5 Mango tilapia *Sarotherodon galilaeus* (Cichliformes, Cichlidae).

with multiple tissue and organ histological alterations associated with heavy metal pollution.[66]

There are also other native fish species, rather than those previously mentioned, that are commonly used in biomonitoring and ecotoxicological studies but not in laboratory-based studies. Although many of them have been reported to be cultivable species but are neither cultured at all, nor actively cultured in the country. Among these species we can mention *Chrysichtys nigrodigitatus*, *Heterotis niloticus*, *Synodontis clarias* as well as *Parachana* sp. The silver catfish *C. nigrodigitatus* is a catfish species of the family Claroteidae (Siluriformes). The adults are found mostly in shallow waters of lakes, are omnivorous and feed on a wide range of food including leaves, phytoplankton, zooplankton, detritus, insects, bivalves, seeds, decapods and smaller fish.[71] They have been recorded in the Qua Iboe River estuary,[72] Ibaka and Ifiayong Rivers,[73] Akwa Ibom State as well as Okemeshi River, Delta State,[55] with evidence of bioaccumulation of heavy metals and PAHs in their tissues and associated histological alterations (see Table 8.1).

Heterotis niloticus, the African bonytongue (Osteoglossiformes, Arapaimidae), is an obligate-air breather, hence, its ability to survive water bodies with limited oxygen concentrations. It is widespread in many of the tropical rivers and lakes of Africa, including Nigeria. The spread is attributed to its ability to utilize varieties of food and its versatile feeding habit which is correlated to its alimentary canal and digestive process.[74] It has been observed in the Lekki Lagoon[75] and Lake Alau, Borno State[66] showing tissue and organ histological alterations in heavy metal-polluted fresh and brackish waters (see Table 8.1). It feeds on small invertebrates, mud, vegetable debris and phytoplanktons.

Parachana obscura, the African obscure snakehead (Anabantiformes, Channidaei), is commonly present in Nigerian water bodies. It is an obligate air-breather found in marginal vegetation and flood plains. It was recorded to bioaccumulate organochlorine pesticides in its body tissues from the Lekki Lagoon.[76] According to Kpogue *et al.*, *P. obscura* is the most common species of the African Channidae and it feeds on insects, other invertebrates, plant detritus, fish and fish remains.[77]

Among other fish species encountered in biomonitoring studies in Nigeria we find the Mandi or red tailed synodontis *Synodontis clarias* (Siluriformes, Mochokidae) from the Lekki Lagoon, Lagos State,[78] Ogun River, Ogun State[79] and Anambra River, Anambra State[70] (see Table 8.1), able to bioaccumulate pyrethroids and showing nuclear abnormalities. Bioluminescent bacteria were observed in the gut of the flathead grey mullet *Mugil cephalus* (Mugiliformes, Mugilidae) from the Oniru beach, Lagos.[80] Organochlorine pesticides (OCPs) were bioaccumulated in the Nile fish *Eleostris soaresi* (Gobiiformes, Eleotridae) from the Lagos Lagoon,[81] polycyclic aromatic hydrocarbons (PAHs) in the Atlantic mudskipper *Periophtalmus barbarus* from the Kporgor and Iko Rivers (Akwa Ibom State) and Sime waters (Rivers State)[82] as well as PAHs bioaccumulated in the Guinean tilapia *Coptodon guineensis* (Cichliformes, Cichlidae) from Ogoni communities landing beaches in Rivers State.[83]

8.4 Review Limitations

The limitations to this review are the use of only one search engine or database, namely Google Scholar (https://scholar.google.com/) for the literature search, though it is recognised to capture most literature abstracts in other specialist databases. We limited the articles extracted to the first 200 results with the potential or having missed some other relevant articles in subsequent result pages. Only articles written in English were included although no article was observed to have been published in any other language apart from English.

8.5 Conclusions and Recommendations

In this review, we have documented that the commonly used native Nigerian fish species in laboratory and field based (biomonitoring) ecotoxicological studies are the African sharptooth catfish *C. gariepinus*, the Nile tilapia *O. niloticus* and the Guinean tilapia *C. guineensis*. Other fish species commonly encountered in biomonitoring studies are the silver catfish *C. nigrodigitatus*, the African bonytongue *H. niloticus*, the Mandi or red tailed synodontis *S. clarias*, among others. These model fish species have been demonstrated as valid models to study biomarkers of exposure and effect of toxicants including pesticides, effluents, heavy metals, hydrocarbons, pharmaceuticals, plant and sediment extracts. They serve as bioindicators of stressors or pollutants in biomonitoring programmes with indices ranging from cellular (nuclear abnormalities, haematological effects) to tissue and organ histological alterations, oxidative stress indices as well as individual/population level effects (behavioural changes). Still, these native model fish species (cultured species) are limited compared to the number of species that have been documented to be culturable or cultivable.[26,27] Furthermore, all of the fish species are mostly freshwater fishes and very few brackish water species. There are hardly any native marine fish species that are used in ecotoxicological studies in Nigeria.

We recommend complementary targeted studies on the culture exploitation of other native Nigerian freshwater, brackish and marine fish species as models in ecotoxicological studies and aquaculture in order to have a more faithful picture of the environmental status. We also endorse the ratification of *C. gariepinus* and *O. niloticus* as globally recognised model fish species for use in ecotoxicological studies as in the case of other temperate fish species like the well-known zebrafish *Danio rerio*.

Acknowledgements

The authors wish to express their gratitude for the assistance of Mr Confidence Kpakol, Master student and Research Assistant in the Environmental Toxicology and Pollution Management programme (2022/2023 session), Department of Zoology, Faculty of Science, University of Lagos in formatting the References list and downloading the articles data file from Google Scholar.

References

1. E. Sanders and S. C. Farmer, Aquatic models: water quality and stability and other environmental factors, *ILAR J.*, 2019, **60**, 141–149.
2. M. Nordberg, D. M. Templeton, O. Andersen and J. H. Duffus, Glossary of terms used in ecotoxicology (IUPAC Recommendations 2009), *Pure Appl. Chem.*, 2009, **81**, 829–970.
3. E. I. Olaniran, T. O. Sogbanmu and J. K. Saliu, Biomonitoring, physico-chemical, and biomarker evaluations of abattoir effluent discharges into the Ogun River from Kara Market, Ogun State, Nigeria, using *Clarias gariepinus*, *Environ. Monit. Assess.*, 2019, **191**, 44.
4. Organisation for Economic Co-operation and Development, *Test No. 203: Fish, Acute Toxicity Test, OECD Guidelines for the Testing of Chemicals, Section 2*, OECD Publishing, Paris, 2019.
5. United States Environmental Protection Agency, *Ecological Effects Test Guidelines: OCSPP 850.1075 - Freshwater and Saltwater Fish Acute Toxicity Test, No. EPA 712-C-16-007*, U.S. EPA, Washington, DC, 2016.
6. L. Norrgren, Fish models for ecotoxicology, *Acta Vet. Scand.*, 2012, **54**(Suppl 1), S14.
7. R. J. Kendall, T. A. Anderson, R. J. Baker, C. M. Bens and J. A. Carr, *Ecotoxicology*, United States Department of Agriculture, National Wildlife Research Center, University of Nebraska, Lincoln, 2001.
8. Á. Borja, M. J. Belzunce, J. M. Garmendia, J. G. Rodríguez, O. Solaun and I. Zorita, Impact of pollutants on coastal and benthic marine communities, in *Ecological Impacts of Toxic Chemicals*, ed. F. Sánchez-Bayo, P. J. van den Brink, and R. M. Mann, Bentham Science Publishers Ltd., Sharjah, U.A.E., 2011, pp. 165–186.
9. Z. Wang, K. W. H. Kwok, G. C. S. Lui, G.-J. Zhou, J.-S. Lee, M. H. W. Lam and K. M. Y. Leung, The difference between temperate and tropical saltwater species' acute sensitivity to chemicals is relatively small, *Chemosphere*, 2014, **105**, 31–43.
10. O. M. Olarinmoye, O. A. Ugwumba and F. O. Awe, in *The Political Ecology of Oil and Gas Activities in the Nigerian Aquatic Ecosystem, The Nigerian Environment*, ed. P. E. Ndimele, Academic Press, New York, 2018, pp. 3–15.
11. O. A. Aribisala, T. O. Sogbanmu and K. A. Kemabonta, Genotoxic, biochemical and histological biomarkers of subacute concentrations of paraquat and glyphosate in Nile Tilapia, *Environ. Anal. Health Toxicol.*, 2022, **37**, e2022012-0.
12. G. De Boeck, E. Rodgers and R. M. Town, Using Ecotoxicology for Conservation: From Biomarkers to Modelling, in *Fish Physiology*, ed. N. A. Frangue, S. J. Cooke, A. P. Farrel, C. J. Brauner, and E. J. Aliason, Academic Press, Oxford, 2022, vol. 39B, pp. 111–174.
13. American Society for Testing and Materials, Standard guide for conducting early life-stage toxicity tests with fishes, in *Annual Book of ASTM Standards*, West Conshohocken, PA, 2000, vol. 11.05. E 1241-98, pp. 528–555.

14. United States Environmental Protection Agency (U.S. EPA), in *Methods for Measuring the Acute Toxicity of Effluents and Receiving Waters to Rreshwater and Marine Organisms*, ed. C. L. Weber, Environmental Monitoring Systems Laboratory, Cincinnati, OH, Office of Research and Development, U.S. Environmental Protection Agency, 4th edn, 1993.
15. B. Anderson, P. Nicely, K. Gilbert, R. Kosaka, J. Hunt and B. Phillips, *Overview of Freshwater and Marine Toxicity Tests: A Technical Tool for Ecological Risk Assessment*, California Environmental Protection Agency (Cal/EPA), Davis, University of California, 2004.
16. L. T. H. Nguyen and C. R. Janssen, Embryo-larval toxicity tests with the African catfish (*Clarias gariepinus*): comparative sensitivity of endpoints, *Arch. Environ. Contam. Toxicol.*, 2002, **42**, 256–262.
17. H. L. Atwood, Q. C. Fontenot, J. R. Tomasso and J. J. Isely, Toxicity of nitrite to Nile tilapia: effect of fish size and environmental chloride, *North Am. J. Aquacult.*, 2001, **63**, 49–51.
18. Worldometers, Nigeria population live, 2023, https://www.worldometers.info/world-population/nigeria-population/, accessed on November 11, 2023.
19. O. Kehinde, O. J. Ramonu, K. O. Babaremu and L. D. Justin, Plastic wastes: environmental hazard and instrument for wealth creation in Nigeria, *Heliyon*, 2020, **6**, e05131.
20. B. Yalwaji, H. O. John-Nwagwu and T. O. Sogbanmu, Plastic pollution in the environment in Nigeria: a rapid systematic review of the sources, distribution, research gaps and policy needs, *Sci. Afr.*, 2022, **16**, e01220.
21. Z. A. Elum, K. Mopipi and A. Henri-Ukoha, Oil exploitation and its socioeconomic effects on the Niger Delta region of Nigeria, *Environ. Sci. Pollut. Res. Int.*, 2016, **23**, 12880–12889.
22. *Environmental Guidelines and Standards for the Petroleum Industry in Nigeria (EGASPIN)*, Table VIII-F3 – Selected test organisms for effluent toxicity tests, Department of Petroleum Resources, Lagos, 3rd edn, 2018, p. 164.
23. B. D. Olaosebikan and A. Raji, *Field Guide to Nigerian Freshwater Fishes (Revised Edition)*, National Institute of Freshwater Fisheries Research (NIFFR), New Bussa, Nigeria, 2013.
24. Food and Agriculture Organization of the United Nations, *The State of World Fisheries and Aquaculture-Meeting the Sustainable Development Goals*, Fisheries and Aquaculture Department, FAO, Rome, 2018.
25. I. Elegbede, A. B. Dauda, N. F. Osho-Abdulgafar, O. E. Sholanke, L. Badmus and I. D. Johnson, in *Encyclopedia of Sustainable Management*, ed. S. Idowu, R. Schmidpeter, N. Capaldi, L. Zu, M. Del Baldo, and R. Abreu, Springer, Cham, 2022, pp. 1–9.
26. E. K. Ajani, A. O. Akinwole and I. A. Ayodele, *Fundamentals of Fish Farming in Nigeria*, Walecrown Publishers, Ibadan, 2011.
27. A. B. Dauda, I. Natrah, M. Karim, M. S. Kamarudin and A. H. Bichi, African catfish aquaculture in Malaysia and Nigeria: Status, trends, and prospects, *Fish. Aquacult. J.*, 2018, **9**, 1.

28. M. J. Page, J. E. McKenzie, P. M. Bossuyt, I. Boutron, T. C. Hoffmann, C. D. Mulrow, L. Shamseer, J. M. Tetzlaff, E. A. Akl, S. E. Brennan, R. Chou, J. Glanville, J. M. Grimshaw, A. Hróbjartsson, M. M. Lalu, T. Li, E. W. Loder, E. Mayo-Wilson, S. McDonald, L. A. McGuinness, L. A. Stewart, J. Thomas, A. C. Tricco, V. A. Welch, P. Whiting and D. Moher, The PRISMA 2020 statement: an updated guideline for reporting systematic reviews, *BMJ*, 2021, **372**, n71.
29. C. N. Amadi, C. Frazzoli and O. E. Orisakwe, Sentinel species for biomonitoring and biosurveillance of environmental heavy metals in Nigeria, *J. Environ. Sci. Health, Part C: Toxicol. Carcinog.*, 2022, **38**, 21–60.
30. A. P. Onyena, O. M. Folorunso, N. Nwanganga, G. J. Udom, O. C. Ekhator, C. Frazzoli, F. Ruggieri, B. Bocca and O. E. Orisakwe, Engaging one health in heavy metal pollution in some selected nigerian Niger Delta cities. A systematic review of pervasiveness, bioaccumulation and subduing environmental health challenges, *Biol. Trace Elem. Res.*, 2024, **202**, 1356–1389.
31. W. M. Bramer, M. L. Rethlefsen, J. Kleijnen and O. H. Franco, Optimal database combinations for literature searches in systematic reviews: A prospective exploratory study, *Syst. Rev.*, 2017, **6**, 245.
32. E. O. Ayotunde, B. O. Offem and A. F. Bekeh, Toxicity of *Carica papaya* Linn: Haematological and piscicidal effect on adult catfish (*Clarias gariepinus*), *J. Fish. Aquat. Sci.*, 2011, **6**, 291–308.
33. J. E. Eyo, C. A. Levi, C. N. Asogwa, E. C. Odii, C. O. Chukwuka, N. Ivoke, U. S. Onoja and C. C. Onyeke, Toxicity and effect of *Carica papaya* seed aqueous extract on liver biomarkers of *Clarias gariepinus*, *Int. J. Indig. Med. Plants*, 2013, **46**, 1301–1307.
34. R. E. Abalaka, M. Y. Fatihu, N. D. G. Ibrahim and S. F. Ambali, Liver histopathological changes in *Clarias gariepinus* exposed to ethanol extract of *Adenium obesum* stem bark, *J. Morphol. Sci.*, 2015, **32**, 22–28.
35. C. M. Ozoagudike and K. A. Bawa-Allah, Comparative assessment of the acute and sub-acute effects of some plant based ichthyotoxins on aquatic organisms: A case study of *Clarias gariepinus*, *Proc. Niger. Acad. Sci.*, 2021, **13**, 36–45.
36. O. D. Owolabi and S. I. Abdulkareem, *Carica papaya* and *Mangifera indica* modulate haematological, biochemical and histological alterations in atrazine-intoxicated fish, *Clarias gariepinus* (Burchell 1822), *J. Basic Appl. Zool.*, 2021, **82**, 42.
37. J. Ekpenyong, S. Iheanacho, F. Asuquo and I. Okey, Ichthyotoxicity of ethanolic extract of *Justicia extensa* on behavioral and histopathological changes in *Clarias gariepinus* (Burchell, 1822), *Environ. Chem. Ecotoxicol.*, 2022, **4**, 211–215.
38. O. A. Agboola and O. O. Fawole, Chronic toxicity of pharmaceutical effluent to *Clarias gariepinus* (Burchell, 1822), *Covenant J. Phys. Life Sci. (CJPL)*, 2014, **1**, 27–42.
39. D. O. Oyeniran, T. O. Sogbanmu and T. A. Adesalu, Antibiotics, algal evaluations and subacute effects of abattoir wastewater on liver function

enzymes, genetic and haematologic biomarkers in the freshwater fish, *Clarias gariepinus*, *Ecotoxicol. Environ. Saf.*, 2021, **212**, 111982.
40. L. Ezemonye and T. E. Ogbomida, Histopathological effects of gammalin 20 on African catfish (*Clarias gariepinus*), *Appl. Environ. Soil Sci.*, 2010, 138019.
41. F. Ajani and A. A. Awogbade, Hematological changes of the African catfish *Clarias gariepinus* (Burchell, 1822) juveniles induced by diuron, *Br. Biotechnol. J.*, 2012, **2**, 247–256.
42. C. D. Nwani, N. Ivoke, D. O. Ugwu, C. Atama, G. C. Onyishi, P. C. Echi and S. A. Ogbonna, Investigation on acute toxicity and behavioral changes in a freshwater African catfish, *Clarias gariepinus* (Burchell, 1822), exposed to organophosphorous pesticide, Termifos®, *Pak. J. Zool.*, 2013, **45**, 959–965.
43. G. E. Odo, J. E. Agwu, J. Madu, C. O. Ezea, E. I. Nnamonu and V. O. Eneje, Histopathological effects of Cyperdicot and vitamin E supplementation on selected organs of *Clarias gariepinus* (Burchell, 1822) reared in a tropical fish farm in Nigeria, *Afr. J. Biotechnol.*, 2016, **15**, 303–314.
44. P. O. Michael, Toxicity effect of atrazine on histology, haematology and biochemical indices of *Clarias gariepinus*, *Int. J. Fish. Aquat. Stud.*, 2018, **6**, 87–92.
45. O. A. Aghoghovwia, P. I. Morgan and S. C. Izah, Behavioral response and acute toxicity of fingerlings of African catfish, *Clarias gariepinus* exposed to paraquat dichloride, *J. Plant Anim. Ecol.*, 2019, **1**, 13–20.
46. E. Ogueji, J. Nwakpa, J. Ekpenyong, M. Olaolu, A. Yaji, I. Okey, G. Amana, C. Elo and U. Agbo, Histological and hematological changes to *Clarias gariepinus* juveniles exposed to acute doses of Emamectin benzoate in a static bioassay, *Ecotoxicology*, 2023, **32**, 569–582.
47. F. E. Olaifa, Bioassay using the water soluble fraction of a Nigerian light crude oil on *Clarias gariepinus* fingerlings, *Niger. J. Physiol. Sci.*, 2012, **27**, 181–187.
48. C. O. Ezike, F. O. Echor, I. Raymond and N. C. Uwadiegwu, Benzo [a] Pyrene-carcinogenic and mutagenic equivalents of *Clarias gariepinus* (Burchell, 1822) to Nigerian-crude and diesel oils, *Int. J. Fish. Aquat. Stud.*, 2019, **7**, 62–67.
49. E. O. Ogueji, C. D. Nwani, S. C. Iheanacho, C. E. Mbah, C. O. Okeke and A. Yaji, Acute toxicity effects of ibuprofen on behaviour and haematological parameters of African catfish *Clarias gariepinus* (Burchell, 1822), *Afr. J. Aquat. Sci.*, 2018, **43**, 293–303.
50. T. O. Sogbanmu, O. T. Fatunsin, F. O. Echebiri, A. A. Otitoloju and K. O. Olayinka, Sawmill activities near the Lagos lagoon, Nigeria: polycyclic aromatic hydrocarbons and embryotoxic evaluations of sediment extracts using *Clarias gariepinus*, *Bull. Environ. Contam. Toxicol.*, 2020, **104**, 809–819.
51. S. O. Ayoola, Toxicity of herbicide glyphosate on Nile tilapia (*Oreochromis niloticus*) juvenile, *Afr. J. Agric. Res.*, 2008, **3**, 825–834.
52. E. T. Soriwei, H. C. Umeokeke, H. N. Amaeze, O. O. Ogunfeitimi and A. S. Labinjo, Dichlorvos and Paraquat induced spatial avoidance

response: A more realistic determinant of population decline of *Oreochromis niloticus*, *Ecotoxicol. Environ. Contam.*, 2021, **16**, 27–34.

53. A. C. Chindah, F. D. Sikoki and I. Vincent-Akpu, Toxicity of an organophosphate pesticide (chloropyrifos) on a common Niger Delta wetland fish -*Tilapia guineensis* (Blecker 1862), *J. Appl. Sci. Environ. Manage.*, 2004, **8**, 11–17.
54. P. B. A. Fai and S. O. Fagade, Acute toxicity of *Euphorbia kamerunica* on *Oreochromis niloticus*, *Ecotoxicol. Environ. Saf.*, 2005, **62**, 128–131.
55. C. R. Ekeanyanwu, C. A. Ogbuinyi and O. F. Etienajirhevwe, Trace metals distribution in fish tissues, bottom sediments and water from Okumeshi River in Delta State, Nigeria, *Ethiop. J. Environ. Stud. Manage.*, 2010, **3**, 12–17.
56. S. O. Ayoola, Acute toxicity and histopathology of Nile tilapia (*Oreochromis niloticus*) fingerlings exposed to aqueous and ethanolic extracts of *Euphorbia poissonii* leaves, *New Clues Sci.*, 2011, **1**, 55–68.
57. A. Tesfahun and S. Alebachew, Food and feeding habits of the Nile tilapia *Oreochromis niloticus* (Linnaeus, 1758) from Ribb reservoir, Lake Tana sub-basin, Ethiopia, *Cogent Food Agric.*, 2023, **9**, 2212457.
58. A. Ibrahim, J.-O. Adetola, A. E. Kolawole and A. A. Yusuf, Natural occurrence of *Diplostomum* spp. in farm-raised African catfish (*Clarias gariepinus*) from Oyo state, Nigeria, *Int. J. Vet. Sci. Med.*, 2016, **4**, 41–45.
59. N. K. Egun, U. R. Okotie and I. P. Oboh, Distribution and toxicological risk assessment of metals in commercial fish species from a lentic ecosystem in Nigeria, *J. Trace Elem. Miner.*, 2023, **6**, 100100.
60. O. A. Arojojoye and A. M. Adeosun, Effect of environmental pollution on oxidative stress biomarkers in African cat fish (*Clarias gariepinus*) from Asejire River in Oyo State, Nigeria, *J. Environ. Occup. Health*, 2016, **5**, 71–76.
61. T. A. Ayandiran and O. O. Fawole, Seasonal variation and length-weight relationship of *Clarias gariepinus* from Oluwa River, Nigeria, *IOSR J. Environ. Sci., Toxicol. Food Technol.*, 2014, **8**, 51–59.
62. T. A. Ayandiran and S. O. Dahunsi, Toxicological assessment of fish (*Clarias gariepinus*) from bitumen-polluted River Oluwa, Nigeria, *Environ. Monit. Assess.*, 2016, **188**, 71.
63. I. A Ololade, N. A. Olodoja, O. O. Ololade, T. D. Saliu, A. B. Alabi, S. B. Obadawo and M. M. Anifowose, Bioaccumulation and toxic potencies of polycyclic aromatic hydrocarbons in freshwater biota from the Ogbese River, Nigeria, *Environ. Monit. Assess.*, 2021, **193**, 8.
64. O. Ogbeide, G. Uhunamure, L. Uwagboe, T. Osakpamwan, M. Glory and A. Chukwuka, Comparative gill and liver pathology of *Tilapia zilli, Clarias gariepinus* and *Neochanna diversus* in Owana River (Nigeria): Relative ecological risks of species in a pesticide-impacted river, *Chemosphere*, 2019, **234**, 1–13.
65. A. T. Anifowoshe, S. O. Oladipo, A. N. Oyinloye, A. Opute, E. O. Odofin, A. Omotola, Y. M. Abdulrahim, K. M. Akinseye, S. I. Abdulkareem and O. A. Iyiola, Induction of oxidative stress and DNA damage in two

common fish species of rivers and reservoirs in Ilorin, Northcentral, Nigeria, *J. Taibah Univ. Sci.*, 2022, **16**, 480–494.
66. G. A. Dimari and S. S. Hati, Interaction profile for As, Cd, Cr and Pb in tissues of fishes (*Tilapia gallier, Clarias lazera* and *Heterotis niloticus*), *Sci. Res. Essays*, 2009, **4**, 894–899.
67. M. Ahmad, F. U. Shagari and A. N. Sani, Fish biodiversity and fishing methods of some water bodies in Katsina State of Nigeria, *Int. J. Fish. Aquat. Stud.*, 2014, **1**, 218–221.
68. A. E. Falaye, E. K. Ajani, O. K. Kareem and A. N. Olanrewaju, Assessment of ichthyofaunal assemblage of Erelu Reservoir, Oyo, Nigeria, *Ecologia*, 2015, **5**, 43–53.
69. U. Friday, O. A. F. Wokoma and A. J. Edoghotu, Levels of bioaccumulation of some heavy metals in fish (*Tilapia zilli*) and their concentration in water and sediment of Owubu Creek, Niger Delta, Nigeria, *Resour. Environ.*, 2013, **3**, 59–64.
70. M. O. Obiakor, J. C. Okonkwo and C. D. Ezeonyejiaku, Genotoxicity of freshwater ecosystem shows DNA damage in preponderant fish as validated by *in vivo* micronucleus induction in gill and kidney erythrocytes, *Mutat. Res.*, 2014, **775–776**, 20–30.
71. I. K. Esenowo, A. A. A. Ugwumba, I. I. Umoh and A. B. Andem, Aspects of the biology of silver catfish (*Chrysichthys nigrodigitatus*) in Nwaniba River, Southeast Nigeria, *Asian J. Biol.*, 2017, **3**, 1–9.
72. J. O. Ekanem, A. Y. Itah and U. Ndubuisi-Nnaji, Microbial diversity, heavy metals and hydrocarbons concentration in some fish species from Qua Iboe River Estuary, Akwa Ibom State, Nigeria, *GSC Adv. Res. Rev.*, 2023, **16**, 242–251.
73. E. I. Akpakpan, E. O. Akpanyung, N. O. Anazodo and U. M. Ekanemesang, Biomarkers of oxidative stress and histopathological studies in fish from Ibaka and Ifiayong Rivers, Akwa Ibom State, Nigeria, *World Appl. Sci. J.*, 2014, **32**, 1209–1218.
74. M. O. Agbugui, H. O. Egbo and F. E. Abhulimen, The biology of the African bonytongue *Heterotis niloticus* (Cuvier, 1829) from the Lower Niger River at Agenebode in Edo State, Nigeria, *Int. J. Zool.*, 2021, **2021**, 1–13.
75. B. Akinsanya, I. O. Ayanda, A. O. Fadipe, B. Onwuka and J. K. Saliu, Heavy metals, parasitologic and oxidative stress biomarker investigations in *Heterotis niloticus* from Lekki Lagoon, Lagos, Nigeria, *Toxicol Rep*, 2020, **7**, 1075–1082.
76. B. Akinsanya, F. Olaleru, O. B. Samuel, E. Akeredolu, P. O. Isibor, O. S. Adeniran, J. K. Saliu and D. I. Akhiromen, Bioaccumulation of organochlorine pesticides, *Procamallanus* sp. (Baylis, 1923) infections, and microbial colonization in African snakehead fish sampled from Lekki Lagoon, Lagos, Nigeria, *Braz. J. Biol.*, 2020, **81**, 1095–1105.
77. D. N. S. Kpogue, G. A. Mensah and E. D. Fiogbe, A review of biology, ecology and prospect for aquaculture of *Parachanna obscura*, *Rev. Fish Biol. Fish.*, 2012, **23**, 41–50.

78. C. G. Alimba, J. K. Saliu and O. A. Ubani-Rex, Cytogenotoxicity and histopathological assessment of Lekki Lagoon and Ogun River in *Synodontis clarias* (Linnaeus, 1758), *Toxicol. Environ. Chem.*, 2015, **97**, 221–234.
79. B. Akinsanya, P. O. Isibor, F. Olaleru, A. Abayomi, E. Akeredolu, M. I. Ohazulike and J. K. Saliu, Bioaccumulation of pyrethroid in parasite *Wenyonia acuminata* (Cestoda: Caryophyllaeidae) and host fish *Synodontis clarias* (Linnacus, 1758) from Lekki Lagoon, Lagos Nigeria, *Braz. J. Biol.*, 2020, **81**, 822–834.
80. O. A. Agwu and O. J. Aguta, Bioluminescent bacterial flora of Nigerian coastal fish species, *Indian J. Mar. Sci.*, 2022, **51**, 26–32.
81. B. Akinsanya, R. Alani, U. D. Ukwa, F. Bamidele and J. K. Saliu, Bioaccumulation and distribution of organochlorine residues across the food web in Lagos Lagoon, Nigeria, *Afr. J. Aquat. Sci.*, 2015, **40**, 403–408.
82. E. O. Nwaichi and S. A. Ntorgbo, Assessment of PAHs levels in some fish and seafood from different coastal waters in the Niger Delta, *Toxicol Rep*, 2016, **3**, 167–172.
83. K. W. Nkpaa, M. O. Wegwu and E. B. Essien, Assessment of polycyclic aromatic hydrocarbons (PAHs) levels in two commercially important fish species from crude oil polluted waters of Ogoniland and their carcinogenic health risks, *J. Environ. Earth Sci.*, 2013, **3**, 128–137.

CHAPTER 9

*Behavioural Endpoints of Nile Tilapia (*Oreochromis niloticus*): Potential Biomarkers for Evaluating the Ecotoxicity of Contaminants of Emerging Concern and Legacy Pollutants*

GUILHERME MALAFAIA[a,b,c*] AND THIARLEN MARINHO DA LUZ[a,b,c]

[a]Laboratory of Toxicology Applied to the Environment, Goiano Federal Institute – Urutaí Campus. Rodovia Geraldo Silva Nascimento, 2,5 km, Zona Rural, Urutaí, GO, Brazil; [b]Post-Graduation Program in Biotechnology and Biodiversity Federal University of Goiás, Goiânia, GO, Brazil; [c]Post-Graduation Program in Ecology, Conservation and Biodiversity Federal University of Uberlândia, Uberlândia, MG, Brazil
*E-mail: guilhermeifgoiano@gmail.com

9.1 Introduction

In recent years, there has been a notable increase in ecotoxicological studies and information on the most varied impacts of chemical and non-chemical stressors on aquatic and terrestrial biota.[1] These trends have been especially

Issues in Toxicology No. 50
Fish Species in Environmental Risk Assessment Strategies
Edited by Guillermo Eli Liwszyc and Marcelo L. Larramendy

Published by the Royal Society of Chemistry, www.rsc.org

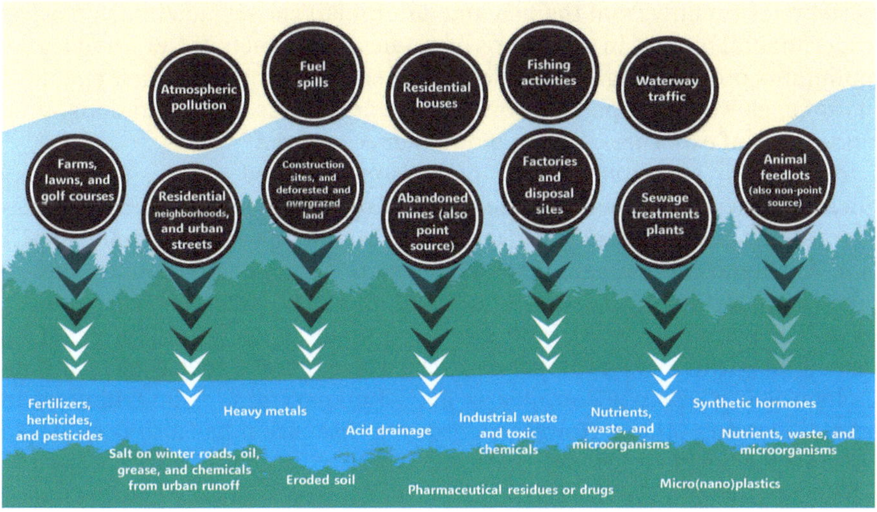

Figure 9.1 Schematic representation of the different sources of aquatic pollution/contamination.

important, given that the diversity and concentration of pollutants and contaminants from various sources have increased proportionately to the growth of product consumption (see Figure 9.1).[2] In recent years, great attention has been paid to contaminants of emerging concern (CECs) that, in general terms, refer to unregulated substances detected in the environment that may present a risk to human health, aquatic life or the environment, and for which scientific understanding of potential risks is evolving.[3] CECs include different types of manufactured chemicals and substances [*e.g.*, agricultural products, micro(nano)particles, pharmaceuticals, industrial chemicals, *etc.*] as well as naturally occurring substances (*e.g.*, algal toxins).[4,5] On the other hand, legacy pollutants are a public health concern due to their environmental persistence, toxicity to ecosystems and biota, and the potential for long-range transport.[6] Examples of legacy pollutants include organochlorine pesticides, polychlorinated biphenyls and polycyclic aromatic hydrocarbons, among others.[7]

Given this, an academic-scientific effort has been noted to develop studies that evaluate how much these contaminants/pollutants affect biodiversity, aiming to provide subsidies that are useful in decision-making and in proposing mitigation or remediation measures for current pollution.[8] Recent reviews on the topic, such as those by Chaturvedi *et al.*,[9] Khan *et al.*,[10] Ahmad *et al.*[11] and Marson *et al.*[12] confirm the growth of research focused on the effects of CECs and legacy pollutants on organisms. Overall, these studies point to using diverse model organisms and a wide range of toxicity biomarkers. To evaluate the possible impacts of current or future levels of aquatic pollution, invertebrates, for example (especially from the orders Cladocera, Amphipoda, Decapoda, Ephemetoptera, Odonata and Diptera) and vertebrates (mainly from the fish and amphibian classes) have been used as experimental models in thousands of studies. The effects of CECs and legacy pollutants on these organisms are as diverse and complex as their chemical natures, as well as dependent on the test organisms, concentrations, time and routes of exposure, and the biomarkers evaluated.

Notably, in vertebrates, the most commonly used toxicity biomarkers include those predictive of biochemical changes (*e.g.*, Freitas *et al.*[13] and Ferreira *et al.*[14]), histological (*e.g.*, Dos-Santos Mendonça *et al.*,[15] Nataraj *et al.*[16] and Attademo *et al.*[17]), genic (*e.g.*, Zhang *et al.*[18] and Agarwal *et al.*[19]), molecular (*e.g.*, Martin-Folgar *et al.*[20] and Khatib *et al.*[21]), cytological (*e.g.*, Lavanya *et al.*[22] and Araújo *et al.*[23]), biometric (*e.g.*, Araújo *et al.*[24] and Filice *et al.*[25]), haematological (*e.g.*, Powolny *et al.*,[26] Sayed *et al.*[27] and Ogueji *et al.*[28]), immunological (*e.g.*, Bernal-González *et al.*,[29] Rutkoski *et al.*[30] and Elmowalid *et al.*[31]), among others. However, a growing discussion has also been noted around the need to expand the list of toxicity biomarkers to predict the effect of environmental stressors on organisms,[32–34] mainly because many of the responses evaluated are only manifested after long periods of exposure. Although this is undoubtedly important, using biomarkers capable of detecting earlier effects and with more far-reaching biological consequences can be crucial for making decisions and applying faster measures to control pollution.

As discussed by Ford *et al.*, the advent of technology and more precise tools for detecting and quantifying subtle changes has increased scientists' interest in using animal behaviour to identify and characterize the impact of CECs and legacy pollutants on biota.[35] This becomes especially relevant since small behavioural changes can be detected even after a short exposure to low concentrations of pollutants/contaminants.[36] Therefore, this particularity supports the classification of behavioural endpoints as sensitive indicators of disturbance,[37,38] which are directly related to fundamental processes that influence the fitness of individuals, as well as the dynamics of their populations.[35]

Despite this and the fact that environmental risk assessment does not exclude the results of behavioural ecotoxicity tests, the behavioural responses of animals when exposed to pollutants/contaminants are rarely included in discussions about the regulation of chemicals.[39] Furthermore, there has been a perception that the results of tests carried out within the scope of ecotoxicological studies are not as reliable as those obtained *via* traditional biomarkers.[40] Part of this is supported by the argument that several tests or protocols used in behavioural assessment have low reproducibility (primarily due to a lack of standardized methods)[35] or that animals' behavioural responses are too complex to be extrapolated to higher levels within the ecosystem structure.[41,42] However, a range of studies conducted over the last decade have demonstrated that a wide variety of fitness-related behaviours are vulnerable to disruption (see reviews by Bertram *et al.*,[37] Melvin *et al.*,[43] Brodin *et al.*,[44] Saaristo *et al.*,[45] Sievers *et al.*,[46] Aulsebrook *et al.*,[47] Jacquin *et al.*,[48] Bownik *et al.*[49] and Grunst *et al.*[50]), reinforcing the validity and importance of behavioural endpoints being incorporated into experimental designs as a complement to more traditionally used biomarkers.

In this context, this chapter presents a synthesis of the endpoints and experimental procedures used to evaluate the behavioural response of Nile tilapia (*Oreochromis niloticus*) to different CECs and legacy pollutants. By focusing on the methodological possibilities to be used in evaluating the behaviour of this Neotropical species, we shed light on the feasibility and reasonableness of using this animal model as a complement or alternative to the use of non-Neotropical fish species in behavioural ecotoxicology, *e.g.*, the zebrafish *Danio rerio* (Cypriniformes, Cyprinidae), the medaka *Oryzias latipes* (Beloniformes Adrianichthyidae), the fathead minnow *Pimephales promela* (Cypriniformes, Cyprinidae) and the three-spined stickleback *Gasterosteus aculeatus* (Scorpaeniformes, Gasterosteidae), among others.

9.2 *Oreochromis niloticus* as a Model System in Ecotoxicological Studies

Oreochromis niloticus (see Figure 9.2A–E) is a fish from the Teleostei Infraclass (Order Perciforme, Family Cichlidae, Subfamily Pseudocrenilabrinae),[51–53] originating from the Nile River basin in East Africa,[54] and which is present on all continents.[55] As highlighted by Bavia *et al.*,[56] the species is considered, among the freshwater fish species already catalogued, to be the one with the greatest migratory capacity to territories around the world, being able to adapt to diverse habitats and biotopes, with a wide range of temperatures (26–28 °C),[57] allowing expansion and cultivation in different regions of the world. From Figure 9.3, it is possible to observe that in regions of Egypt, the species was widely disseminated (naturally or introduced by man) from the 1940s to the mid-to-late 1990s.

Biologically, the species reaches sexual maturity just a few months after birth (3–6 months, depending on temperature),[58] has an omnivorous feeding habit and has been bred naturally in rivers, lakes, estuaries and ponds. Furthermore, *O. niloticus* presents fast, high resistance to stress, growth rates, tolerance to extreme environmental conditions and diseases, trophic plasticity and feeding at low trophic levels and a high ability to reproduce in captivity.[59] Such characteristics give the species a label as an ideal candidate for aquaculture all over the globe.[60] According to the Food and Agriculture Organization (FAO),[61] *O. niloticus* is considered the fifth most cultivated aquatic species globally, reaching 27 million tons with a value of 52 billion dollars during the last decade. However, it is essential to highlight that the species does not only stand out in the global aquaculture sector. *O. niloticus* has also been considered a good model for ecotoxicological investigations of different pollutants (microplastics,[62–73] nanoplastics,[74,75] heavy metals,[76–82] pharmaceuticals,[83–95] pesticides,[96–107] metallic nanomaterials,[108–118] industrial chemicals,[119–129] persistent organic pollutants,[130–133] polychlorinated biphenyls[134–136] and polycyclic aromatic hydrocarbons[137–144]), predominantly using predictive biomarkers of biochemical,

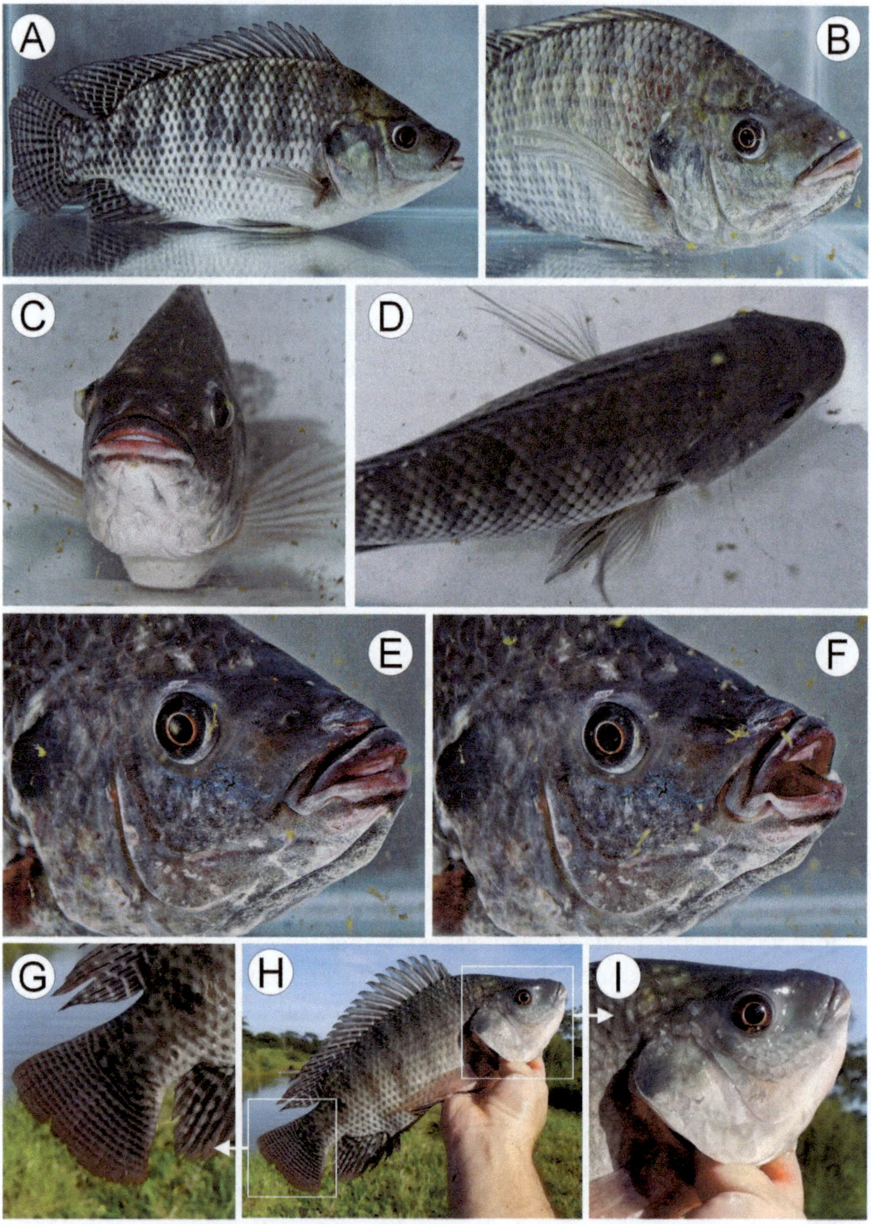

Figure 9.2 (A–I) Representative images of the external morphology of Nile tilapia *Oreochromis niloticus*.

histopathological, mutagenic, genotoxic, cytotoxic, morphotoxic, immunotoxic and haematotoxic effects.

On the other hand, some studies have also evaluated the effect of CECs and legacy pollutants on the behaviour of *O. niloticus*, exposing individuals of different stages of development (larvae, juveniles and adults) to various

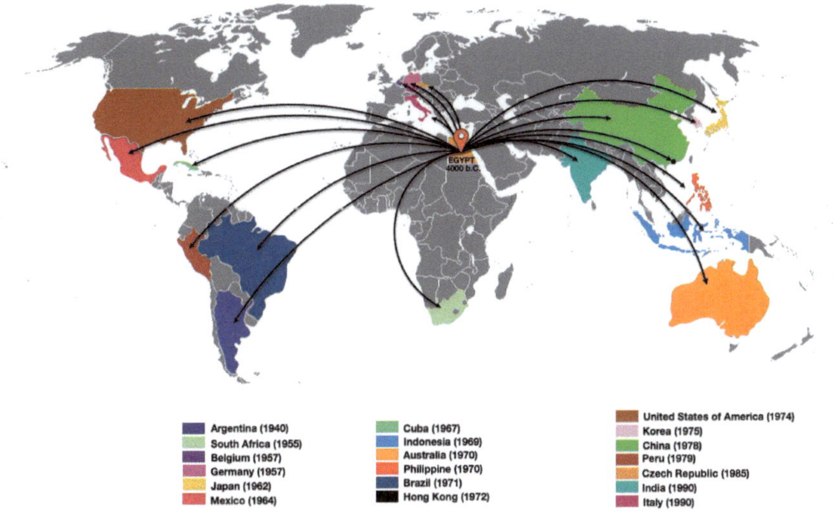

Figure 9.3 Global spreading of Nile tilapia *Oreochromis niloticus* from Egypt. The worldwide spreading of the species (represented by black arrows) started in the 1940s and extended principally until the 1990s. The coloured boxes below represent some countries where Nile tilapia was introduced in the corresponding year. Reproduced from ref. 56 with permission from Elsevier, Copyright 2022.

xenobiotics at varying concentrations. Table 9.1 summarizes the main studies involving the behavioural assessment of *O. niloticus* exposed to CECs or legacy pollutants, which will be mentioned throughout this chapter.

9.3 *Oreochromis niloticus* and General Considerations About Animal Behaviour

Like all cichlids, *O. niloticus* exhibits a wide range of behaviours which have diverse biological functions and varying levels of complexity. Such behaviours include, for example, those related to parental care, aggressive behaviour (almost always related to territoriality), advanced social structuring with stable dominance hierarchies, and courtship (having chemical communication, chemical alarm cues and sonic behaviour).[164,165] As highlighted by Olivares-Rubio and Arce,[166] such behaviours are essential for reproduction, predation, anti-predatory response, cognition and neurotransmission. Therefore, when altered, they can harm individuals and their populations.

That said, before presenting the possibilities of methods and techniques that can be used in behavioural ecotoxicology involving *O. niloticus*, it is crucial to recognize two important aspects: (i) the behavioural repertoire of the species displayed in its natural environments may not be observed under conditions in laboratory tests and (ii) careful analyses and interpretations of results obtained in laboratory experiments are necessary. The first aspect

Table 9.1 Summary of some of the main studies that used behavioural endpoints of Nile tilapia *Oreochromis niloticus* in evaluating the ecotoxicity of contaminants of emerging concern or legacy pollutants.

References	CECs and LPs[a]	Concentrations	Time exposure	Development stage	Tests or assessment procedures
Al-Kahem[145]	Trivalent chromium ($CrCl_3$)	3 and 6 mg L^{-1}	96 h	Juvenile	Visual and behavioural observation
Gül[146]	Chlorpyrifos-methyl	1.0, 1.2, 1.3, 1.5 and 1.7 mg L^{-1}	96 h	Larva	Visual and behavioural observation
Benli and Köksal[147]	Ammonium chloride (NH_4Cl)	Larvae: 0.018–0.0075 g L^{-1} Fingerlings: 0.18–0.037 g L^{-1}	24 h	Larva and fingerling	Visual and behavioural observation
Yildirim *et al.*[148]	Deltamethrin	12 µg L^{-1}	45 min	Juvenile	Visual and behavioural observation
Almeida *et al.*[149]	Cadmium chloride ($CdCl_2$)	0.75 mg L^{-1}	15 days	Juvenile	Aggressive behaviour test
Pessoa *et al.*[150]	Carbofuran	69.9, 140, 297 and 397 µg L^{-1}	96 h	Larva	Test of swimming activity
Ezeonyejiaku *et al.*[151]	Copper sulphate ($CuSO_4{\cdot}5H_2O$)	50, 60, 70, 80, 100 and 120 mg	96 h	Adult	Visual and behavioural observation
El-Sayed *et al.*[152]	Pendimethalin	0.5 to 6.5 mg L^{-1}	96 h	Adult	Visual and behavioural observation
Boscolo *et al.*[153]	Diuron, 3,4-dichloroaniline (DCA), and 3,4-dichlorophenyl-*N*-methylurea (DCPMU)	100 ng L^{-1}	10 days	Adult	Aggressive behaviour test
Campos *et al.*[108]	Zinc oxide nanoparticles (ZnONPs)	760 and 76 000 µg L^{-1}	72 h	Juvenile	Open field test Light/Dark transition test Eating behaviour test Anti-predatory response test
Julinta *et al.*[154]	Oxytetracycline	80, 240, 400 and 800 mg kg^{-1}	30 days	Juvenile	Visual and behavioural observation

Guedegba *et al.*[97]	Lambda-cyhalothrin, acetamiprid, Acer 35EC, and mixture	Farag	96 h	Juvenile	Visual and behavioural observation
Batalhão *et al.*[155]	Methylphenidate (MPH)	20 and 100 ng L^{-1}	2 and 4 days	Juvenile	Mirror test
Sheethal *et al.*[156]	Nickel dichloride ($NiCl_2$)	46.98 mg L^{-1}	96 h	Juvenile	Visual and behavioural observation
Zeid *et al.*[157]	Organic methyl mercury (MeHg)	0.5, 1, 1.5 and 2 mg kg^{-1} diet MeHg chloride (MeHgCl)	30 days	Juvenile	Visual and behavioural observation
El-Saadony *et al.*[158]	Biological zinc nanoparticles (BIO-ZnONPs)	0.1 and 1 mg L^{-1}	70 days	Juvenile	Visual and behavioural observation
Farag *et al.*[159]	Bifenthrin	2, 4, 6, 8 and 10 μg L^{-1}	96 h	Juvenile	Visual and behavioural observation
Mahi *et al.*[160]	Lead nitrate [$Pb(NO_3)_2$]	5.20, 10.40 and 20.80 mg L^{-1}	30 days	Juvenile	Visual and behavioural observation
Rahman *et al.*[161]	Lambada-cyhalothrin and methomyl	Various concentrations (alone or in combinations)	60 days	Juvenile	Visual and behavioural observation
Farag *et al.*[162]	Thallium (Tl)	1/10th of 96 h LC50 = 41.9 μg L^{-1}	60 days	Juvenile	Visual and behavioural observation
Ibrahim *et al.*[163]	Mancozeb (MAZ)	4, 8, 12, 16 and 20 mg L^{-1}	96 h	Juvenile	Visual and behavioural observation

[a]CECs: contaminants of emerging concern; LPs: legacy pollutants; LC_{50}: is the amount of a substance suspended in the air required to kill 50% of a test animal during a predetermined observation.

draws our attention to the need to keep in mind that a set of factors can interfere with the animal's behavioural response in the laboratory, and, in ecotoxicology, behavioural changes should not always be interpreted as being the sole cause of exposure to pollutants/contaminants tested or that similar responses may occur in polluted/contaminated natural environments. As discussed by Dawkins,[167] in laboratory studies, it is common and highly recommended that we have a keen ability to control the greatest number of unwanted variables that can bias the results (in this case, the animal's behavioural response). In ecotoxicological studies involving fish, for example, it is common for researchers to distribute test animals into replicates, considering strict control of co-variables, such as number of individuals, sex, age, biomass, colour, genotype as well as size/volume of tanks or aquariums, ambient luminosity, water temperature, composition of their diets, oxygenation and other physical–chemical parameters of the water. Undoubtedly, attention and care to these aspects constitute the cornerstone of a good scientific procedure and contribute to increasing the statistical power of the proposed design.

However, the influence of these procedures on the behavioural response of test animals must be a point to be considered, both during the preliminary discussions to define the experimental design and during the analysis and interpretation of the results obtained post-exposure to contaminants/pollutants. The division of animals from their social groups (especially in the case of gregarious species) when causing a stressful condition may, for example, be more preponderant for the exhibition of behavioural abnormalities than exposure to the pollutants/contaminants. Therefore, understanding which and how many factors interact in the experimental context is particularly important for studying animal behaviour, as it rarely has a single "cause."[168,169] In other words, it is necessary to consider that the behaviour exhibited by the animal, whether studied in natural contexts or under laboratory conditions, will depend on a series of factors that include how other animals behave around it, to signals from within the body in terms of neural, biochemical, endocrine, sensory, *etc.*

In this context, it is worth remembering a substantial legacy left by Karl von Frisch, Konrad Lorenz and Nikolaas Tinbergen (awarded the Nobel Prize in Physiology or Medicine 1973),[170] which refers to the importance of taking into account that the scientific study of animal behaviour has four basic components, reflected in the following questions: What has caused an animal (*i.e.*, what is the cause) to exhibit a specific behaviour? Can this behaviour be changed as the animal grows and develops? What would be the evolutionary background implicit in the behaviour exhibited by the animal? And finally, how does a particular behaviour help the animal to survive and reproduce successfully? Such questions obviously instigate important reflections as researchers. However, they also clarify that this chapter does not question the (complex) biological meaning of the results obtained in the studies mentioned in Table 9.1. The following items present behavioural endpoints, methods or procedures used in behavioural ecotoxicology involving *O. niloticus* in an informative context for future studies in the area. Therefore, there is no intention to make a value judgment about how efficient a given method, procedure, or technique is in reflecting the species' innate behaviours, nor how much the results of studies can or should be extrapolated to the real world.

9.4 Behavioural Endpoints Recorded by the Focal-animal Sampling Technique

When we return to the studies summarized in Table 9.1, we will see that the behaviour of *O. niloticus* (larvae, juveniles or adults) was evaluated based on procedures based on visual observation of the test individual during or after the respective exposure periods. Although the procedural details adopted in the observation of the animals were not provided by some authors, the information described by Guedegba *et al.*,[97] Gül,[146] Yildirim *et al.*,[148] Zeid *et al.*[157] and Rahman *et al.*[161] provide the methodological possibilities that can be adopted. Gül[146] and Yildirim *et al.*[148] reported the behavioural response of *O. niloticus* larvae and juveniles (respectively) conducted at 1–8 h and every 12 h during acute toxicity tests. Similarly, Guedegba *et al.*[97] evaluated the possible behavioural abnormalities of *O. niloticus* juveniles at an initial 1–5 h followed by every 12 h during the acute bioassay. On the other hand, recordings of the behavioural activities of *O. niloticus* exposed to the organometallic cation methyl mercury $[CH_3Hg]^+$ occurred for 15 min through 8 hours per week during an experimental period of 30 days.[157] Rahman *et al.*[161] reported that, for 60 days, fish behaviour was documented with a stopwatch and a video camera at the same time for 4 days weekly (at 8 am).

Regardless of the experimental design, period and frequency of the animals' behavioural recording, these authors adopted the focal-animal sampling technique. As discussed by Altmann,[171] this technique requires that observers sample the behaviours of only one individual at a time, even though other individuals from the same group may be seen during the assessments. In the focal-animal sampling technique, individuals are observed for a predetermined period in which all observed behaviours are duly recorded. In studies that used this technique to evaluate the behavioural response of *O. niloticus* during or after exposure to CESs or legacy pollutants, the categories presented in Table 9.2 were the most frequently reported, the following items providing a more specific view of each of them.

9.4.1 Erratic Movement (Erratic Swimming/Locomotion)

Erratic movements, also called erratic swimming/locomotion, encompass the display of complex behaviours characterized by sudden changes in direction or speed and repeated rapid darting.[172] The fictional three-dimensional (3D) temporal reconstruction presented by Cachat *et al.*[172] reveals that episodes of high speed, rapid movements, high angular velocity, high mobility and steep curve angles are typical characteristics of the erratic movements of a given

Table 9.2 Behavioural categories reported in studies that evaluated the response of Nile tilapia *Oreochromis niloticus* to different contaminants of emerging concern or legacy pollutants.

Recorded behaviours	References
Erratic movement (erratic swimming/locomotion)	Al-Kahem,[145] Gül,[146] Benli and Köksal,[147] Ezeonyejiaku *et al.*,[151] El-Sayed *et al.*,[152] Julinta *et al.*,[154] Guedegba *et al.*,[97] Sheethal *et al.*,[156] Zeid *et al.*,[157] El-Saadony *et al.*,[158] Farag *et al.*,[159] Mahi *et al.*[160] and Rahman *et al.*[161]
Changes in operculum movement	Guedegba *et al.*,[97] Al-Kahem,[145] Yildirim *et al.*,[148] El-Sayed *et al.*,[152] and Sheethal *et al.*[156]
Lost equilibrium in water	Guedegba *et al.*,[97] Gül,[146] Benli and Köksal,[147] Yildirim *et al.*[148] and Rahman *et al.*[161]
Surfacing behaviour	El-Saadony *et al.*[158] and Ibrahim *et al.*[163]
Comfort behaviours	Farag *et al.*[159] and Farag *et al.*[162]
Aggressive behaviour	Al-Kahem,[145] Zeid *et al.*,[157] El-Saadony *et al.*,[158] Farag *et al.*,[159] Farag *et al.*[162] and Ibrahim *et al.*[163]
Feeding behaviour	Julinta *et al.*,[154] Zeid *et al.*,[157] El-Saadony *et al.*,[158] Mahi *et al.*[160] and Rahman *et al.*[161]

fish. This type of behavioural abnormality has been recorded in *O. niloticus* exposed to various pollutants/contaminants, including, among others, lambda-cyhalothrin,[97] trivalent chromium,[145] chlorpyrifos-methyl,[146] ammonium chloride,[147] copper sulphate,[151] pendimethalin,[152] oxytetracyckine,[154] nickel dichloride,[156] organic methyl mercury,[157] biological zinc nanoparticles,[158] bifenthrin[159] and lead nitrate.[160] Although the mechanisms of action of these pollutants/contaminants may have different characteristics in the animals evaluated, according to Kalueff *et al.*,[173] erratic movements are commonly evoked by acute stressors and reflect a general state of anxiety-like behaviour.

9.4.2 Changes in Operculum Movement

Opercular movements are provided by the fish's gill breathing mechanisms, which can be easily visualized and evaluated using slow-mode video recordings. As discussed by Tantarpale *et al.*[174] and Dan-kishiya *et al.*,[175] changes in opercular movements can be mediated by different physiological mechanisms (activated or inactivated by stress conditions), constituting a measure of the fish's oxygen consumption and metabolism.

In the study by Guedegba *et al.*,[97] for example, the increased opercular activity in *O. niloticus* juveniles exposed to lambda-cyhalothrin was reported to be associated with a hypoxia condition caused by the chemical. As discussed by Val *et al.*,[176] when fish cannot escape from hypoxic stress, hyperventilation may be evoked to compensate for the low oxygen supply. Furthermore, opercular movements can be influenced by various psychotropic medications, while their increase may be associated with anxiety-like behaviour,[177] their reduction may be associated with a reduction in the animals' general activity during episodes of resting/sleep.[178] An interesting tool that can be used to evaluate possible changes in fish's opercular movements is the AEFishBIT tri-axial accelerometer, proposed by Ferrer *et al.*[179] Although this tool has not yet been used in *O. niloticus*, it will allow any changes in opercular movements (opening and closing) to be recorded in an integrated way with the speed of operculum movement, providing more detailed information on the effects of pollutants/contaminants.

9.4.3 Lost Equilibrium in Water

Loss of equilibrium in water – recorded in *O. niloticus* larvae exposed to different concentrations of ammonia,[147] fingerlings exposed to deltamethrin[148] and juveniles fed a taurine-supplemented diet[161] – can be defined as the inability of fish to maintain an upright position within the water column.[180] Gül[145] reported that the loss of equilibrium observed in *O. niloticus* larvae exposed to chlorpyrifos-methyl was described as the larvae swimming to the water surface, followed by vertical decay to the bottom of the aquarium. Studies involving the species[181] and other representatives of freshwater fish[182–186] have revealed that maintaining the ability to retain a balanced upright position in the water is also associated with hypoxic conditions, as well as thermal changes,[187] whether or not related to pollutants/contaminants. Furthermore, the results of the study by Schneider *et al.*[188] involving sub-adult largemouth bass *Micropterus salmoides* (Perciformes, Centrarchidae) show that loss of equilibrium can also occur during exposure to high carbon dioxide.

9.4.4 Surfacing Behaviour

Surfacing behaviour or surface touching refers to the swimming of fish close to the water surface. El-Saadony *et al.*[158] and Ibrahim *et al.*,[163] for example, observed that this behaviour was significantly increased in *O. niloticus* juveniles exposed to mycogenic zinc nanoparticles and mancozeb fungicide, respectively. As discussed by different authors, this behaviour may be associated with a hypoxia condition (to avoid the negative effect of hypoxia on their metabolic scope)[189–192] or buoyancy dysregulation, characterized by an inability to remain at a constant elevation associated, commonly, with changes in the swim bladder or other peripheral systems.[173,193]

On the other hand, it is important to be careful when interpreting the results of surfacing behaviour analyses since their frequency may depend on the time of day the assessments are carried out. As highlighted by Eveson *et al.*,[194] a fish's time at the surface varies enormously during the day (morning, afternoon and night). In the species studied by these authors – the Southern bluefin tuna *Thunnus maccoyii* (Scombriformes, Scombridae) for example, – the proportion of time fish spend at the surface tends to be high during most days and either very high (>90%) or very low (<10%) during most nights.

9.4.5 Comfort Behaviours

Comfort behaviour includes a variety of behavioural categories that can be observed in *O. niloticus*, the most frequent of which include chafing or scratching, resting and schooling. The chafing or scratching category is recorded when animals exhibit movements that appear to represent attempts to alleviate possible irritation on the skin surface or dislodge foreign objects from the skin.[195] This behaviour can be recorded when we observe fish rubbing their bodies against the bottom of the aquarium and/or against walls, equipment, substrates, *etc.* The resting behaviour (resting), characterized by the immobility and setting of fish on the aquarium base under normal conditions, reflects a period of "suspended animation" (sleep-like) that can perform restorative functions in fish that sleep performs. However, studies involving the zebrafish *D. rerio* have suggested that changes in the levels of melatonin and its receptors, as well as in the interactions between melatonin and the hypocretin/orexin system, can induce significant changes in the display of rest behaviour.[196–199]

On the other hand, schooling behaviour is characterized by forming a relatively polarized group (school) in which multiple fish swim together, presenting tight relationships between individuals and closely coordinating their behaviour.[200,201] According to Larsson[202] and Larsson and Abbott,[203] fish can present a range of schooling types, depending on factors such as age, physiological state, motivation, habitat type and environmental conditions. However, it is essential to highlight that schooling behaviour differs significantly from shoal behaviour. As discussed by Pavlov and Kasumyan,[204] shoals represent simple spatial aggregations of fish, which can be attracted by certain stimuli, including light, temperature, food and/or olfactory stimuli, among others. In this type of behaviour, the aggregation of individuals into shoals occurs individually, and fish do not exhibit mutual attraction; they even lack unitary orientation and synchronization of movements in response to external stimuli.

In the studies by Farag *et al.*,[159,162] the authors reported changes in the comfort behaviours of *O. niloticus*. At the same time, Farag *et al.*[159] reported that *O. niloticus* exhibited changes in their comfort behaviours in response to bifenthrin insecticide whereas a reduction in comfort behaviours was recorded in animals exposed to thallium trace metal.[162]

9.4.6 Aggressive Behaviour

Aggressive behaviour constitutes a category widely explored in studies involving *O. niloticus* due to its aggressive nature.[205–209] Using the focal-animal sampling technique, it is possible to evaluate changes in the aggressiveness of *O. niloticus*, recording patterns that include approach, chasing, mouth pushing, biting, butting and fin tugging. As described by Zeid *et al.*,[157] the approach is considered to be the movement of a given fish towards another and chasing, the frequency with which vigorous swims are directed at the pursued fish. On the other hand, the frequency with which two fish have their mouths open facing each other is recorded as "mouth pushing," and "biting" is characterized by the contact (multiple or not) of a fish's mouth in some regions of another fish. Butting behaviour is recorded when the focal fish attacks the genital papilla of another fish, using its snout, and fin tugging refers to the specific act of biting the ends of the other fish. Among the studies that used the aggressive behaviour of *O. niloticus* as a toxicity biomarker, those by Al-Kahem,[145] Zeid *et al.*,[157] El-Saadony *et al.*,[158] Farag *et al.*[159] and Farag *et al.*[162] can be mentioned. These authors reported changes in the aggressive behaviour of these animals after exposure to trivalent chromium, organic methyl mercury, mycogenic zinc nanoparticles, bifenthrin and thallium, respectively.

9.4.7 Feeding Behaviour

The feeding behaviour of *O. niloticus* constitutes another endpoint frequently explored in studies involving exposure to contaminants of emerging concern or legacy pollutants. In biological terms, the species has versatile feeding behaviour, being characterized as a generalist and opportunistic omnivore that feeds on algae, plankton, detritus, small invertebrates and bacterial films, among others.[210–212] As discussed by Temesgen *et al.*,[211] the composition of the diet of this species can present significant variations depending on the temporal and spatial conditions of the environment, which justifies classifying feeding behaviour as a critical factor for the survival of individuals in natural environments (or in cultivation systems).

One of the approaches adopted to evaluate the effect of pollutants/contaminants on the feeding behaviour of *O. niloticus* refers to that adopted by Julinta *et al.*[154] This approach is based on a scoring system (five-point ordinal scale) initially proposed by Bowker *et al.*[213] when studying rainbow trout *Oncorhynchus mykiss* (Salmoniformes, Salmonidae). In this system, a qualitative score ranging from 0 to 4 is assigned to the test fish after food has been offered, with a score of 0 being assigned when no feed is consumed and 5 when approximately 100% of the feed is voraciously consumed. Although the evaluation period and the amount of food offered may vary depending on the experimental design, this system helps us infer how much pollutants/contaminants interfere with the animals' interest in food.

9.5 Behaviours of *Oreochromis niloticus* Recorded From Behavioural Tests

In addition to the focal-animal sampling technique, the behavioural assessment of *O. niloticus* after exposure to pollutants/contaminants can also occur through behavioural tests. Although, in general, the lack of universal standardization of the procedures adopted in these tests can undermine the validity of different experimental designs and lead to inconsistency in the patterns of animal responses documented among studies, the flexibility of these tests can be seen as a positive attribute.[214] Using pre-defined and planned procedures, such tests can provide data that researchers can easily interpret which is important in behavioural ecotoxicology. Furthermore, the wide accessibility of computational software tools contributes to the more precise quantification of behaviour patterns observable during the behavioural assessment of animals. In this sense, some tests have been used to identify and characterize the effects of pollutants/contaminants on the behaviour of *O. niloticus*, such as those described in the studies by Campos *et al.*[108] and Boscolo *et al.*,[153] which serve as a starting point for the development or adaptation of tests applicable to the behavioural ecotoxicology of the species.

The aggressive behaviour test was used in the study by Boscolo *et al.*[153] to evaluate the possible effect of exposure of *O. niloticus* male adults to diuron and its metabolites, which consisted of pairing the animals and recording aggressive interactions (frequency and types of behaviours) for 20 min. In the study, the animals were paired in a single aquarium (different from the exhibition aquariums) containing the pollutants and their respective concentrations, and the identification of individuals during the test was carried out using visually implanted elastomer ties. Although details about the dimensions and volume of the aquarium used in the test as well as the general conditions of the room where the test was carried out (including temperature, luminosity, acoustic insulation, *etc.*), were not provided in the paper, the decision to use an aquarium single-family home for both individuals constituted an interesting approach to exclude the effects of previous residence. As discussed by Chellappa *et al.*,[215] Harwood *et al.*[216] and Jones *et al.*,[217] when two individuals are confronted, the one that has spent some time in an area (resident) is more likely to invest time and energy in agonistic behaviours when new individuals (intruders) are introduced into their aquariums.

In the work of Campos *et al.*,[108] multiple tests were conducted on *O. niloticus* juveniles to evaluate the possible impacts of zinc oxide nanoparticles (ZnO NPs) after 72 h of exposure on the animals' behaviour. At the individual level, each animal of each experimental group was sequentially subjected to the tests on the same day and aquarium (see Figures 9.4 and 9.5). Furthermore, the authors highlighted that all tests were performed in specific rooms equipped with sound insulation, temperature and light control. Table 9.3 presents a summary of the tests carried out by Campos *et al.* [open field (see Figure 9.4A), light/dark transition (see Figure 9.4B–D), eating behaviour (see Figure 9.5A and B), and anti-predatory response tests (see Figure 9.5C and D)] and the following sections provide an overview of the principles and objectives of each test.[108]

9.5.1 Open Field Test

The open field test, originally developed as a test for emotionality in rodents,[218] has currently been used in the behavioural assessment of several other animal groups, including other orders of mammals,[219–221] as well

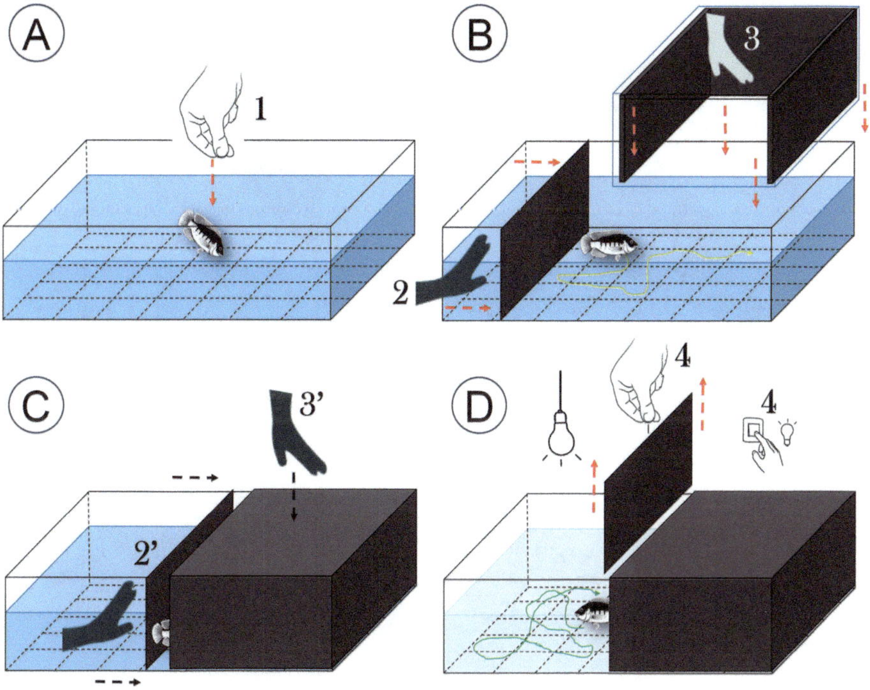

Figure 9.4 Schematic drawings and sequential order of the open-field and light-dark transition tests. (A) Representation of the apparatus used in the open field test; (B and C) sequential procedures performed shortly after the open field test and (D) representation of the light–dark transition test. The numbers highlighted in the figure indicate the sequence of the performed procedures. Equal numbers indicate that events happened simultaneously, and numbers followed by (') indicate a more advanced sequence of the same procedure. Colours are merely illustrative. Reproduced from ref. 108 with permission from Elsevier, Copyright 2019.

as amphibians,[222–224] birds,[225–229] reptiles[230–232] and fish.[233–235] Such a test is often used to measure (simultaneously) exploratory behaviour, total locomotor activity and anxiety-like behaviour, which are determined by the conflict between the motivation to explore a new area and the aversion to open, unprotected and bright places. When animals are introduced into the apparatus for the first time, there is a tendency to remain close to the walls (*i.e.*, thigmotaxis behaviour) exhibiting a natural aversion to illuminated areas. On the other hand, animals have an innate drive to explore a perceived threatening stimulus.

9.5.2 Light/Dark Transition Test

The light/dark transition test was also initially developed for behavioural assessments in murine models.[236] However, in recent years, it has also been – with adaptations – widely used for behavioural assessment in different

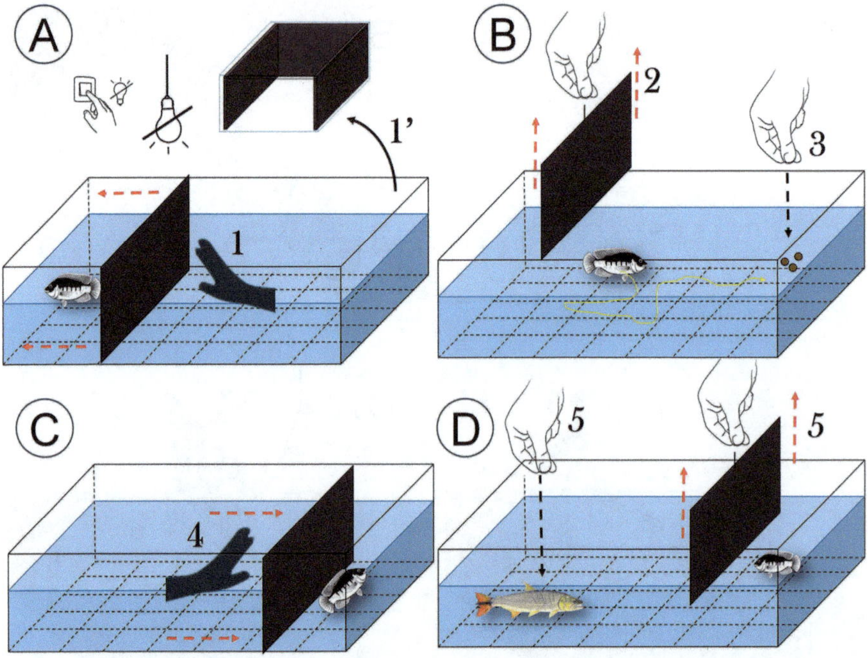

Figure 9.5 Schematic drawings, sequential food behaviour order, and anti-predatory response tests (individual condition). (A and B) Representation of the sequential dynamics adopted in the food behaviour test. (C) Previous organization of the apparatus used in the (D) anti-predatory response test. Numbers in the figure indicate the sequence in which the procedures were performed. Equal numbers indicate that events happened simultaneously, and numbers followed by (') indicate a more advanced sequence of the same procedure. Colours are merely illustrative. Reproduced from ref. 108 with permission from Elsevier, Copyright 2019.

species of fish, such as the zebra fish *D. rerio*,[237–240] the goldfish *Carassius auratus* (Cypriniformes, Cyprinidae),[241] the medaka *O. latipes*[242] and the guppy *Poecilia reticulata* (Cyprinodontiformes, Poeciliidae),[243] among others. This test makes it possible to evaluate the swimming activity of fish in response to a visual stimulus (change from light to dark and *vice versa*). As discussed by Maeda *et al.*, fish commonly show low locomotor activity in light, followed by a substantial increase in activity after moving to darkness as fish feel safer in dark environments.[238]

9.5.3 Eating Behaviour Test

The eating behaviour test conducted by Campos *et al.*[108] was like that of Domingues *et al.*,[244] in which commercial-grade pellets were used as a food stimulus. This test allows us to infer, like observations that can be recorded *via* the focal-animal sampling technique, the effects of pollutants/

Table 9.3 Brief summary of the apparatus, experimental conditions and biomarkers evaluated in the behavioural tests conducted by Campos *et al.*[108] after exposure of Nile tilapia *Oreochromis niloticus* exposed to zinc oxide nanoparticles.

Open field test	
Apparatus	Glass aquarium with opaque and white sides
Dimensions	40 cm length × 35 cm width × 17.8 cm height
Volume of water used	10 L dechlorinated water
Assessment time	3 min
Behavioural biomarkers evaluated	Total number of quadrant crossings (at the bottom of the aquarium), travelled distance, and the anxiety index
Light/dark transition test	
Apparatus	Glass aquarium with opaque and white sides, containing a pre-moulded black structure that covered half of the aquarium (top and sides)
Dimensions	40 cm length × 35 cm width × 17.8 cm height
Volume of water used	10 L dechlorinated water
Outdoor lighting	Intense light provided by a 6 watt LED illuminator
Assessment time	3 min
Behavioural biomarkers evaluated	Latency (s) for a first trip to the clear compartment, frequency of trips and dwelling time (s) in the light compartment, and locomotor activity of the animals (measured based on the total quadrant crossings at the bottom of the aquarium, proportional to the length of stay in the lit compartment)
Eating behaviour test	
Apparatus	Glass aquarium with opaque and white sides
Dimensions	40 cm length × 35 cm width × 17.8 cm height
Volume of water used	10 L dechlorinated water
Food source	Three spherical feed pellets (6.2 mg ± 1.5 mg/each), totalling 18.6 mg ± 3.8 mg of feed, 143 mm ± 13.19 mm diameter
Assessment time	3 min
Behavioural biomarkers evaluated	Latency (s) to capture the first feed pellet, distance (cm), and intake score (ranging from 0 to 3) equivalent to the number of feed pellets of tilapia during the test
Anti-predatory response test	
Apparatus	Glass aquarium with opaque and white sides
Dimensions	40 cm length × 35 cm width × 17.8 cm height
Volume of water used	10 L dechlorinated water
Predatory stimulus	*Salminus brasiliensis* (10–11 cm long and had 6.3 g of body biomass)
Assessment time	3 min
Behavioural biomarkers evaluated	Frequency of predator avoidance, explosive escape and territorialism behaviours

contaminants on the animals' feeding motivation. However, it is necessary to consider that fish's eating behaviour can depend on several factors, including the pellets' size, shape, colour and palatability (in terms of texture and flavour). Controlling these variables in an experimental design is important so that adequate interpretations of the animals' responses can be made.

9.5.4 Anti-predatory Response Test

The anti-predatory response of fish has been widely evaluated using a series of protocols. In zebra fish *D. rerio*, for example, anti-predatory behaviour has already been evaluated using computer-animated images,[245–247] indirect contact (only visualization of the predator)[248] and alarm substance.[249,250] However, a limiting aspect of many of these protocols refers to the fact that the test fish does not come into direct contact with its potential predator, making it impossible for all (natural) predatory stimuli to be offered to the individual being evaluated (visual, olfactory and sound). In this sense, the anti-predatory response test adopted by Campos *et al.*[108] explored the animals' innate defensive abilities of *O. niloticus* juveniles (unexposed and exposed to ZnO NPs) when individually confronted by a potential predator (the golden dorado *Salminus brasiliensis* (Characiformes, Characidae) (see Figure 9.5C and D)), which constitutes a more realistic approach. As discussed by Apfelbach *et al.*,[251] approaches like this are important and useful for a more holistic understanding of the biological mechanisms affected by pollutants/contaminants (neural circuits, biochemical responses, genes, *etc.*) that modulate fear and anxiety responses. Furthermore, Hendrie *et al.* highlight that naturalistic approaches allow us to differentiate more precisely what is physiologically "normal" from what is considered "abnormal" (pathological and aberrant).[252]

9.6 Final Considerations and Recommendations for Future Research

Throughout this chapter, we address different aspects involving behavioural endpoints in ecotoxicological studies, focusing on the different methodological possibilities that can be used in the behavioural assessment of *O. niloticus* after exposure to different pollutants/contaminants. Naturally, using a non-conventional species such as *O. niloticus* in behavioural ecotoxicology does not overcome the countless possibilities and biomarkers widely used in evaluating traditional species, such as *D. rerio*. Therefore, there are still extensive paths to be covered, which include a more significant appropriation of the behaviour of *O. niloticus* in the designs of ecotoxicological studies, the need for standardization or harmonization of evaluation protocols and particular aspects of the behaviour of individuals of this species, which need be incorporated into the analyses and interpretations of results. Furthermore, some pertinent issues addressed in this chapter are presented below, which aim to contribute to expanding the use of behavioural endpoints in ecotoxicology.

- ✓ Although there is a consensus that pollutants/contaminants affect the behaviour of organisms (at different levels), it is important that future studies are designed to go beyond the identification and characterization of the behavioural changes induced by xenobiotics. Equally important is that we advance in understanding the toxicity mechanisms implicit in the changes, allowing more appropriate generalizations about the impact of behavioural changes in the context of the animals' natural lives. To this end, the ecological/environmental relevance of the results reported in studies must be, whenever possible, explained and emphasized, therefore providing more applicable subsidies to risk assessors and environmental managers.
- ✓ For some time now, ecotoxicologists have been highlighting the importance of standardization/harmonization of behavioural assessments, especially to ensure the validity (quality and reproducibility) of different experimental designs and to obtain greater consistency in the response patterns documented between studies.[253–255] This will also enable reliable comparisons of data from different regions, species, *etc.* In this sense, considering that the use of *O. niloticus* in behavioural ecotoxicology is on the rise,[166] procedural details must be provided in studies, as the greater the detail of the procedures adopted, the more studies will be reproducible and greater consistency in behavioural response patterns will be observed. Therefore, it is recommended to pay special attention to information related to the dimensions and characteristics of the apparatus used in the evaluations, the water used during the behavioural evaluations (in terms of physical–chemical quality, volume and the presence or absence of xenobiotics), lighting conditions and temperature of the place where the behaviours were recorded, the test individuals (size, biomass, sex and age) and the statistical design (in terms of number of individuals and existence of replicates or pseudoreplicates). Special attention to these aspects will be important to consolidate behavioural ecotoxicology as an area not restricted to secondary observations from studies initially designed to determine standard toxicological biomarkers.
- ✓ To provide more environmentally realistic input for environmental risk assessments, it is recommended that behavioural biomarkers be evaluated in conjunction with traditional/standard ecotoxicological tests/procedures (*i.e.*, mortality, biochemistry, morphology, *etc.*). Furthermore, it is suggested that approaches to assessing the impacts of pollutants/contaminants involve laboratory (controlled) studies, semi-realistic ones (*e.g.*, micro- and mesocosms) and field studies.
- ✓ Finally, we note in this chapter that metals and pesticides were the most studied types of pollutants in the behaviour of *O. niloticus*. However, as discussed by Araújo *et al.*,[23] aquatic environments receive a vast diversity

of substances and chemical products from different sources. Therefore, this reinforces the need for future studies to test the effects of a greater variety of xenobiotics on the behaviour of individuals of this species, including CECs (*e.g.*, micro(nano)particles, pharmaceuticals, industrial chemicals, *etc.*), naturally occurring substances (*e.g.*, algal toxins) and legacy pollutants such as polychlorinated biphenyls and polycyclic aromatic hydrocarbons.

Acknowledgements

The author thanks the National Council for Scientific and Technological Development (CNPq/Brazil) for the financial support needed to conduct this research. Malafaia G. is a CNPq Research Productivity Fellow – Level 1D (Proc. #308854/2021-7).

References

1. R. A. Griffin, J. D. McCuaig and T. A. Blewett, in *Ecophysiology of the European Green Crab (Carcinus Maenas) and Related Species*, ed. D. Weihrauch, and I. J. McGaw, Academic Press, Cambridge, 2024, pp. 249–260.
2. Y. Yu, S. Wang, P. Yu, D. Wang, B. Hu, P. Zheng and M. Zhang, A bibliometric analysis of emerging contaminants (ECs) (2001– 2021): Evolution of hotspots and research trends, *Sci. Total Environ.*, 2024, **907**, 168116.
3. N. V. Raman, A. Dubey, E. Millar, V. Nava, B. Leoni and I. Gallego, Monitoring contaminants of emerging concern in aquatic systems through the lens of citizen science, *Sci. Total Environ.*, 2023, **874**, 162527.
4. X. Wu, J. Ren, Q. Xu, Y. Xiao, X. Li and Y. Peng, Priority screening of contaminant of emerging concern (CECs) in surface water from drinking water sources in the lower reaches of the Yangtze River based on exposure-activity ratios (EARs), *Sci. Total Environ.*, 2023, **856**, 159016.
5. Q. Shi, Y. Xiong, P. Kaur, N. M. Dennis, D. Schlenk and J. Gan, Strategy to minimize plant accumulation of contaminants of emerging concern: Alternation between treated wastewater and freshwater irrigation, *Environ. Sci. Technol. Lett.*, 2023, **10**, 302–307.
6. M. C. O. Souza, B. A. Rocha, J. A. Adeyemi, M. Nadal, J. L. Domingo and F. Barbosa Jr, Legacy and emerging pollutants in Latin America: A critical review of occurrence and levels in environmental and food samples, *Sci. Total Environ.*, 2022, **848**, 157774.
7. J. Cantoni, Z. Kalantari and G. Destouni, Legacy contributions to diffuse water pollution: Data-driven multi-catchment quantification for nutrients and carbon, *Sci. Total Environ.*, 2023, **879**, 163092.
8. C. Rudén, J. Adams, M. Ågerstrand, T. C. Brock, V. Poulsen, C. E. Schlekat and T. R. Henry, Assessing the relevance of ecotoxicological studies for regulatory decision making, *Integr. Environ. Assess. Manage.*, 2017, **13**, 652–663.

9. P. Chaturvedi, P. Shukla, B. S. Giri, P. Chowdhary, R. Chandra, P. Gupta and A. Pandey, Prevalence and hazardous impact of pharmaceutical and personal care products and antibiotics in environment: A review on emerging contaminants, *Environ. Res.*, 2021, **194**, 110664.
10. S. Khan, M. Naushad, M. Govarthanan, J. Iqbal and S. M. Alfadul, Emerging contaminants of high concern for the environment: Current trends and future research, *Environ. Res.*, 2022, **207**, 112609.
11. A. Ahmad, S. B. Kurniawan, S. R. S. Abdullah, A. R. Othman and H. A. Hasan, Contaminants of emerging concern (CECs) in aquaculture effluent: Insight into breeding and rearing activities, alarming impacts, regulations, performance of wastewater treatment unit and future approaches, *Chemosphere*, 2022, **290**, 133319.
12. E. O. Marson, C. E. Paniagua, O. G. Júnior, B. R. Gonçalves, V. M. Silva, I. A. Ricardo and A. G. Trovó, A review toward contaminants of emerging concern in Brazil: Occurrence, impact and their degradation by advanced oxidation process in aquatic matrices, *Sci. Total Environ.*, 2022, **836**, 155605.
13. Í. N. Freitas, A. V. Dourado, A. P. da Costa Araújo, S. S. de Souza, T. M. da Luz, A. T. B. Guimarães and G. Malafaia, Toxicity assessment of SARS-CoV-2-derived peptides in combination with a mix of pollutants on zebrafish adults: a perspective study of behavioral, biometric, mutagenic, and biochemical toxicity, *Sci. Total Environ.*, 2023, **858**, 159838.
14. R. O. Ferreira, A. T. B. Guimarães, T. M. da Luz, A. S. de Lima Rodrigues, A. R. M. T. Islam, M. M. Rahman and G. Malafaia, First report on the toxicity of SARS-CoV-2, alone and in combination with polyethylene microplastics in neotropical fish, *Sci. Total Environ.*, 2023, **882**, 163617.
15. J. Dos-Santos Mendonça, J. C. N. de Almeida, L. G. Vieira, L. Q. L. Hirano, A. L. Q. Santos, D. V. Andrade and M. E. Beletti, Mutagenicity, hepatotoxicity, and neurotoxicity of glyphosate and fipronil commercial formulations in Amazon turtles neonates (*Podocnemis expansa*), *Sci. Total Environ.*, 2023, **898**, 165529.
16. B. Nataraj, D. Hemalatha, G. Malafaia, K. Maharajan and M. Ramesh, "Fishcide" effect of the fungicide difenoconazole in freshwater fish (*Labeo rohita*): A multi-endpoint approach, *Sci. Total Environ.*, 2023, **857**, 159425.
17. A. M. Attademo, L. M. Curi, A. P. C. Boccioni, C. E. Barrios, P. M. Peltzer, M. F. Simoniello and J. M. Ríos, Microplastics and plastic additives as contaminants of emerging concern: a multi-biomarker approach using *Rhinella arenarum* tadpoles, *Environ. Adv.*, 2023, 100444.
18. H. Zhang, P. Si, Q. Kong and J. Ma, Transcriptome reveals the toxicity and genetic response of zebrafish to naphthenic acids and benzo [a] pyrene at ambient concentrations, *Ecotoxicol. Environ. Saf.*, 2023, **253**, 114700.
19. M. Agarwal, A. Sharma, R. A. Kagoo and A. Bhargava, Interactions between genes altered during cardiotoxicity and neurotoxicity in

zebrafish revealed using induced network modules analysis, *Sci. Rep.*, 2023, **13**, 6257.

20. R. Martin-Folgar, M. Torres-Ruiz, M. de Alba, A. I. Cañas-Portilla, M. C. González and M. Morales, Molecular effects of polystyrene nanoplastics toxicity in zebrafish embryos (*Danio rerio*), *Chemosphere*, 2023, **312**, 137077.
21. I. Khatib, O. Horyn, O. Bodnar, O. Lushchak, P. Rychter and H. Falfushynska, Molecular and biochemical evidence of the toxic effects of terbuthylazine and malathion in zebrafish, *Animals*, 2023, **13**, 1029.
22. M. Lavanya, R. Krishnamoorthy, M. A. Alshuniaber, S. Manoharadas, C. P. Palanisamy, V. P. Veeraraghavan and R. Padmini, Formulation, characterization and evaluation of gelatin-syringic acid/zinc oxide nanocomposite for its effective anticancer, antioxidant and anti-inflammatory activities, *J. King Saud Univ., Sci.*, 2023, **35**, 102909.
23. A. P. C. Araújo, T. M. da Luz, M. A. I. Ahmed, M. M. Ali, M. M. Rahman, B. Nataraj and G. Malafaia, Toxicity assessment of polyethylene microplastics in combination with a mix of emerging pollutants on *Physalaemus cuvieri* tadpoles, *J. Environ. Sci.*, 2023, **127**, 465–482.
24. A. P. C. Araújo, N. F. S. de Melo, A. G. de Oliveira Junior, F. P. Rodrigues, T. Fernandes, J. E. de Andrade Vieira and G. Malafaia, How much are microplastics harmful to the health of amphibians? A study with pristine polyethylene microplastics and *Physalaemus cuvieri*, *J. Hazard. Mater.*, 2020, **382**, 121066.
25. M. Filice, F. R. Reinero, M. C. Cerra, C. Faggio, F. L. Leonetti, P. Micarelli and S. Imbrogno, Contamination by trace elements and oxidative stress in the skeletal muscle of *Scyliorhinus canicula* from the Central Tyrrhenian Sea, *Antioxidants*, 2023, **12**, 524.
26. T. Powolny, R. Scheifler, F. Raoul, M. Coeurdassier and C. Fritsch, Effects of chronic exposure to toxic metals on haematological parameters in free-ranging small mammals, *Environ. Pollut.*, 2023, **317**, 120675.
27. A. E. D. H. Sayed, S. K. Idriss, S. K. Abdel-Ghaffar and A. A. Hussein, Haemato-biochemical, mutagenic, and histopathological changes in *Oreochromis niloticus* exposed to BTX, *Environ. Sci. Pollut. Res.*, 2023, **30**, 59301–59315.
28. E. Ogueji, J. Nwakpa, J. Ekpenyong, M. Olaolu, A. Yaji, I. Okey and U. Agbo, Histological and hematological changes to *Clarias gariepinus* juveniles exposed to acute doses of emamectin benzoate in a static bioassay, *Ecotoxicology*, 2023, 1–14.
29. K. G. Bernal-González, C. E. Covantes-Rosales, M. R. Camacho-Pérez, U. Mercado-Salgado, V. W. Barajas-Carrillo, D. A. Girón-Pérez and M. I. Girón-Pérez, Organophosphate-pesticide-mediated immune response modulation in Invertebrates and Vertebrates, *Int. J. Mol. Sci.*, 2023, **24**, 5360.
30. C. F. Rutkoski and E. A. de-Almeida, in *Toxicology of Amphibian Tadpoles*, ed. E. A. Almeida, and J. S. Freitas, CRC Press, Boca Raton, 2023, pp. 152–161.

31. G. A. Elmowalid, W. A. Ghonimi, H. M. Abd Allah, H. Abdallah, A. El-Murr and A. M. Abdelwahab, β-1, 3-glucan improved the health and immunity of juvenile African catfish (*Clarias gariepinus*) and neutralized the histological changes caused by lead and fipronil pollutants, *BMC Vet. Res.*, 2023, **19**, 1–13.
32. S. Lomartire, J. C. Marques and A. M. Gonçalves, Biomarkers based tools to assess environmental and chemical stressors in aquatic systems, *Ecol. Indic.*, 2021, **122**, 107207.
33. N. Iftikhar, R. Zafar and I. Hashmi, Multi-biomarkers approach to determine the toxicological impacts of sulfamethoxazole antibiotic on freshwater fish *Cyprinus carpio*, *Ecotoxicol. Environ. Saf.*, 2022, **233**, 113331.
34. Y. Chen, M. Duan, X. Xu and C. Wu, Multi-biomarkers hazard assessment of microplastics with different polymers by acute embryo test and chronic Larvae test with zebrafish (*Danio rerio*), *Aquat. Toxicol.*, 2023, 106595.
35. A. T. Ford, M. Ågerstrand, B. W. Brooks, J. Allen, M. G. Bertram, T. Brodin, Z. Dang, S. Duquesne, R. Sahm, F. Hoffmann and H. Hollert, The role of behavioral ecotoxicology in environmental protection, *Environ. Sci. Technol.*, 2021, **55**, 5620–5628.
36. T. Rahman and U. Candolin, Linking animal behavior to ecosystem change in disturbed environments, *Front. Ecol. Evol.*, 2022, **10**, 893453.
37. M. G. Bertram, J. Martin, E. S. McCallum, L. A. Alton, J. A. Brand, B. W. Brooks and T. Brodin, Frontiers in quantifying wildlife behavioural responses to chemical pollution, *Biol. Rev.*, 2022, **97**, 1346–1364.
38. S. J. Cooke, H. L. Auld, K. Birnie-Gauvin, C. K. Elvidge, M. L. Piczak, W. M. Twardek and A. M. Muir, On the relevance of animal behavior to the management and conservation of fishes and fisheries, *Environ. Biol. Fishes*, 2023, **106**, 785–810.
39. M. Ågerstrand, K. Arnold, S. Balshine, T. Brodin, B. W. Brooks, G. Maack, E. S. McCallum, G. Pyle, M. Saaristo and A. T. Ford, Emerging investigator series: use of behavioural endpoints in the regulation of chemicals, *Environ. Sci.: Processes Impacts*, 2020, **22**, 49–65.
40. J. M. Pérez-Iglesias, L. Franco-Belussi, C. R. de Arcaute, N. C. Bach, S. Soloneski, P. González and J. C. Brodeur, Ecotoxicological methodologies to evaluate biomarkers at different scales in Neotropical anurans, *J. Visualized Exp.*, 2023, **194**, e64520.
41. G. Pyle and A. Ford, Behaviour revised: contaminant effects on aquatic animal behaviour, *Aquat. Toxicol.*, 2017, **182**, 226–228.
42. J. B. Legradi, C. Di Paolo, M. H. Kraak, H. G. Van der Geest, E. L. Schymanski, A. J. Williams, M. M. Dingemans, R. Massei, W. Brack, X. Cousin and M. L. Begout, An ecotoxicological view on neurotoxicity assessment, *Environ. Sci. Eur.*, 2018, **30**, 1–34.
43. S. D. Melvin and S. P. Wilson, The utility of behavioral studies for aquatic toxicology testing: a meta-analysis, *Chemosphere*, 2013, **93**, 2217–2223.
44. T. Brodin, S. Piovano, J. Fick, J. Klaminder, M. Heynen and M. Jonsson, Ecological effects of pharmaceuticals in aquatic systems—impacts

through behavioural alterations, *Philos. Trans. R. Soc. London, Ser. B*, 2014, **369**, 20130580.

45. M. Saaristo, T. Brodin, S. Balshine, M. G. Bertram, B. W. Brooks, S. M. Ehlman, E. S. McCallum, A. Sih, J. Sundin, B. B. Wong and K. E. Arnold, Direct and indirect effects of chemical contaminants on the behaviour, ecology and evolution of wildlife, *Proc. R. Soc. B*, 2018, **285**, 20181297.
46. M. Sievers, R. Hale, K. M. Parris, S. D. Melvin, C. M. Lanctot and S. E. Swearer, Contaminant-induced behavioural changes in amphibians: A meta-analysis, *Sci. Total Environ.*, 2019, **693**, 133570.
47. L. C. Aulsebrook, M. G. Bertram, J. M. Martin, A. E. Aulsebrook, T. Brodin, J. P. Evans, M. D. Hall, M. K. O'Bryan, A. J. Pask, C. R. Tyler and B. B. Wong, Reproduction in a polluted world: implications for wildlife, *Reproduction*, 2020, **160**, 13–23.
48. L. Jacquin, Q. Petitjean, J. Côte, P. Laffaille and S. Jean, Effects of pollution on fish behavior, personality, and cognition: some research perspectives, *Front. Ecol. Evol.*, 2020, **8**, 86.
49. A. Bownik and D. Wlodkowic, Applications of advanced neuro-behavioral analysis strategies in aquatic ecotoxicology, *Sci. Total Environ.*, 2021, **772**, 145577.
50. A. S. Grunst, M. L. Grunst and J. Fort, Contaminant-by-environment interactive effects on animal behavior in the context of global change: Evidence from avian behavioral ecotoxicology, *Sci. Total Environ.*, 2023, **879**, 163169.
51. C. Linnaeus, *Systema Naturae per Regna Tria Naturae, Secundum Classes, Ordines, Genera, Species, cum Characteribus, Differentiis, Synonymis, Locis*, Laurentii Salvii, Stockholm, 1758.
52. G. L. Grammer, W. T. Slack, M. S. Peterson and M. A. Dugo, Nile tilapia *Oreochromis niloticus* (Linnaeus, 1758) establishment in temperate Mississippi, USA: multi-year survival confirmed by otolith ages, *Aquat. Invasions*, 2012, **7**, 367–376.
53. R. Froese and D. Pauly, FishBase, World Wide Web electronic publication, version (02/2023), https://www.fishbase.org, accessed on November 2023.
54. P. D. Tibihika, M. Curto, E. Alemayehu, H. Waidbacher, C. Masembe, P. Akoll and H. Meimberg, Molecular genetic diversity and differentiation of Nile tilapia (*Oreochromis niloticus*, L. 1758) in East African natural and stocked populations, *BMC Evol. Biol.*, 2020, **20**, 1–20.
55. T. Ishikawa, T. Shimose and K. Tachihara, Life history of an invasive and unexploited population of Nile tilapia (*Oreochromis niloticus*) and geographical variation across its native and non-native ranges, *Environ. Biol. Fishes*, 2013, **96**, 603–616.
56. L. Bavia, L. E. Santiesteban-Lores, M. C. Carneiro and M. M. Prodocimo, Advances in the complement system of a teleost fish, Oreochromis niloticus, *Fish Shellfish Immunol.*, 2022, **123**, 61–74.
57. J. N. Leonard and P. V. Skov, Capacity for thermal adaptation in Nile tilapia (*Oreochromis niloticus*): Effects on oxygen uptake and ventilation, *J. Therm. Biol.*, 2022, **105**, 103206.

58. F. Duponchelle and J. Panfili, Variations in age and size at maturity of female Nile tilapia, *Oreochromis niloticus*, populations from man-made lakes of Côte d'Ivoire, *Environ. Biol. Fishes*, 1998, **52**, 453–465.
59. A. F. M. El-Sayed and K. Fitzsimmons, From Africa to the world—The journey of Nile tilapia, *Rev. Aquacult.*, 2023, **15**, 6–21.
60. J. C. R. Dorantes-de-la-O, A. N. Maeda-Martínez, L. D. Espinosa-Chaurand and R. Garza-Torres, Bioeconomic modelling in tilapia aquaculture: A review, *Rev. Aquac.*, 2023, **15**(4), 1765–1777.
61. FAO, El estado mundial de la pesca y la acuicultura, 2022, https://www.fao.org/3/cc0461es/cc0461es.pdf, accessed on December 2023.
62. J. Ding, S. Zhang, R. M. Razanajatovo, H. Zou and W. Zhu, Accumulation, tissue distribution, and biochemical effects of polystyrene microplastics in the freshwater fish red tilapia (*Oreochromis niloticus*), *Environ. Pollut.*, 2018, **238**, 1–9.
63. S. Zhang, J. Ding, R. M. Razanajatovo, H. Jiang, H. Zou and W. Zhu, Interactive effects of polystyrene microplastics and roxithromycin on bioaccumulation and biochemical status in the freshwater fish red tilapia (*Oreochromis niloticus*), *Sci. Total Environ.*, 2019, **648**, 1431–1439.
64. M. Hamed, H. A. Soliman, A. G. Osman and A. E. D. H. Sayed, Assessment the effect of exposure to microplastics in Nile Tilapia (*Oreochromis niloticus*) early juvenile: I. Blood biomarkers, *Chemosphere*, 2019, **228**, 345–350.
65. M. Hamed, H. A. Soliman, A. G. Osman and A. E. D. H. Sayed, Antioxidants and molecular damage in Nile Tilapia (*Oreochromis niloticus*) after exposure to microplastics, *Environ. Sci. Pollut. Res.*, 2020, **27**, 14581–14588.
66. M. Hamed, H. A. Soliman, A. E. Badrey and A. G. Osman, Microplastics induced histopathological lesions in some tissues of tilapia (*Oreochromis niloticus*) early juveniles, *Tissue Cell*, 2021, **71**, 101512.
67. R. F. Ismail, N. E. Saleh and A. E. D. H. Sayed, Impacts of microplastics on reproductive performance of male tilapia (*Oreochromis niloticus*) pre-fed on *Amphora coffeaeformis*, *Environ. Sci. Pollut. Res.*, 2021, **28**, 68732–68744.
68. R. Dong, C. Zhou, S. Wang, Y. Yan and Q. Jiang, Probiotics ameliorate polyethylene microplastics-induced liver injury by inhibition of oxidative stress in Nile tilapia (*Oreochromis niloticus*), *Fish Shellfish Immunol.*, 2022, **130**, 261–272.
69. F. Zhang, D. Li, Y. Yang, H. Zhang, J. Zhu, J. Liu and X. Wang, Combined effects of polystyrene microplastics and copper on antioxidant capacity, immune response and intestinal microbiota of Nile tilapia (*Oreochromis niloticus*), *Sci. Total Environ.*, 2022, **808**, 152099.
70. J. Hasan, M. A. Siddik, A. K. Ghosh, S. B. Mesbah, M. A. Sadat and M. Shahjahan, Increase in temperature increases ingestion and toxicity of polyamide microplastics in Nile tilapia, *Chemosphere*, 2023, **327**, 138502.

71. M. P. Mtega, M. J. Mihale and K. F. Kilulya, Evaluation of toxicological risks and effects of microplastics on Nile tilapia (*Oreochromis niloticus*) under *in vitro* laboratory conditions, *East Afr. J. Biophys. Comput. Sci.*, 2023, **4**, 54–68.
72. L. S. Do-Nascimento, S. L. de Oliveira, C. C. da Costa, M. F. Aracati, L. F. Rodrigues, I. Charlie-Silva and M. A. de Andrade Belo, Deleterious effects of polypropylene microplastic ingestion in Nile tilapia (*Oreochromis niloticus*), *Bull. Environ. Contam. Toxicol.*, 2023, **111**, 13.
73. X. Hu, L. J. Meng, H. D. Liu, Y. S. Guo, W. C. Liu, H. X. Tan and G. Z. Luo, Impacts of Nile Tilapia (*Oreochromis niloticus*) exposed to microplastics in bioflocs system, *Sci. Total Environ.*, 2023, **901**, 165921.
74. T. Hao, Y. Gao, Z. C. Li, X. X. Zhou and B. Yan, Size-dependent uptake and depuration of nanoplastics in tilapia (*Oreochromis niloticus*) and distinct intestinal impacts, *Environ. Sci. Technol.*, 2023, **57**, 2804–2812.
75. X. Wu, M. Li, S. Yang, J. Dong, W. Pan, Y. Ning and D. Liang, Liver metabolic dysregulation induced by polypropylene nano-and microplastics in Nile tilapia using internal extractive electrospray ionization mass spectrometry, *Anal. Chem.*, 2023, **95**(20), 7863–7871.
76. F. Karayakar, Ö. Yurt, B. Cicik and M. Canli, Accumulation and elimination of cadmium by the Nile tilapia (*Oreochromis niloticus*) in differing temperatures and responses of oxidative stress biomarkers, *Bull. Environ. Contam. Toxicol.*, 2022, **109**, 1126–1134.
77. M. F. Abdelzaher, A. M. Azab, M. Authman and W. M. Shaban, Effects of lead and cadmium accumulation on survival and growth of the Nile tilapia (*Oreochromis niloticus*), *Egypt. J. Aquat. Biol. Fish.*, 2022, **26**(6), 155–171.
78. N. A. El-Fahla, M. E. Mohallal, H. N. Gad El-Hak, A. A. Dessouki and H. Abdelrazek, Effect of lead toxicity on the kidney of Nile tilapia: amelioration by Beta-MOS®, *Egypt. Acad. J. Biol. Sci. D, Histol. Histochem.*, 2022, **14**, 31–45.
79. E. Zahran, F. Ahmed, Z. Hassan, I. Ibrahim, A. A. Khaled, D. Palić and M. G. El Sebaei, Toxicity Evaluation, Oxidative, and Immune Responses of Mercury on Nile Tilapia: Modulatory Role of Dietary Nannochloropsis oculata, *Biol. Trace Elem. Res.*, 2024, **202**(4), 1752–1766.
80. B. Yulianto, A. Soegianto, M. Affandi and C. M. Payus, The impact of various periods of mercury exposure on the osmoregulatory and blood gas parameters of tilapia (*Oreochromis niloticus*), *Emerging Contam.*, 2023, **9**, 100244.
81. R. Gulzar, Z. Riaz, Q. U. A. Gillani, A. Mehreen, F. Jameel and R. Nawaz, Bioaccumulation and influence of cadmium chloride on histology of muscles and gills in Nile tilapia (*Oreochromis niloticus*), *J. Surv. Fish. Sci.*, 2023, 992–1000.
82. N. M. Abu-Elala, M. S. Khattab, H. O. AbuBakr, S. Helmy, A. Hesham, N. A. Younis and M. F. El Basuini, Neem leaf powder (*Azadirachta indica*) mitigates oxidative stress and pathological alterations triggered by lead toxicity in Nile tilapia (*Oreochromis niloticus*), *Sci. Rep.*, 2023, **13**, 9170.

83. F. Gröner, A. Ziková and W. Kloas, Effects of the pharmaceuticals diclofenac and metoprolol on gene expression levels of enzymes of biotransformation, excretion pathways and estrogenicity in primary hepatocytes of Nile tilapia (*Oreochromis niloticus*), *Comp. Biochem. Physiol. C Toxicol. Pharmacol.*, 2015, **167**, 51–57.
84. P. K. Pandey, M. N. Ajima, K. Kumar, N. Poojary and S. Kumar, Evaluation of DNA damage and physiological responses in Nile tilapia, *Oreochromis niloticus* (Linnaeus, 1758) exposed to sub-lethal diclofenac (DCF), *Aquat. Toxicol.*, 2017, **186**, 205–214.
85. S. M. Limbu, L. Zhou, S. X. Sun, M. L. Zhang and Z. Y. Du, Chronic exposure to low environmental concentrations and legal aquaculture doses of antibiotics cause systemic adverse effects in Nile tilapia and provoke differential human health risk, *Environ. Int.*, 2018, **115**, 205–219.
86. R. G. Botelho, C. A. Christofoletti, J. E. Correia, Y. Ansoar, R. A. D. Olinda and V. L. Tornisielo, Genotoxic responses of juvenile tilapia (*Oreochromis niloticus*) exposed to florfenicol and oxytetracycline, *Chemosphere*, 2015, **132**, 206–212.
87. C. C. Mattioli, B. M. Chiste, N. A. Takeshita, C. M. Jonsson, V. L. Ferracini and H. Hisano, Acute toxicity and risk assessment of florfenicol for Nile tilapia larvae, *Bull. Environ. Contam. Toxicol.*, 2020, **105**, 721–727.
88. Y. H. Liu, Y. Z. Lv, Z. Huang, Y. F. Guan, J. W. Huang, J. L. Zhao and G. G. Ying, Uptake, elimination, and toxicokinetics of selected pharmaceuticals in multiple tissues of Nile tilapia (*Oreochromis niloticus*) exposed to environmentally relevant concentrations, *Ecotoxicol. Environ. Saf.*, 2021, **226**, 112874.
89. S. M. Lan, T. M. D. Briggs, H. E. Obanya, N. H. Amaeze and A. A. Otitoloju, Toxicity evaluation of pharmaceutical wastewater to the Nile tilapia (*Oreochromis niloticus*), *J. Environ. Prot.*, 2021, **12**, 296–309.
90. M. N. Ajima, K. Kumar, N. Poojary and P. K. Pandey, Sublethal diclofenac induced oxidative stress, neurotoxicity, molecular responses and alters energy metabolism proteins in Nile tilapia, *Oreochromis niloticus*, *Environ. Sci. Pollut. Res.*, 2021, **28**, 44494–44504.
91. M. N. Ajima, K. Kumar, N. Poojary and P. K. Pandey, Oxidative stress biomarkers, biochemical responses and Na^+-K^+-ATPase activities in Nile tilapia, *Oreochromis niloticus* exposed to diclofenac, *Comp. Biochem. Physiol., Part C: Toxicol. Pharmacol.*, 2021, **240**, 108934.
92. B. Dey, T. J. Abraham, J. Singha, A. Roy, S. Karmakar, P. Kumar Patil and U. Roy, Histopathological changes and tissue residue concentrations of monosex Nile tilapia (*Oreochromis niloticus*, L) fries exposed to oxytetracycline, *Aquacult. Int.*, 2022, **30**, 2113–2128.
93. A. Bardhan, T. J. Abraham, J. Singha, T. K. Sar, R. Rajisha, E. K. N. Krishna and P. K. Patil, Histopathological aberrations and oxidative stress responses in Nile tilapia *Oreochromis niloticus* as influenced by dietary florfenicol and its metabolites, *Aquaculture*, 2022, **559**, 738447.
94. A. Bardhan, T. J. Abraham, J. Singha, R. Rajisha, E. K. N. Krishna, S. K. Panda and P. K. Patil, Impacts of oral florfenicol medication and

residues on the kidney and liver of Nile tilapia *Oreochromis niloticus* (L.), *Vet. Sci.*, 2023, **10**, 36.

95. P. Vijitkul, M. Kongsema, T. Toommakorn and V. Bullangpoti, Investigation of genotoxicity, mutagenicity, and cytotoxicity in erythrocytes of Nile tilapia (*Oreochromis niloticus*) after fluoxetine exposure, *Toxicol. Rep*, 2022, **9**, 588–596.
96. M. M. Hussein, H. A. Elsadaawy, A. El-Murr, M. M. Ahmed, A. M. Bedawy, H. A. Tukur and I. M. Saadeldin, Endosulfan toxicity in Nile tilapia (*Oreochromis niloticus*) and the use of lycopene as an ameliorative agent, *Comp. Biochem. Physiol., Part C: Toxicol. Pharmacol.*, 2019, **224**, 108573.
97. N. L. Guedegba, I. Imorou Toko, P. T. Agbohessi, B. S. Zoumenou, C. Douny, S. N. Mandiki and P. Kestemont, Comparative acute toxicity of two phytosanitary molecules, lambda-cyhalothrin and acetamiprid, on Nile Tilapia (*Oreochromis Niloticus*) juveniles, *J. Environ. Sci. Health B*, 2019, **54**, 580–589.
98. F. M. Abdelhamid, G. E. Elshopakey and A. E. Aziza, Ameliorative effects of dietary *Chlorella vulgaris* and β-glucan against diazinon-induced toxicity in Nile tilapia (*Oreochromis niloticus*), *Fish Shellfish Immunol.*, 2020, **96**, 213–222.
99. M. Abdel-Tawwab, H. A. El-Saadawy, H. I. El-Belbasi, S. A. Abd El-Hameed and A. A. Attia, Dietary spirulina (*Arthrospira platenesis*) mitigated the adverse effects of imidacloprid insecticide on the growth performance, haemato-biochemical, antioxidant, and immune responses of Nile tilapia, *Comp. Biochem. Physiol., Part C: Toxicol. Pharmacol.*, 2021, **247**, 109067.
100. K. C. Kanu, N. H. Amaeze and A. A. Otitoloju, Estimation of pesticide mixture interaction in Nile tilapia (*Oreochromis niloticus*) using survival analysis, *Afr. J. Aquat. Sci.*, 2022, **47**, 66–77.
101. K. C. Kanu, A. A. Otitoloju and N. H. Amaeze, Survival analysis of African catfish and Nile tilapia briefly exposed to complex pesticide mixtures, *J. Basic Appl. Zool.*, 2022, **83**, 1–15.
102. M. A. Hossain, L. Sutradhar, T. R. Sarker, S. Saha and M. M. Iqbal, Toxic effects of chlorpyrifos on the growth, hematology, and different organs histopathology of Nile tilapia, *Oreochromis niloticus*, *Saudi J. Biol. Sci.*, 2022, **29**, 103316.
103. M. R. Fouad, A. F. El-Aswad and M. I. Aly, Acute toxicity, biochemical and histological of fenitrothion and thiobencarb on fish Nile tilapia (*Oreochromis niloticus*), *Nusantara Biosci.*, 2022, **14**, 217–226.
104. M. Amin, M. Yousuf, M. Attaullah, N. Ahmad, M. N. Azra, M. Lateef and M. Naeem, Cholinesterase activity as a potential biomarker for neurotoxicity induced by pesticides in vivo exposed *Oreochromis niloticus* (Nile tilapia): Assessment tool for organophosphates and synthetic pyrethroids, *Environ. Technol.*, 2023, **44**, 2148–2156.
105. A. D. Abdelmagid, A. M. Said, E. A. Abd El-Gawad, S. A. Shalaby and M. A. Dawood, Glyphosate-induced liver and kidney dysfunction, oxidative stress, immunosuppression in Nile tilapia, but ginger showed a protection role, *Vet. Res. Commun.*, 2023, **47**, 445–455.

106. I. Taufik, A. Ibrahim, M. S. Syawal, E. Setiadi and E. Supriyono, Lethal toxicity LC50 of permethrin insecticide and its physiological and histopathological effect on common carp and Nile tilapia, *IOP Conf. Ser.: Earth Environ. Sci.*, 2022, **1221**, 012019.
107. E. A. Kamal, N. K. Abdelkhalek, A. E. Hassan and M. El-Adl, Metabolomic analyses, toxicity biomarkers and histopathological changes in the liver of Nile tilapia exposed to diazinon toxicity, *J. Adv. Vet. Res.*, 2023, **13**, 1161–1165.
108. R. P. Campos, T. Q. Chagas, T. G. da Silva Alvarez, C. Mesak, J. E. de Andrade Vieira, C. F. C. Paixão and G. Malafaia, Analysis of ZnO nanoparticle-induced changes in *Oreochromis niloticus* behavior as toxicity endpoint, *Sci. Total Environ.*, 2019, **682**, 561–571.
109. W. A. Mansour, N. R. Abdelsalam, M. Tanekhy, A. A. Khaled and A. T. Mansour, Toxicity, inflammatory and antioxidant genes expression, and physiological changes of green synthesis silver nanoparticles on Nile tilapia (*Oreochromis niloticus*) fingerlings, *Comp. Biochem. Physiol., Part C: Toxicol. Pharmacol.*, 2021, **247**, 109068.
110. H. M. Abdel-Latif, M. Shukry, O. I. El Euony, M. Mohamed Soliman, A. E. Noreldin, H. A. Ghetas and M. A. Khallaf, Hazardous effects of SiO_2 nanoparticles on liver and kidney functions, histopathology characteristics, and transcriptomic responses in Nile Tilapia (*Oreochromis niloticus*) Juveniles, *Biology*, 2021, **10**, 183.
111. S. Veisi, M. Sarkheil, S. A. Johari and O. Safari, Dietary supplementation with melatonin: influence on growth performance, oxidative stress status, and amelioration of silver nanoparticles-induced toxicity in Nile tilapia (*Oreochromis niloticus*), *Trop. Anim. Health Prod.*, 2021, **53**, 314.
112. N. E. Ismael, S. A. Abd El-hameed, A. M. Salama, M. A. Naiel and H. M. Abdel-Latif, The effects of dietary clinoptilolite and chitosan nanoparticles on growth, body composition, haemato-biochemical parameters, immune responses, and antioxidative status of Nile tilapia exposed to imidacloprid, *Environ. Sci. Pollut. Res.*, 2021, **28**, 29535–29550.
113. H. A. Soliman, M. Hamed and A. E. D. H. Sayed, Investigating the effects of copper sulfate and copper oxide nanoparticles in Nile tilapia (*Oreochromis niloticus*) using multiple biomarkers: the prophylactic role of Spirulina, *Environ. Sci. Pollut. Res.*, 2021, **28**, 30046–30057.
114. M. M. Mabrouk, A. T. Mansour, A. F. Abdelhamid, K. M. Abualnaja, A. Mamoon, W. S. Gado and H. F. Ayoub, Impact of aqueous exposure to silver nanoparticles on growth performance, redox status, non-specific immunity, and histopathological changes of Nile Tilapia, *Oreochromis niloticus*, challenged with *Aeromonas hydrophila*, *Aquacult. Rep.*, 2021, **21**, 100816.
115. H. S. Hamed, R. M. Amen, A. H. Elelemi, H. H. Mahboub, H. Elabd, A. M. Abdelfattah and A. K. Ismail, Effect of dietary *Moringa oleifera* leaves nanoparticles on growth performance, physiological, immunological responses, and liver antioxidant biomarkers in Nile tilapia

(*Oreochromis niloticus*) against Zinc oxide nanoparticles toxicity, *Fishes*, 2022, **7**, 360.
116. M. N. Goda, A. A. Shaheen and H. S. Hamed, Potential role of dietary parsley and/or parsley nanoparticles against zinc oxide nanoparticles toxicity induced physiological, and histological alterations in Nile tilapia, *Oreochromis niloticus*, *Aquac. Rep.*, 2023, **28**, 101425.
117. C. Zheng, M. R. Farag, M. Alagawany, S. El-Kassas, M. M. Azzam, A. Di Cerbo and E. Wagih, Dietary supplementation of quercetin nanoparticles enhances the growth performance hematological and immunological responses and resistance to *Aeromonas hydrophila* infection in Nile tilapia (*Oreochromis niloticus*) exposed to silver nanoparticles toxicity, *Aquac. Rep.*, 2023, **33**, 101780.
118. M. R. Farag, H. G. Abo-Al-Ela, M. Alagawany, M. M. Azzam, M. T. El-Saadony, S. Rea and D. S. Nouh, Effect of quercetin nanoparticles on hepatic and intestinal enzymes and stress-related genes in Nile tilapia fish exposed to silver nanoparticles, *Biomedicines*, 2023, **11**, 663.
119. M. Prabakaran, C. Binuramesh, D. Steinhagen and R. D. Michael, Immune response in the tilapia, *Oreochromis mossambicus* on exposure to tannery effluent, *Ecotoxicol. Environ. Saf.*, 2007, **68**, 372–378.
120. W. M. AbdEl-Rahim, W. K. Khalil and M. G. Eshak, Evaluation of the gene expression changes in Nile tilapia (*Oreochromis niloticus*) as affected by the bio-removal of toxic textile dyes from aqueous solution in small-scale bioreactor, *The Environmentalist*, 2010, **30**, 242–253.
121. A. Zagal and B. Mazmanci, Oxidative stress response in Nile tilapia (*Oreochromis niloticus*) exposed to textile mill effluent, *Toxicol. Ind. Health*, 2011, **27**, 81–85.
122. K. B. Workagegn, Toxicity evaluation of wastewater treatment plant of textile effluent using fish: Nile tilapia *Oreochromis niloticus*, *Int. J. Fish. Aquacult.*, 2013, **3**, 43–48.
123. N. J. Julius, F. O. Arimoro, A. O. Edegbene, N. Danbaba, A. G. Jiya and M. D. Usman, Effects of tannery waste on Nile tilapia (*Oreochromis niloticus*), *J. Biosci.*, 2014, **2**, 63–66.
124. A. Weldetinsae, M. Dawit, A. Getahun, H. S. Patil, E. Alemayehu, M. Gizaw and D. Abera, Aneugenicity and clastogenicity in freshwater fish *Oreochromis niloticus* exposed to incipient safe concentration of tannery effluent, *Ecotoxicol. Environ. Saf.*, 2017, **138**, 98–104.
125. M. Abdel-Tawwab and H. S. Hamed, Effect of bisphenol A toxicity on growth performance, biochemical variables, and oxidative stress biomarkers of Nile tilapia, *Oreochromis niloticus* (L.), *J. Appl. Ichthyol.*, 2018, **34**, 1117–1125.
126. M. Muliari, I. Zulfahmi, Y. Akmal, N. W. K. Karja, C. Nisa, K. A. Sumon and M. M. Rahman, Toxicity of palm oil mill effluent on the early life stages of Nile tilapia (*Oreochromis niloticus*, Linnaeus 1758), *Environ. Sci. Pollut. Res.*, 2020, **27**, 30592–30599.
127. S. M. Lan, T. M. D. Briggs, H. E. Obanya, N. H. Amaeze and A. A. Otitoloju, Toxicity Evaluation of Pharmaceutical Wastewater to the Nile Tilapia (*Oreochromis niloticus*), *J. Environ. Prot.*, 2021, **12**, 296–309.

128. M. M. Abdellatif, N. M. Sabry, S. Ibrahim, S. M. Bassem, W. K. Khalil and F. K. Abdel-Gawad, Evaluation of mitigation role of L-phenylalanine-based low-molecular-weight gelator against oil pollution-induced Nile tilapia oxicity, *Gels*, 2023, **9**, 848.
129. M. Banaee, A. A. Badr, C. R. Multisanti, B. N. Haghi and C. Faggio, The toxicity effects of the individual and combined exposure of methyl tert-butyl ether (MTBE) and tire rubber powder (RP) on Nile tilapia fish (*Oreochromis niloticus*), *Comp. Biochem. Physiol., Part C: Toxicol. Pharmacol.*, 2023, **274**, 109759.
130. E. Deribe, B. O. Rosseland, R. Borgstrøm, B. Salbu, Z. Gebremariam, E. Dadebo and O. M. Eklo, Bioaccumulation of persistent organic pollutants (POPs) in fish species from Lake Koka, Ethiopia: the influence of lipid content and trophic position, *Sci. Total Environ.*, 2011, **410**, 136–145.
131. O. R. Ibor, A. O. Adeogun, A. V. Chukwuka and A. Arukwe, Gross pathology, physiological and toxicological responses in relation to metals and persistent organic pollutants (POPs) burden in tilapia species from Ogun River, Nigeria, *Mar. Environ. Res.*, 2017, **129**, 245–257.
132. H. M. El-Khayat, H. S. Gaber and H. E. Flefel, Experimental studies on the toxicity of certain heavy metals and persistent organic pollutants on the Nile tilapia health, *Egypt. J. Aquat. Biol. Fish.*, 2022, **26**, 321–346.
133. K. Chinnadurai, P. Prema, V. Veeramanikandan, K. R. Kumar, V. H. Nguyen, N. Marraiki and P. Balaji, Toxicity evaluation and oxidative stress response of fumaronitrile, a persistent organic pollutant (POP) of industrial waste water on tilapia fish (*Oreochromis mossambicus*), *Environ. Res.*, 2022, **204**, 112030.
134. C. H. O. Kyu-Seok, Changes of hepatic microsomal cytochrome P450 monooxygenase system in Nile tilapia, *Oreochromis niloticus* exposed to PCBs, *J. Fish. Aquat. Sci.*, 2001, **34**, 194–198.
135. C. K. Seok, M. E. Young and K. J. Chan, Changes of haematological constituents in the Nile tilapia, *Oreochromis niloticus* exposed to HCB and PCBs, *J. Fish. Aquat. Sci.*, 2002, **35**, 110–114.
136. Y. Zheng, L. P. Qiu, S. L. Meng, L. M. Fan, C. Song, D. D. Li and J. Z. Chen, Effect of polychlorinated biphenyls on oxidation stress in the liver of juvenile GIFT, *Oreochromis niloticus*, *Genet. Mol. Res.*, 2016, **15**, 1–9.
137. Y. Zheng, L. Qiu, L. Fan, C. Song, S. Meng and J. Chen, Effect of polychlorinated biphenyls on osmoregulatory response and apoptosis in GIFT tilapia, Oreochromis niloticus, *Genet. Mol. Res.*, 2016, **15**(3), DOI: 10.4238/gmr.15038620.
138. M. S. Shailaja and C. D'Silva, Evaluation of impact of PAH on a tropical fish, *Oreochromis mossambicus* using multiple biomarkers, *Chemosphere*, 2003, **53**, 835–841.
139. A. Pathiratne and C. K. Hemachandra, Modulation of ethoxyresorufin O-deethylase and glutathione S-transferase activities in Nile tilapia (*Oreochromis niloticus*) by polycyclic aromatic hydrocarbons containing

two to four rings: implications in biomonitoring aquatic pollution, *Ecotoxicology*, 2010, **19**, 1012–1018.
140. J. Costa, M. Ferreira, L. Rey-Salgueiro and M. A. Reis-Henriques, Comparison of the waterborne and dietary routes of exposure on the effects of Benzo (a) pyrene on biotransformation pathways in Nile tilapia (*Oreochromis niloticus*), *Chemosphere*, 2011, **84**, 1452–1460.
141. M. Martinez-Porchas, M. Henández-Rodrнgues, J. Dávila-Ortiz, V. Villa-Cruz and J. R. Ramos-Enríquez, A preliminary study about the effect of benzo [a] pyrene (BaP) injection on the thermal behavior and plasmatic parameters of Nile tilapia (*Oreochromis niloticus* L.) acclimated to different temperatures, *Pan-Am. J. Aquat. Sci.*, 2011, **6**, 76–85.
142. T. O. Sogbanmu, A. O. Osibona, O. A. Oguntunde and A. A. Otitoloju, Biomarkers of toxicity in *Clarias gariepinus* exposed to sublethal concentrations of polycyclic aromatic hydrocarbons, *Afr. J. Aquat. Sci.*, 2018, **43**, 281–292.
143. N. A. Albornoz-Abud, G. F. Canul-Marín, I. Chan-Cuá, E. Hernández-Núñez, M. A. Cañizares-Martínez, D. Valdés-Lozano and R. C. Colli-Dula, Gene expression analysis on growth, development and toxicity pathways of male Nile tilapia (*Oreochromis niloticus*), after acute and sub-chronic benzo (α) pyrene exposures, *Comp. Biochem. Physiol., Part C: Toxicol. Pharmacol.*, 2021, **250**, 109160.
144. M. Quintanilla-Mena, J. Vega-Arreguin, M. Del Río-García, V. Patiño-Suárez, S. Peraza-Echeverria and C. Puch-Hau, The effect of benzo [a] pyrene on the gut microbiota of Nile tilapia (*Oreochromis niloticus*), *Appl. Microbiol. Biotechnol.*, 2021, **105**, 7935–7947.
145. A. Gül, Investigation of acute toxicity of chlorpyrifos-methyl on Nile tilapia (Oreochromis niloticus L.) larvae, *Chemosphere*, 2005, **59**(2), 163–166.
146. A. Gül, Investigation of acute toxicity of chlorpyrifos-methyl on Nile tilapia (*Oreochromis niloticus* L.) larvae, *Chemosphere*, 2005, **59**, 163–166.
147. A. Ç. K. Benli and G. Köksal, The acute toxicity of ammonia on tilapia (*Oreochromis niloticus* L.) larvae and fingerlings, *Turk. J. Vet. Anim. Sci.*, 2005, **29**, 339–344.
148. M. Z. Yildirim, A. Ç. K. Benlı, M. Selvı, A. Özkul, F. Erkoc and O. Kocak, Acute toxicity, behavioral changes, and histopathological effects of deltamethrin on tissues (gills, liver, brain, spleen, kidney, muscle, skin) of Nile tilapia (*Oreochromis niloticus* L.) fingerlings, *Environ. Toxicol.*, 2006, **21**, 614–620.
149. J. A. Almeida, R. E. Barreto, E. L. Novelli, F. J. Castro and S. E. Moron, Oxidative stress biomarkers and aggressive behavior in fish exposed to aquatic cadmium contamination, *Neotrop. Ichthyol.*, 2009, **7**, 103–108.
150. P. C. Pessoa, K. H. Luchmann, A. B. Ribeiro, M. M. Veras, J. R. M. B. Correa, A. J. Nogueira and P. S. M. Carvalho, Cholinesterase inhibition and behavioral toxicity of carbofuran on *Oreochromis niloticus* early life stages, *Aquat. Toxicol.*, 2011, **105**, 312–320.

151. C. D. Ezeonyejiaku, M. O. Obiakor and C. Ezenwelu, Toxicity of copper sulphate and behavioral locomotor response of tilapia (*Oreochromis niloticus*) and catfish (*Clarias gariepinus*) species, *Turk. J. Vet. Anim. Sci.*, 2011, **1**, 130–134.
152. Y. S. El-Sayed, D. H. Samak, I. Y. Abou-Ghanema and M. K. Soliman, Physiological and oxidative stress biomarkers in the freshwater monosex Nile tilapia, *Oreochromis niloticus* L., exposed to pendimethalin-based herbicide, *Environ. Toxicol.*, 2015, **30**, 430–438.
153. C. N. P. Boscolo, T. S. B. Pereira, I. G. Batalhão, P. L. R. Dourado, D. Schlenk and E. A. de Almeida, Diuron metabolites act as endocrine disruptors and alter aggressive behavior in Nile tilapia (*Oreochromis niloticus*), *Chemosphere*, 2018, **191**, 832–838.
154. R. B. Julinta, T. J. Abraham, A. Roy, J. Singha, G. Dash, P. Mali and K. A. Kumar, Effect of oxytetracycline-dosing on the growth, safety and intestinal histology of Nile tilapia, *Oreochromis niloticus* (L.) juveniles, *Int. J. Curr. Microbiol. Appl. Sci.*, 2019, **8**, 2708–2724.
155. I. G. Batalhão, D. Lima, A. P. M. Russi, C. N. P. Boscolo, D. G. H. Silva, T. S. B. Pereira and E. A. de Almeida, Effects of methylphenidate on the aggressive behavior, serotonin and dopamine levels, and dopamine-related gene transcription in brain of male Nile tilapia (*Oreochromis niloticus*), *Fish Physiol. Biochem.*, 2019, **45**, 1377–1391.
156. K. U. Sheethal, T. C. Mohan, G. Gowda and H. Nayak, Acute toxicology of nickel to Nile tilapia, *Oreochromis niloticus* (Linnaeus, 1758), *J. Entomol. Zool. Stud.*, 2020, **8**, 1804–1806.
157. E. H. A. Zeid, B. A. Khalifa, E. N. Said, A. H. Arisha and R. M. Reda, Neurobehavioral and immune-toxic impairments induced by organic methyl mercury dietary exposure in Nile tilapia *Oreochromis niloticus*, *Aquat. Toxicol.*, 2021, **230**, 105702.
158. M. T. El-Saadony, F. M. Alkhatib, S. O. Alzahrani, M. E. Shafi, S. E. Abdel-Hamid, T. F. Taha and N. H. Ahmed, Impact of mycogenic zinc nanoparticles on performance, behavior, immune response, and microbial load in *Oreochromis niloticus*, *Saudi J. Biol. Sci.*, 2021, **28**, 4592–4604.
159. M. R. Farag, H. K. Mahmoud, S. A. El-Sayed, S. Y. Ahmed, M. Alagawany and S. M. Abou-Zeid, Neurobehavioral, physiological and inflammatory impairments in response to bifenthrin intoxication in *Oreochromis niloticus* fish: Role of dietary supplementation with *Petroselinum crispum* essential oil, *Aquat. Toxicol.*, 2021, **231**, 105715.
160. T. F. Mahi, G. Chowdhury, M. A. Hossain, A. K. Baishnab, P. Schneider and M. M. Iqbal, Assessment of lead (Pb) toxicity in juvenile Nile tilapia, *Oreochromis niloticus* — growth, behaviour, erythrocytes abnormalities, and histological alterations in vital organs, *Toxics*, 2022, **10**, 793.
161. A. N. A. Rahman, A. A. R. Mohamed, N. Dahran, M. F. Farag, L. S. Alqahtani, M. A. Nassan and N. I. El-Naseery, Appraisal of sub-chronic exposure to lambada-cyhalothrin and/or methomyl on the behavior and hepato-renal functioning in *Oreochromis niloticus*: Supportive role of taurine-supplemented feed, *Aquat. Toxicol.*, 2022, **250**, 106257.

162. M. R. Farag, Y. A. Attia, L. S. Al Solami, F. Bovera, A. Nizza and M. Alagawany, Behavioral, physiological, and inflammatory responses of *Oreochromis niloticus* fish exposed to thallium and/or supplementation with *Astragalus membranaceus* polysaccharides, *Aquaculture*, 2022, **553**, 738100.
163. R. E. Ibrahim, M. A. Elbealy, G. A. Salem, A. A. Abdelwarith, E. M. Younis, E. Wagih and A. N. A. Rahman, Acute mancozeb-fungicide exposure induces neuro-ethology disruption, health disorders, and immune-oxidative dysfunction in Nile tilapia (*Oreochromis niloticus*), *Aquat. Toxicol.*, 2023, **261**, 106630.
164. T. Keller-Costa, A. V. Canário and P. C. Hubbard, Chemical communication in cichlids: a mini-review, *Gen. Comp. Endocrinol.*, 2015, **221**, 64–74.
165. U. Lopez, J. Gautrais, I. D. Couzin and D. Theraulaz, From behavioural analyses to models of collective motion in fish schools, *Interface Focus*, 2012, **2**, 693–707.
166. H. F. Olivares-Rubio and E. Arce, Effects of chemical pollution on the behaviour of cichlid fish, *Environ. Biol. Fishes*, 2023, **106**, 1149–1176.
167. M. S. Dawkins, *Observing Animal Behaviour: Design and Analysis of Quantitative Data*, Oxford University Press, Oxford, 2007.
168. L. A. Dugatkin, *Principles of Animal Behavior*, University of Chicago Press, Chicago, 2020.
169. G. Fehlmann, M. J. O'riain, I. FÜrtbauer and A. J. King, Behavioral causes, ecological consequences, and management challenges associated with wildlife foraging in human-modified landscapes, *BioScience*, 2021, **71**, 40–54.
170. Nobel Prize 1973, The Nobel Prize in Physiology or Medicine 1973, https://www.nobelprize.org/prizes/medicine/1973/summary/, accessed on November 2023.
171. J. Altmann, Observational study of behavior: sampling methods, *Behaviour*, 1974, **49**, 227–266.
172. J. Cachat, A. Stewart, E. Utterback, P. Hart, S. Gaikwad, K. Wong and A. V. Kalueff, Three-dimensional neurophenotyping of adult zebrafish behavior, *PLoS One*, 2011, **6**, e17597.
173. A. V. Kalueff, M. Gebhardt, A. M. Stewart, J. M. Cachat, M. Brimmer, J. S. Chawla, H. Schneider, and The Zebrafish Neuroscience Research H. Consortium, Towards a comprehensive catalog of zebrafish behavior 1.0 and beyond, *Zebrafish*, 2013, **10**, 70–86.
174. V. T. Tantarpale, S. H. Rathod and S. Kapil, Temperature stress on opercular beats and respiratory rate of freshwater fish *Channa punctatus*, *Int. J. Sci. Res. Publ.*, 2012, **2**, 1–5.
175. A. S. Dan-kishiya, J. R. Solomon and U. A. Alhaji, Influence of temperature on the respiratory rate of Nile Tilapia, *Oreochromis niloticus* (Pisces: Cichlidae) in the laboratory, *Cuad. Inv. UNED Res. J.*, 2016, **8**, 27–30.
176. A. L. Val, M. N. P. Silva and V. M. F. Almeida-Val, Hypoxia adaptation in fish of the Amazon: a never-ending task, *S. Afr. J. Zool.*, 1998, **33**, 107–114.

177. J. M. Wilson, R. M. Bunte and A. J. Carty, Evaluation of rapid cooling and tricaine methanesulfonate (MS222) as methods of euthanasia in zebrafish (*Danio rerio*), *J. Am. Assoc. Lab. Anim. Sci.*, 2009, **48**, 785–789.
178. I. V. Zhdanova, S. Y. Wang, O. U. Leclair and N. P. Danilova, Melatonin promotes sleep-like state in zebrafish, *Brain Res.*, 2001, **903**, 263–268.
179. M. A. Ferrer, J. A. Calduch-Giner, M. Díaz, J. Sosa, E. Rosell-Moll, J. S. Abril and J. Pérez-Sánchez, From operculum and body tail movements to different coupling of physical activity and respiratory frequency in farmed gilthead sea bream and European sea bass. Insights on aquaculture biosensing, *Comput. Electron. Agric.*, 2020, **175**, 105531.
180. R. D. Calfee, H. J. Puglis, E. E. Little, W. G. Brumbaugh and C. A. Mebane, Quantifying fish swimming behavior in response to acute exposure of aqueous copper using computer assisted video and digital image analysis, *JoVE*, 2016, **108**, e53477.
181. J. H. Bergstedt, T. Pfalzgraff and P. V. Skov, Hypoxia tolerance and metabolic coping strategies in *Oreochromis niloticus*, *Comp. Biochem. Physiol., Part C: Toxicol. Pharmacol.*, 2021, **257**, 110956.
182. B. G. Borowiec, K. L. Darcy, D. M. Gillette and G. R. Scott, Distinct physiological strategies are used to cope with constant hypoxia and intermittent hypoxia in killifish (*Fundulus heteroclitus*), *J. Exp. Biol.*, 2015, **218**, 1198–1211.
183. K. L. Gilmore, Z. A. Doubleday and B. M. Gillanders, Prolonged exposure to low oxygen improves hypoxia tolerance in a freshwater fish, *Conserv. Physiol.*, 2019, **7**, coz058.
184. D. McPhee, J. R. Watson, D. J. Harding, A. Prior, J. H. Fawcett, C. E. Franklin and R. L. Cramp, Body size dictates physiological and behavioural responses to hypoxia and elevated water temperatures in Murray cod (*Maccullochella peelii*), *Conserv. Physiol.*, 2023, **11**, coac087.
185. L. Shi, A. Zhang, H. Liu and H. Wang, Deletion of the *foxO*$_4$ gene increases hypoxia tolerance in zebrafish, *Int. J. Mol. Sci.*, 2023, **24**, 8942.
186. W. Heinrichs-Caldas, H. Ikert, V. M. F. Almeida-Val and P. M. Craig, Sex matters: Gamete-specific contribution of microRNA following parental exposure to hypoxia in zebrafish, *Comp. Biochem. Physiol., Part D: Genomics Proteomics*, 2023, **47**, 101090.
187. H. Guo, L. Whitehouse, R. Danzmann and B. Dixon, Effects of juvenile thermal preconditioning on the heat-shock, immune, and stress responses of rainbow trout upon a secondary thermal challenge, *Comp. Biochem. Physiol., Part A: Mol. Integr. Physiol.*, 2023, **280**, 111413.
188. E. V. Schneider, C. T. Hasler and C. D. Suski, Swimming performance of a freshwater fish during exposure to high carbon dioxide, *Environ. Sci. Pollut. Res.*, 2019, **26**, 3447–3454.
189. A. G. Ponniah and T. J. Pandian, Surfacing activity and food utilization in the air-breathing fish *Polyacanthus cupanus* exposed to constant Po2, *Hydrobiologia*, 1977, **53**, 221–227.
190. P. Domenici, C. Lefrançois and A. Shingles, Hypoxia and the antipredator behaviours of fishes, *Philos. Trans. R. Soc. London, Ser. B*, 2007, **362**, 2105–2121.

191. L. J. Chapman and D. J. Mckenzie, in *Fish Physiology*, ed. J. G. Richards, A. P. Farrell, and C. J. Brauner, Academic Press, Cambridge, 2009, vol. 27, pp. 25–77.
192. M. Mandic, K. A. Sloman and J. G. Richards, Escaping to the surface: a phylogenetically independent analysis of hypoxia-induced respiratory behaviors in sculpins, *Physiol. Biochem. Zool.*, 2009, **82**, 730–738.
193. G. N. Robertson, B. W. Lindsey, T. C. Dumbarton, R. P. Croll and F. M. Smith, The contribution of the swimbladder to buoyancy in the adult zebrafish (*Danio rerio*): a morphometric analysis, *J. Morphol.*, 2018, **269**, 666–673.
194. J. P. Eveson, T. A. Patterson, J. R. Hartog and K. Evans, Modelling surfacing behaviour of southern bluefin tuna in the Great Australian Bight, *Deep Sea Res., Part II*, 2018, **157**, 179–189.
195. S. G. Reebs, *Oxygen and Fish Behaviour*, Université de Moncton, Moncton, 2009.
196. I. V. Zhdanova, Sleep in zebrafish, *Zebrafish*, 2006, **3**, 215–226.
197. P. Aranda-Martínez, J. Fernández-Martínez, Y. Ramírez-Casas, C. Rodríguez-Santana, I. Rusanova, G. Escames and D. Acuña-Castroviejo, Chronodisruption and loss of melatonin rhythm, associated with alterations in daily motor activity and mitochondrial dynamics in parkinsonian zebrafish, are corrected by melatonin treatment, *Antioxidants*, 2023, **12**, 954.
198. L. De Oliveira-Mello, D. Baronio and P. Panula, Zebrafish embryonically exposed to valproic acid present impaired retinal development and sleep behavior, *Autism Res.*, 2023, **16**, 1877–1890.
199. J. Zhang, J. Feng and C. Feng, Analysis of the underlying mechanism of Ziziphi Spinosae Semen for treating anxiety disorder in a zebrafish sleep deprivation model, *Pharm. Sci. Adv.*, 2024, **2**, 100019.
200. E. Shaw and H. Topoff, 17, Fish in Schools, in *The Natural History Reader in Animal Behavior*, Columbia University Press, New York Chichester, West Sussex, 1987, pp. 162–167.
201. T. J. Pitcher, in *Encyclopedia of Ocean Sciences: Marine Biology*, ed. J. H. Steele, S. A. Thorpe, and K. K. Turekian, Elsevier, New York, 2001, pp. 337–349.
202. M. Larsson, Why do fish school?, *Curr. Zool.*, 2012, **58**(1), 116–128.
203. M. Larsson and B. W. Abbott, Is the capacity for vocal learning in vertebrates rooted in fish schooling behavior?, *Evol. Biol.*, 2018, **45**, 359–373.
204. D. S. Pavlov and A. O. Kasumyan, Patterns and mechanisms of schooling behavior in fish: A review, *J. Ichthyol.*, 2000, **40**, S163–S231.
205. M. P. Carrieri and G. L. Volpato, Does snatching frequency really indicate food ingestion in the Nile tilapia?, *Physiol. Behav.*, 1991, **50**, 489–492.
206. E. Gonçalves-de-Freitas, F. B. Teresa, F. S. Gomes and P. C. Giaquinto, Effect of water renewal on dominance hierarchy of juvenile Nile tilapia, *Appl. Anim. Behav. Sci.*, 2008, **111**, 187–195.
207. R. E. Barreto, G. L. Volpato, C. de Brito Faturi, P. C. Giaquinto, E. Gonçalves de Freitas and M. Fernandes de Castilho, Aggressive behaviour traits predict physiological stress responses in Nile tilapia (*Oreochromis niloticus*), *Mar. Freshwater Behav. Physiol.*, 2009, **42**, 109–118.

208. E. Gonçalves-de-Freitas, M. C. Bolognesi, A. C. D. S. Gauy, M. L. Brandão, P. C. Giaquinto and M. Fernandes-Castilho, Social behavior and welfare in Nile tilapia, *Fishes*, 2019, **4**, 23.
209. A. Waheed, H. Naz, M. Wajid and M. S. Khan, Impact of isolation on growth performance, behavior, and stress responses in Nile tilapia, *Oreochromis niloticus*, *Lat. Am. J. Aquat. Res.*, 2023, **51**, 483–490.
210. A. Tesfahun and M. Temesgen, Food and feeding habits of Nile tilapia *Oreochromis niloticus* (L.) in Ethiopian water bodies: A review, *Int. J. Fish. Aquat. Stud.*, 2018, **6**, 43–47.
211. M. Temesgen, A. Getahun, B. Lemma and G. P. Janssens, Food and feeding biology of Nile tilapia (*Oreochromis niloticus*) in Lake Langeno, Ethiopia, *Sustainability*, 2022, **14**, 974.
212. S. C. Debnath, J. McMurtrie, B. Temperton, J. Delamare-Deboutteville, C. V. Mohan and C. R. Tyler, Tilapia aquaculture, emerging diseases, and the roles of the skin microbiomes in health and disease, *Aquacult. Int.*, 2023, **31**, 2945–2976.
213. J. D. Bowker, D. Carty and M. P. Bowman, The safety of SLICE (0.2% emamectin benzoate) administered in feed to fingerling rainbow trout, *North. Am. J. Aquacult.*, 2013, **75**, 455–462.
214. S. D. Melvin, M. A. Petit, M. C. Duvignacq and J. P. Sumpter, Towards improved behavioural testing in aquatic toxicology: Acclimation and observation times are important factors when designing behavioural tests with fish, *Chemosphere*, 2017, **180**, 430–436.
215. S. Chellappa, M. E. Yamanoto, M. S. R. F. Cacho and F. A. Huntingford, Prior residence, body size and the dynamics of territorial disputes between male freshwater angelfish, *J. Fish Biol.*, 1999, **55**, 1163–1170.
216. A. J. Harwood, S. W. Griffiths, N. B. Metcalfe and J. D. Armstrong, The relative influence of prior residency and dominance on the early feeding behaviour of juvenile Atlantic salmon, *Anim. Behav.*, 2003, **65**, 1141–1149.
217. M. Jones, M. E. Alexander, S. Lightbody, D. Snellgrove, P. Smith, S. Bramhall and K. A. Sloman, Influence of social enrichment on transport stress in fish: a behavioural approach, *Appl. Anim. Behav. Sci.*, 2023, **262**, 105920.
218. C. S. Hall, Emotional behavior in the rat. I. Defecation and urination as measures of individual differences in emotionality, *J. Comp. Psychol.*, 1934, **18**, 385–403.
219. I. Olivas and A. Villagrá, Assessing fear in rabbits. Preliminary results, in *2012 IX International Livestock Environment Symposium (ILES IX)*, American Society of Agricultural and Biological Engineers, 2012, p. 3.
220. A. Haigh, J. Y. Chou and K. O'Driscoll, Variations in the behavior of pigs during an open field and novel object test, *Front. Vet. Sci.*, 2020, **7**, 607.
221. R. Choudhury, A. Middelkoop, J. E. Bolhuis and M. Kleerebezem, Exploring the association between microbiota and behaviour in suckling piglets, *Sci. Rep.*, 2022, **12**, 12322.
222. A. P. C. Araújo and G. Malafaia, Can short exposure to polyethylene microplastics change tadpoles' behavior? A study conducted with

Neotropical tadpole species belonging to order anura (*Physalaemus cuvieri*), *J. Hazard. Mater.*, 2020, **391**, 122214.
223. S. L. Spruiell Eldridge, J. F. Teetsel, R. A. Torres, C. H. Ulrich, V. V. Shah, D. Singh and A. K. Sater, A focal impact model of traumatic brain injury in *Xenopus* tadpoles reveals behavioral alterations, neuroinflammation, and an astroglial response, *Int. J. Mol. Sci.*, 2022, **23**, 7578.
224. S. Castellano, L. Falbo, D. Seglie and O. Friard, Anti-predator behavior in two brown frogs: differences in the mean behaviors and in the structure of animal personality variation, *Behav. Ecol. Sociobiol.*, 2023, **77**, 98.
225. C. Mesak, D. M. dos Reis Sampaio, R. de Oliveira Ferreira, B. de Oliveira Mendes, A. S. de Lima Rodrigues and G. Malafaia, The effects of predicted environmentally relevant concentrations of ZnO nanoparticles on the behavior of *Gallus gallus domesticus* (Phasianidae) chicks, *Environ. Pollut.*, 2018, **242**, 1274–1282.
226. D. M. R. Sampaio, F. N. Estrela, B. de Oliveira Mendes, D. da Costa Estrela, M. F. Montalvão, C. Mesak and G. Malafaia, Ingestion of tannery effluent as a risk factor to the health of birds: A toxicological study using *Coturnix coturnix japonica* as a model system, *Sci. Total Environ.*, 2019, **681**, 275–291.
227. X. Zhang, X. Wang, W. Wang, R. Xu, C. Li and F. Zhang, Effects of personality traits on the food-scratching behaviour and food intake of Japanese quail (*Coturnix japonica*), *Animals*, 2021, **11**, 3423.
228. I. Tiemann, S. Becker, W. Büscher and V. Meuser, Exploring animal genetic resources of the domestic chicken and their behavior in the open field, *J. Appl. Poult. Res.*, 2022, **31**, 100237.
229. C. Li, X. Zhang, L. Cheng, B. Zhang and F. Zhang, Food patch use of Japanese quail (*Coturnix japonica*) varies with personality traits, *Front. Zool.*, 2023, **20**, 30.
230. H. C. Morlock, Behavior following ablation of the dorsal cortex of turtles, *Brain Behav. Evol.*, 1972, **5**, 256–263.
231. S. Allard, G. Fuller, L. Torgerson-White, M. D. Starking and T. Yoder-Nowak, Personality in zoo-hatched Blanding's turtles affects behavior and survival after reintroduction into the wild, *Front. Psychol.*, 2019, **10**, 2324.
232. S. Miller, A. Derenne, S. Ellis-Felege and T. Rhen, Incubation temperature and satiety influence general locomotor and exploratory behaviors in the common snapping turtle (*Chelydra serpentina*), *Physiol. Behav.*, 2020, **220**, 112875.
233. K. R. Scatterty, T. Pitman, T. Eckersley, R. Schmaltz and T. J. Hamilton, Zebrafish aversion to infrasound in an open field test, *Front. Behav. Neurosci.*, 2023, **16**, 493.
234. K. Lu, X. Jia, J. Wu, Q. Wang and X. F. Liang, Neuropeptide Y receptor Y2 (npy2r) deficiency reduces anxiety and increases food intake in Japanese medaka (*Oryzias latipes*), *Front. Cell Dev. Biol.*, 2023, **11**, 1273006.
235. A. Corral-Lopez, A. Kotrschal, A. Szorkovszky, M. Garate-Olaizola, J. Herbert-Read, W. van der Bijl and N. Kolm, Evolution of schooling drives

changes in neuroanatomy and motion characteristics across predation contexts in guppies, *Nat. Commun.*, 2023, **14**, 6027.
236. J. Crawley and F. K. Goodwin, Preliminary report of a simple animal behavior model for the anxiolytic effects of benzodiazepines, *Pharmacol., Biochem. Behav.*, 1980, **13**, 167–170.
237. C. Maximino, T. Marques de Brito, C. A. G. D. M. Dias, A. Gouveia Jr and S. Morato, Scototaxis as anxiety-like behavior in fish, *Nat. Protoc.*, 2010, **5**, 209–216.
238. H. Maeda, N. Fukushima and A. Hasumi, Standardized method for the assessment of behavioral responses of zebrafish larvae, *Biomedicines*, 2021, **9**, 884.
239. A. C. Haigis, R. Ottermanns, A. Schiwy, H. Hollert and J. Legradi, Getting more out of the zebrafish light dark transition test, *Chemosphere*, 2022, **295**, 133863.
240. L. M. Lovin, K. R. Scarlett, A. N. Henke, J. L. Sims and B. W. Brooks, Experimental arena size alters larval zebrafish photolocomotor behaviors and influences bioactivity responses to a model neurostimulant, *Environ. Int.*, 2023, **177**, 107995.
241. F. Bano, P. Gupta, P. Agarwal and M. Serajuddin, Black/White preference and novel tank test to evaluate anxiety in adult goldfish, *Carassius auratus*, *Trends Biosci.*, 2018, **11**, 2493–2498.
242. S. Ansai, H. Hosokawa, S. Maegawa and M. Kinoshita, Chronic fluoxetine treatment induces anxiolytic responses and altered social behaviors in medaka, *Oryzias latipes*, *Behav. Brain Res.*, 2016, **303**, 126–136.
243. K. Volkova, N. R. Caspillo, T. Porseryd, S. Hallgren, P. Dinnetz, H. Olsén and I. P. Hällström, Transgenerational effects of 17α-ethinyl estradiol on anxiety behavior in the guppy, Poecilia reticulate, *Gen. Comp. Endocrinol.*, 2015, **223**, 66–72.
244. I. Domingues, R. Oliveira, A. M. V. M. Soares and M. J. B. Amorim, Effects of ivermectin on *Danio rerio*: a multiple endpoint approach: behaviour, weight and subcellular markers, *Ecotoxicology*, 2016, **25**, 491–499.
245. R. Gerlai, Y. Fernandes and T. Pereira, Zebrafish (*Danio rerio*) responds to the animated image of a predator: towards the development of an automated aversive task, *Behav. Brain Res.*, 2019, **201**, 318–324.
246. R. M. Luca and R. Gerlai, In search of optimal fear inducing stimuli: differential behavioral responses to computer animated images in zebrafish, *Behav. Brain Res.*, 2012, **226**, 66–76.
247. T. Scerbina and R. Gerlai, Acute administration of ethanol and of a D1-receptor antagonist affects the behavior and neurochemistry of adult zebrafish, *Biomedicines*, 2022, **10**, 2878.
248. L. J. G. Barcellos, F. Ritter, L. C. Kreutz, R. M. Quevedo, L. B. da Silva, A. C. Bedin and L. Cericato, Whole-body cortisol increases after direct and visual contact with a predator in zebrafish, *Danio rerio*, *Aquaculture*, 2007, **272**, 774–778.
249. N. Speedie and R. Gerlai, Alarm substance induced behavioral responses in zebrafish (*Danio rerio*), *Behav. Brain Res.*, 2008, **188**, 168–177.

250. M. I. Johnson, K. A. Hanson, J. M. Simon, A. Shastri and B. D. Wisenden, Ring a bell? Variable recognition of a multicomponent auditory stimulus associated with predation risk by zebrafish responding to full and partial matches, *Behaviour*, 2023, **160**, 409–426.
251. R. Apfelbach, C. D. Blanchard, R. J. Blanchard, R. A. Hayes and I. S. McGregor, The effects of predator odors in mammalian prey species: a review of field and laboratory studies, *Neurosci. Biobehav. Rev.*, 2005, **29**, 1123–1144.
252. C. A. Hendrie, S. M. Weiss and D. Eilam, Exploration and predation models of anxiety: evidence from laboratory and wild species, *Pharmacol., Biochem. Behav.*, 1996, **54**, 13–20.
253. M. O. Parker, Adult vertebrate behavioural aquatic toxicology: reliability and validity, *Aquat. Toxicol.*, 2016, **170**, 323–329.
254. S. D. Melvin, M. A. Petit, M. C. Duvignacq and J. P. Sumpter, Towards improved behavioural testing in aquatic toxicology: Acclimation and observation times are important factors when designing behavioural tests with fish, *Chemosphere*, 2017, **180**, 430–436.
255. L. Rejo, J. M. Malgouyres, E. Bonnafé and C. Vignet, Optimization and calibration of behavioural tests on different species of planaria for ecotoxicological studies, *Environ. Toxicol. Pharmacol.*, 2023, **101**, 104189.

CHAPTER 10

Commonly Caught Fish Species as Potential Indicators of Pollution Along Lake Nyasa (Lake Malawi), Tanzania

ALINANUSWE J. MWAKALESI[a*] AND
ALEX N. NEHEMIA[b]

[a]Department of Chemistry and Physics, College of Applied and Natural Sciences, Sokoine University of Agriculture, P.O. Box 3038, Tanzania;
[b]Department of Bioscience, College of Applied and Natural Sciences, Sokoine University of Agriculture, P.O. Box 3038, Tanzania
*E-mail: mwakalesi@sua.ac.tz

10.1 Introduction

Lakes are estimated to contain about 90% of all surface freshwater found on the planet and are regarded as important freshwater reservoirs on the Earth.[1] Thus, they provide support for a wide range of aquatic ecosystems, human livelihoods and economic development. However, lakes are recipients of numerous pollutants produced by a variety of human activities due to the rapid economic development of society. Among the water bodies, lakes show magnified pollution levels due to their availability for a wide range of uses. They usually reflect the practical levels of pollution taking place in the

Issues in Toxicology No. 50
Fish Species in Environmental Risk Assessment Strategies
Edited by Guillermo Eli Liwszyc and Marcelo L. Larramendy

Published by the Royal Society of Chemistry, www.rsc.org

surrounding areas. Consequently, they have been the subject of research on the effects caused by anthropogenic activities.

Lake Nyasa is the third largest and most southern of the lakes in East Africa's Eastern Rift Valley. It is also the world's ninth largest and third deepest freshwater lake. The lake has a length of between 560 and 580 km, a maximum width of 75 km, an average depth of 292 m, a maximum depth of 706 m and an approximate average volume of 8000 km^3. Tanzania's Lake Nyasa catchment is estimated to be 27 500 km long.[2] Tanzania, Malawi and Mozambique share the Lake with different names, *i.e.*, Nyasa in Tanzania, Malawi in Malawi and Niassa in Mozambique. The people from these three countries benefit from the ecosystem services such as tourism, agriculture, fishing and wildlife sectors provided by the lake.[3]

This lake is reported to contain the largest number of freshwater fish species in the world, which is estimated to be between 800–1000.[4] Consequently, UNESCO declared the lake to be among the World Heritage Sites in 1984.[4] Fish species found in Lake Nyasa include, among others, the lake salmon or mpasa *Opsaridium microlepis* (Cypriniformes, Cyprinidae), the ray-finned fish or sanjika *Opsaridium microcephalum* (Cypriniformes, Cyprinidae), the dwarf sanjika *O. tweddleorum* (Cypriniformes, Cyprinidae), the ningwi or redeye labeo *Labeo cylindricus* (Cypriniformes, Cyprinidae), the African sharptooth catfish *Clarias gariepinus* (Siluriformes, Clariidae), the somo *Marcuseniuis livingstonii* (Osteoglossiformes, Mormyridae) and the tiny lampeye killifish *Micropanchax johnstonnii* as well as other cichlid fish species, most of which are threatened by environmental degradation and other human activities.[3] The majority of fish species in the lake are listed by the International Union for Conservation of Nature (IUCN) in the Red List of Threatened Species[5,6] as either near threatened, vulnerable, endangered, critically endangered or extinct, presumably as a result of environmental degradation caused by human activities that significantly contribute to pollution of the fish habitats.[7–10] The effects of pollution can pass through the food chain and produce adverse health effects in various organisms including humans. Therefore, it is critical to monitor the possible threats brought on by pollution using fish species that are found in the area. This chapter looks into the potential effects of water pollution on fish species using genetic indices as well as the bioaccumulation of heavy metals in species that are frequently fished around Lake Nyasa. The estimates of potential risk caused by the pollution to human health are also described.

10.2 Common Sources of Water Pollution Along Lake Nyasa

10.2.1 Coal Mining

One of the economic activities carried out near the Kiwira and Ruhuhu rivers is coal mining. A significant amount of water flow is contributed by these rivers to Lake Nyasa. Coal mining activities are mainly conducted at Kiwira

Figure 10.1 Coal mining activities taking place near Kiwira river that empties its water into Lake Nyasa (March, 2024).

(see Figure 10.1) and Ngaka, located in the regions of Mbeya and Ruvuma, respectively. The movement of waste from the mines to the lakes by water can contaminate the water in the lake. Acid drainage can also result from coal mining when sulfur-containing rocks are exposed to oxygen and water, which forms sulfuric acid.[11] The slurry produced during the washing of coal can leak into water and cause sedimentation of chemical precipitates. Coal mining has been a source of heavy metal pollution that causes loss of aquatic ecosystems. Fish are one of the aquatic species on which coal mining can have a major impact on because solid coal particles stick to their gill lamellae, changing the permeability of oxygen and breathing processes.[12] The effects of reduced fish growth due to the contamination caused by coal mining are known.[13,14] The adverse effects of heavy metal pollution on the success of reproduction are expected to alter the rate of population growth.

10.2.2 Agriculture

Most people who live near Lake Nyasa engage in farming as their livelihood. Rice is among the crops that are highly grown in the area (see Figure 10.2). Most farmers are now obliged to farm in marginal areas that were previously uncultivated, like wetlands and steep hill slopes, as a result of the recent population growth in rural areas. Agricultural activities performed in the area are normally associated with excessive applications of chemicals such as pesticides and fertilizers that may end up in the lake. The soil naturally contains different types of chemicals such carbon (C), nitrogen (N), phosphorus (P) and silicon (Si) that can accumulate in the lake and cause negative effects on aquatic organisms. The chemicals may adversely affect the water quality and ecosystem of the lake. Additionally, a large portion of the land is exposed to several types of erosion due to agriculture. Eroding soil particles are carried

Figure 10.2 Area planted with rice along the Lake Nyasa catchments, Tanzania (March, 2024).

into the lake by this process as well. Previous studies have indicated that the input of sediments and nutrients depends on the size of agricultural land in the sub-basin.[15] The higher amounts of sediments carried by tributaries from dense catchments stem from areas with extensive agricultural activity.

10.2.3 Burning of Trees

Burning of trees is another human activity that may significantly affect the fish industry of Lake Nyasa documented to contribute substantially to the input of nutrients like N and P from atmospheric deposition.[16] The burning of forests in the area surrounding Lake Nyasa has been taking place for numerous reasons, including the preparation of land for cultivation, opening woodland for agriculture, promotion of early growth of grasses for grazing animals, traditions, hunting and accidental fires. Consequently, the increased inputs of nutrients due to chemical deposition are expected. Materials produced after burning can be transported into the lake, raising concentrations of elements like P, Ca, Mg and K. Increased amounts of P, N and C for burned areas compared to unburned areas have been reported.[17] The atmospheric input of nutrients is likely to significantly impact fish production because the activity is performed by most people living in rural areas.

10.2.4 Deforestation or Clear-cutting of Trees

Deforestation is among the activities that significantly contribute to the deterioration of the fishery industry in Lake Nyasa. The activity exposes a considerable part of the land to erosion that deposits materials in the lake, leading to the input of chemicals. Deforestation results from a significant reliance on forest resources as a cheap source of fuel and building materials. The rapid increase in population densities, especially in rural areas, also catalyzes forest loss. This increases the area of land cleared for agricultural activity. Disturbances in forests, particularly those inside water catchment areas, modify the hydrological system, leading to a rise in soil erosion, sedimentation and deterioration of water quality.[18] In addition to lowering water quality, it raises temperature and increases runoff pollutants. Additionally, it impacts the surrounding biodiversity and aquatic life, including fish.[19] Thus, water quality and fish production are likely to be highly affected by the activity due to the high dependence of the rural population on the forest for sustenance and income.

10.3 Reduced Genetic Diversity in the Lake Salmon or Mpasa *Opsaridium microlepis* as a Potential Indicator of the Impact of Water Pollution

The catchments along Lake Nyasa are subjected to anthropogenic activities that include agricultural practices, coal mining, burning of trees and deforestation. The activities significantly affect water and sediment quality. Thus, the quality of sediment and water needs to be periodically monitored to reduce or control the impacts of pollution on various organisms found in the lake. The contamination of aquatic environments causes stresses to aquatic organisms such as fish, frogs, insects and plants, as well as the benthos because they depend on the quality of water for their survival.[19] Thus, aquatic organisms' diversity, structure and distribution can change depending on the level of contamination. It has been indicated that fish, among the aquatic organisms, can serve as valid bioindicators of contamination.[20] Fish cannot escape from the effects caused by the pollution of aquatic systems because they entirely depend on water for their survival. Consequently, fish are more highly sensitive to numerous pollutants in an aquatic ecosystem than invertebrates.[21]

Opsaridium microlepis (see Figure 10.3) is an endemic cyprinid fish that spawns in rivers and is among the fish that fishermen commonly catch in Lake Nyasa.[22,23] The *O. microlepis* spawning run takes place during and after rains, peaking in May and June. They spawn in the early morning hours of the day over a gravel substratum, frequently in very shallow water.[24] Once in the lake, the fry remain close to the river mouth for a considerable amount of time until going offshore in their second year, when they reach a length of roughly 150–200 mm.[23,24] Due to a sharp population decline, *O. microlepis* has

Figure 10.3 *Opsaridium microlepis*, an endemic cyprinid fish of Lake Nyasa.

Table 10.1 Chemical parameters of *Opsaridium microlepis* samples collected from different locations along Lake Nyasa, Tanzania.

		Coordinates			Chemical parameters		
Sampling sites	Status	Longitude	Latitude	Sample size (N)	TN (mg L^{-1})	TP (mg L^{-1})	TDS (ppm)
Kafyofo	Relatively undisturbed	33.9520° E	9.6880° S	15	8.87 ± 2.139	0.052 ± 0.026	46 ± 2.52
Buloma	Relatively undisturbed	33.9570° E	9.5880° S	16	1.87 ± 0.404	0.052 ± 0.037	54 ± 3.51
Katumba	Disturbed	34.0353° E	9.4928° S	16	18.67 ± 1.762	0.072 ± 0.004	45 ± 1.53
Lituhi	Disturbed	34.5817° E	10.4555° S	15	19.60 ± 1.400	0.106 ± 0.019	122 ± 2.00
Ruhuhu	Relatively undisturbed	34.5761° E	10.5392° S	19	5.13 ± 0.404	0.021 ± 0.013	40 ± 1.00
Mbambabay	Disturbed	34.7897° E	11.3111° S	18	13.53 ± 1.762	0.112 ± 0.018	261 ± 0.58

been red-listed as a vulnerable species by the IUCN.[10] Due to rising overfishing, ongoing habitat degradation and habitat loss caused by siltation from soil erosion, water abstraction and farming operations, it is predicted that the species' abundance will drop by 30% in ten years.[10] *Opsaridium microlepis*' sensitivity to human activity may suggest that this species has the potential to be utilized as an indicator of pollution in their natural environments.

Opsaridium microlepis collected from disturbed locations displayed elevated amounts of total nitrogen (TN), total phosphorus (TP) and total dissolved salts (TDS) than those found at sites around Lake Nyasa that were comparatively undisturbed (see Table 10.1). The disturbed sites with elevated TN, TP and TDS were consistently found to have low genetic diversity in the form of haplotype diversity (h), nucleotide diversity (π),[25] observed heterozygosity (Ho) and unbiased expected heterozygosity (uHe) (see Figure 10.4). The samples of *O. microlepis* from disturbed sites along Lake Nyasa also

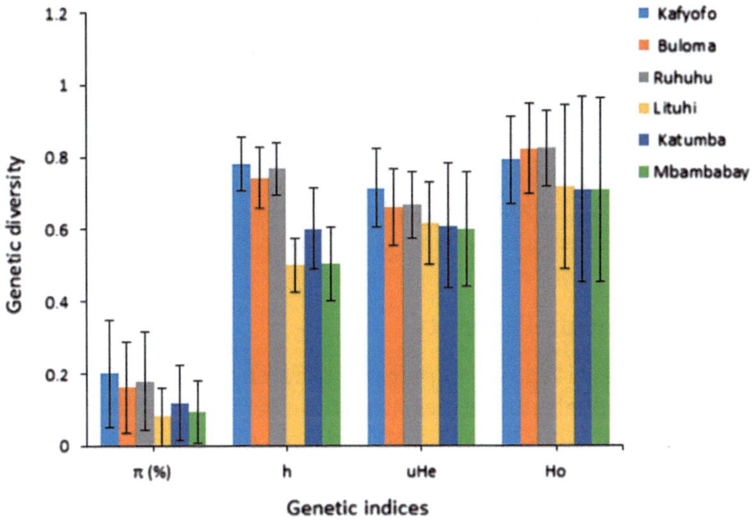

Figure 10.4 Genetic indices of *Opsaridium microlepis* measured in samples collected from disturbed and relatively undisturbed sites along Lake Nyasa, Tanzania. In the figure, π%, *h*, uHe and Ho indicate nucleotide diversity, haplotype diversity, unbiased expected heterozygosity and observed heterozygosity, respectively.

showed a smaller effective population size (0.037667 ± 0.035119) compared to samples taken from relatively undisturbed areas (0.004667 ± 0.003215).[25]

These results corroborate information indicating that this fish species is extremely vulnerable to contamination in the lake. Similar findings have been shown in Tanzania's Lake Victoria, where areas exhibiting high levels of TDS, TP and TN demonstrated low levels of species richness and diversity in the fish community structure.[26] The findings also show that disturbance to fish habitats can have a significant effect on the fish community. Fish species are at risk due to decreased genetic diversity since it limits their capacity to adapt to the environment.[27] One of the biggest challenges to the biodiversity of freshwater fish is habitat fragmentation, which is likely to have genetic effects due to disruptions in migration and population declines.[28] Among the genetic indices affected by water pollution are haplotype diversity, nucleotide diversity and heterozygosity.[29] The indices can serve as potential parameters for indication of the effects associated with pollution of the aquatic environment.

10.3.1 Haplotype Diversity

The likelihood that two randomly selected haplotypes differ from each other is known as haplotype diversity and is controlled by a variety of processes, including mutation, recombination, marker ascertainment and demography.[30] The samples of *O. microlepis* collected along Lake Nyasa indicate consistently low haplotype diversity in disturbed sites compared to relatively undisturbed sites. A similar study on the effect of anthropogenic disturbance on the genetic structure of a native Brazilian Neotropical fish species revealed that the ten spotted livebearer fish *Cnesterodon decemmaculatus* (Cyprinodontifores, Poeciliidae) served as an excellent bioindicator of environmental quality because most of the haplotypes were not able to adapt to stressful conditions.[31] Thus, the haplotype diversity may serve as a potential parameter that indicates the effects of pollution in the aquatic environment.

10.3.2 Nucleotide Diversity

Nucleotide diversity is the mean pairwise difference among all possible pairs of sampled individuals. It is a measure of the degree of polymorphism present within a population. It represents a simple measure of genetic diversity which can accurately be estimated even with very few samples of specimens of a species. Fish with declining population size are expected to have reduced nucleotide diversity.[32] The significant decline in species has been observed to be caused by overexploitation, low productivity and water pollution as well.[33] The samples of *O. microlepis* collected along Lake Nyasa in disturbed sites showed low effective population size and low nucleotide diversity compared to relatively undisturbed sites. In agreement, reduced genetic variations in the Nile tilapia *Oreochromis niloticus* (Cichliformes, Cichlidae) due to water pollution were also reported for samples collected from Riyadh Lake, Saudi Arabia.[34]

10.3.3 Observed Heterozygosity (Ho)

Variation in alleles is essential for a species to survive and enables organisms to adapt to shifting environmental conditions. The proportion of individuals in a population who are heterozygous at a given number of loci is known as the population's observed heterozygosity and it is influenced by both the level of inbreeding and the amount of genetic variation in the population.[33] The Ho was found to be low in samples of *O. microlepis* collected in disturbed sites along Lake Nyasa compared to that from relatively undisturbed sites. Reduction in Ho attributed to enormous pollution due to industrial activities, sand mining and the release of agricultural runoffs has been reported.[35]

10.3.4 Expected Heterozygosity

It is the probability that an individual will be heterozygous at a given locus and is estimated from allele frequencies.[36] Natural populations subjected to random environmental stress can drastically shrink in terms of the populations' size and result in the loss of a random number of alleles, which lowers genetic heterozygosity and variability: a process known as genetic erosion. Additionally, it may have a selective effect, resulting in varying rates of mortality for different genotypes, reducing the frequency of the less resistant ones, then altering the genotypic frequencies in the populations in question.[37] Thus, the expected heterozygosity can be one of the parameters indicating the potential effect of pollution in the aquatic ecosystem.

10.4 Heavy Metal Contamination and Health Risk Assessment

Not only can aquatic contamination have a negative genetic impact on fish populations, but pollutants like heavy metals can also accumulate in the fish bodies and cause adverse effects through the food chain. The accumulation of heavy metals in tissues and organs of various fish species depends on eating behaviour, time of exposure and age. Fish are among organisms that depend entirely on the health of the aquatic system for their survival. As a result, they are more sensitive to pollution of aquatic systems compared to other aquatic organisms. The accumulation of heavy metals in fish can be used to assess the pollution of the aquatic systems. Fish have therefore been employed as a potential indicator of the risks associated with the accumulation of heavy metals in the aquatic environment.[38] The majority of fish species' samples gathered around Lake Nyasa landing locations to evaluate the effects of heavy metal pollution showed that the concentrations of the heavy metals Fe, Zn, Cu and Ni in the gills were comparatively greater than in their muscles (see Table 10.2). These higher concentrations of the heavy metals in gills than muscles are likely to be caused by the higher exposure of gills to the contaminated water than muscles.[39] Due to their frequent contact with the outside world and primary site of heavy metal absorption, gills are the initial target of waterborne pollutants and are highly susceptible to heavy metal accumulation.[40] It has been revealed that the concentration of

Table 10.2 Heavy metals accumulated in gills and muscles (mg kg^{-1}) of commonly caught fish species from Lake Nyasa. Reproduced from ref. 53 with permission from Springer Nature, Copyright 2024[a]

Fish species	Tissue	Fe	Zn	Cu	Ni	Cr
Clarias gariepinus	Gills	19.95 ± 11.53	1.11 ± 0.49	5.59 ± 1.49	2.04 ± 0.36	1.33 ± 0.63
	Muscle	9.83 ± 2.25	8.01 ± 2.42	1.19 ± 0.34	1.94 ± 0.16	1.26 ± 0.57
	Gills	9.66 ± 2.03	8.95 ± 3.16	3.19 ± 1.08	1.77 ± 0.40	BDL
	Muscle	7.24 ± 0.47	0.92 ± 0.63	7.24 ± 0.47	1.27 ± 0.42	BDL
	Gills	9.66 ± 1.52	8.41 ± 0.55	0.67 ± 0.19	1.70 ± 0.40	1.36 ± 0.35
	Muscle	6.51 ± 1.01	6.78 ± 0.97	0.59 ± 0.23	6.51 ± 1.01	1.42 ± 0.18
Raiamas moorii	Gills	1.82 ± 0.38	10.91 ± 2.34	3.76 ± 1.29	2.62 ± 0.97	BDL
	Muscle	7.94 ± 2.69	6.67 ± 2.24	2.02 ± 0.67	1.64 ± 0.63	BDL
Oreochromis niloticus	Gills	13.32 ± 2.82	8.86 ± 2.16	3.37 ± 1.63	2.6 ± 1.01	BDL
	Muscle	7.68 ± 0.53	6.92 ± 1.81	3.05 ± 2.38	1.52 ± 0.52	BDL

[a]BDL: Below detectable limit.

Cd accumulated in the gills of the goldfish *Carassius auratus* (Cypriniformes, Cyprinidae) increased with an enhancement in the concentration of particulate Cd.[40] This indicates that the higher concentration in gills was likely to be due to the accumulation of particulate matter. The accumulation of the particles is caused by a mucus layer in the gills that contains a polyanionic matrix with different affinities for metal ions.[41]

The metals are known to accumulate in the fish muscles after their amounts exceed the body's defence barriers.[42] Thus, the heavy metals accumulated in the fish gills are likely to be a possible good indicator of heavy metal pollution in aquatic environments. However, the measurement of the heavy metal concentrations in the fish muscles is important as well because it is the part of the fish that is highly consumed by humans. Therefore, it is also a good tool for the estimation of human exposure to pollutants from aquatic environments.[43] Consequently, most studies involving the use of fish as bioindicators for risk assessments involve the use of fish muscles.[43,44] The obtained concentrations of the heavy metals can be compared with the standard concentration provided by health organizations like the World Health Organization (WHO).[45] The concentrations of the heavy metals from the common fish species caught by fishermen in Lake Nyasa were below the thresholds provided by WHO. Thus, it is an indication that risks associated with the consumption of the fish from Lake Nyasa are at minimal levels.

10.4.1 Estimated Daily Intake of Heavy Metals

Estimated daily intake of heavy metals (EDI) is a possible indicator used for the assessment of risks caused by under or over-consumption of a particular nutrient. It is also used by health professionals to provide nutritional recommendations. The values of EDI are calculated using eqn (10.1) and normally compared with the recommended dose (RD) using the EDI–RD ratio.[46] Thus, if the value of EDI is less or equal to RD, the potential risk is minimal; however, values greater than RD are a likely indicator that the risk associated with the consumption of a particular metal is high. The estimated daily intake is calculated according to eqn (10.1) as follows:[47]

$$\mathrm{EDI} = C \times \frac{\mathrm{FIR}}{\mathrm{BW}} \tag{10.1}$$

where C is the average concentration of heavy metals in fish tissues (mg kg^{-1}) and FIR is the food ingestion rate or the daily average consumption of fish (g day^{-1}) per capita for the Tanzania population (7.6 g day^{-1}) and BW is the average Tanzanian body weight (75 kg).[48] Children's EDI was computed by taking half of an adult's daily consumption and one-third of an adult's body weight.[49]

The findings (see Tables 10.3 and 10.4) for the common fish species caught by fishermen in Lake Nyasa indicated that the consumption of the fish species was less likely to result in negative health effects. However, the consumption

Table 10.3 Risk assessment for adult Tanzanian population using estimated daily intake (EDI) of metals from gills and muscle of commonly caught fish by fishermen from Lake Nyasa. Adapted from ref. 53 with permission from Springer Nature, Copyright 2024.

Species	Tissue	Fe	Zn	Cu	Ni	Cr
Clarias gariepinus	Gills	2.0216	0.11248	0.566453	0.20672	0.134773
	Muscle	0.996107	0.81168	0.120587	0.196587	0.12768
Bagrus meridionalis	Gills	0.97888	0.906933	0.323253	0.17936	
	Muscle	0.733653	0.093227	0.733653	0.128693	
Opsaridium microlepis	Gills	0.97888	0.852213	0.067893	0.172267	0.137813
	Muscle	0.65968	0.68704	0.059787	0.65968	0.143893
Raiamas moorii	Gills	0.184427	1.105547	0.381013	0.265493	
	Muscle	0.804587	0.675893	0.204693	0.166187	
Oreochromis niloticus	Gills	1.34976	0.897813	0.341493	0.263467	
	Muscle	0.77824	0.701227	0.309067	0.154027	

Table 10.4 Risk assessment for Tanzanian children population using the estimated daily intake (EDI) of metals from gills and muscle of commonly caught fish by fishermen from Lake Nyasa. Adapted from ref. 53 with permission from Springer Nature, Copyright 2024.

Species	Tissue	Fe	Zn	Cu	Ni	Cr
Clarias gariepinus	Gills	3.0324	0.16872	0.84968	0.31008	0.20216
	Muscle	1.49416	1.21752	0.18088	0.29488	0.19152
Bagrus meridionalis	Gills	1.46832	1.3604	0.48488	0.26904	
	Muscle	1.10048	0.13984	1.10048	0.19304	
Opsaridium microlepis	Gills	1.46832	1.27832	0.10184	0.2584	0.20672
	Muscle	0.98952	1.03056	0.08968	0.98952	0.21584
Raiamas moorii	Gills	0.27664	1.65832	0.57152	0.39824	
	Muscle	1.20688	1.01384	0.30704	0.24928	
Oreochromis niloticus	Gills	2.02464	1.34672	0.51224	0.3952	
	Muscle	1.16736	1.05184	0.4636	0.23104	

of some fish species such as the African sharptooth catfish *C. gariepinus*, the Lake Rukwa minnow *Raiamas moorii* (Cypriniformes, Cyprinidae) and the Nile tilapia *O. niloticus* is likely to result in negative health effects in both adults and children. The health effects are largely linked to the consumption of some elements such as Fe and Zn. However, these elements are known to be required by the body for metabolic reactions but excessive amounts can cause negative effects. As an example, Fe is known to be essential in the prevention of anaemia and thus is required in sufficient amounts. However, the deficiency of the element is a common problem in women and children.[50,51] On the other hand, an excessive amount of this metal in the human body is normally associated with heart disease and type 2 diabetes.[52] A recently conducted research also revealed that excess Fe can lead to chronic liver diseases.[38]

10.5 Conclusion

The chapter presented the potential use of frequently caught fish species from Lake Nyasa as bioindicators of pollution that is likely to be a result of human activities. The effects of pollution can be magnified if proper and periodic assessments of risks associated with the pollution are not properly done. The sampling and analyses of pollutants accumulated in the sediments is always difficult. However, pollution induces stresses that can be detected through the molecular analysis of the fish. Thus, parameters such as genetic diversity, haplotype diversity (h), nucleotide diversity (π), observed heterozygosity (Ho) and unbiased expected heterozygosity can be used as potential indicators of the effects of pollution in aquatic systems.

It is also possible for a pollutant contained in the aquatic systems to accumulate in the body tissues of the fish. The concentration of persistent and non-biodegradable heavy metals in fish increases with eating behaviour, age or rate of exposure. Thus, the concentration of heavy metals may vary from one fish species to another depending on the factors. The heavy metals accumulated in the body of the fish can be used to estimate potential risks that can be caused by pollution to human health.

The toxicity of a pollutant highly depends on the chemical form in which it is present in the aqueous environment. For example, organic mercury is more toxic compared to inorganic mercury due to its ability to penetrate through various blood barriers and cause neurological damage. The chemical characteristics of a toxicant vary from one another. Thus, to estimate potential risks associated with a certain pollutant, the analysis of a specific chemical species is required. The use of fish as a potential bioindicator can be significantly more relevant, if these aspects are borne in mind during the chemical speciation analysis.

Acknowledgements

We express our sincere gratitude to the Sokoine University of Agriculture Research and Innovation Support (SUARIS) (Project Name: SUARIS1, Funding Number: DCOP06) for the financial support that made this work possible. Without their invaluable support, this project could not have been realized. We distinctively acknowledge SUARIS for their noteworthy support.

References

1. L. Fu, Y. Shen, H. Zhang and X. Wang, Effects of water diversion projects on water environment in Chaohu Lake, *Limnologica*, 2022, **97**, 126026.
2. Y. Yihdego and A. Paffard, Hydro-engineering solution for a sustainable groundwater management at a cross border region: Case of Lake Nyasa/Malawi basin, Tanzania, *Int. J. Geotech. Eng.*, 2016, 7, 23.
3. S. J. Nindi, Changing livelihoods and the environment along Lake Nyasa, Tanzania, *Afr. Study Monogr., Suppl.*, 2007, **36**, 71–93.

4. G. M. S. Chavula, M. S. M'Balaka, E. Gondwe, M. Ngochera, J. S. Halafo, A. H. Shechonge, N. P. Gabagambi, A. S. Mbonde, D. Banda, N. Muyumbana, T. J. Lawrence, M. Z. Ajode, S. Smith, C. Jonasse, G. Z. Kanyerere and B. P. Ngatunga, Lake Malawi/Niassa/Nyasa basin: Status, challenges, and research needs, *J. Great Lakes Res.*, 2023, **49**, 102241.
5. J. Day, N. Dudley, M. Hockings, G. Holmes, D. d. A. Laffoley, S. Stolton and S. M. Wells, *Guidelines for Applying the IUCN Protected Area Management Categories to Marine Protected Areas*, International Union for Conservation of Nature, Switzerland, 2012.
6. C. Sayer, A. Palmer-Newton and W. Darwall, *Conservation Priorities for Freshwater Biodiversity in the Lake Malawi/Nyasa/Niassa Catchment*, International Union for Conservation of Nature, Switzerland, 2019.
7. E. C. Torell, D. M. Jamu, G. Z. Kanyerere, L. Chiwaula, J. Nagoli, P. Kambewa, A. Brooks and P. Freeman, Assessing the economic impacts of post-harvest fisheries losses in Malawi, *World Dev. Perspect.*, 2020, **19**, 100224.
8. D. Jamu, M. Banda, F. Njaya and R. E. Hecky, Challenges to sustainable management of the lakes of Malawi, *J. Great Lakes Res.*, 2011, **37**, 3–14.
9. M. S. M'balaka, E. Kaunda, G. Kanyerere, D. Jamu and A. Msukwa, Changes in fish species diversity, size structure and distribution in the trawlable demersal zones of Lake Malawi, Malawi, *Environ. Dev. Sustainability*, 2024, DOI: 10.1007/s10668-024-04649-8.
10. D. Kassam, W. Changadeya, H. Stima, W. Jere and E. Kaunda, Morphological and genetic variability among Mpasa (*Opsaridium microlepis* Günther, 1864) populations from the inflow rivers of Lake Malawi, *Int. J. Fish. Aquacult.*, 2017, **9**, 52–58.
11. B. S. Acharya and G. Kharel, Acid mine drainage from coal mining in the United States – An overview, *J. Hydrol.*, 2020, **588**, 125061.
12. G. M. Hughes, Respiratory responses to hypoxia in fish, *Am. Zool.*, 2015, **13**, 475–489.
13. K. L. E. Berry, *Effects of Coal Contamination on Tropical Marine Organisms*, James Cook University, 2017.
14. M. O. Tretyakova, A. I. Vardavas, C. I. Vardavas, E. I. Iatrou, P. D. Stivaktakis, T. I. Burykina, Y. O. Mezhuev, A. M. Tsatsakis and K. S. Golokhvast, Effects of coal microparticles on marine organisms: A review, *Toxicol. Rep.*, 2021, **8**, 1207–1219.
15. R. E. Hecky, H. A. Bootsma and M. L. Kingdon, Impact of land use on sediment and nutrient yields to Lake Malawi/Nyasa (Africa), *J. Great Lakes Res.*, 2003, **29**, 139–158.
16. M. N. Waters, A. P. Metz, J. M. Smoak and H. Turner, Chronic prescribed burning alters nutrient deposition and sediment stoichiometry in a lake ecosystem, *Ambio*, 2019, **48**, 672–682.
17. Z. Song, Z. Li, Y. Luo and Y. Liu, Allocation strategies of carbon, nitrogen and phosphorus following a gradient of wildfire severities, *J. Plant Ecol.*, 2021, **15**, 347–358.

18. X. Kong, S. Ghaffar, M. Determann, K. Friese, S. Jomaa, C. Mi, T. Shatwell, K. Rinke and M. Rode, Reservoir water quality deterioration due to deforestation emphasizes the indirect effects of global change, *Water Res.*, 2022, **221**, 118721.
19. L. F. Leon, D. C. Lam, C. McCrimmon and D. A. Swayne, Watershed management modelling in Malawi: application and technology transfer, *Environ. Modell. Software*, 2003, **18**, 531–539.
20. A. Chovanec, R. Hofer and F. Schiemer, Fish as Bioindicators, in *Bioindicators and Biomonitors*, ed. B. A. Markert, A. M. Breure, and H. G. Zechmeister, Elsevier, 2003, vol. 6, pp. 639–676.
21. L. Weltje, P. Simpson, M. Gross, M. Crane and J. R. Wheeler, Comparative acute and chronic sensitivity of fish and amphibians: A critical review of data, *Environ. Toxicol. Chem.*, 2013, **32**, 984–994.
22. H. Sungani, B. Ngatunga, S. Koblmüller, T. Mäkinen, P. Skelton and M. Genner, Multiple colonisations of the Lake Malawi catchment by the genus *Opsaridium* (Teleostei: Cyprinidae), *Mol. Phylogenet. Evol.*, 2017, **107**, 256–265.
23. H. Sungani, B. Ngatunga and M. Genner, Migratory behaviour shapes spatial genetic structure of cyprinid fishes within the Lake Malawi catchment, *Freshwater Biol.*, 2016, **61**, 1062–1074.
24. D. Tweddle, Breeding behaviour of the mpasa, *Opsaridium microlepis* (Gunther) (Pisces: Cyprinidae), in Lake Malawi, *J. Limnol. Soc. South. Afr.*, 1983, **9**, 23–28.
25. A. Nehemia and A. Mwakalesi, Genetic diversity and structure of *Opsaridium microlepis* along Lake Nyasa, Tanzania, *Tanzania J. Agric. Sci.*, 2023, **22**, 278–283.
26. I. Naigaga, H. Kaiser, W. J. Muller, L. Ojok, D. Mbabazi, G. Magezi and E. Muhumuza, Fish as bioindicators in aquatic environmental pollution assessment: A case study in Lake Victoria wetlands, Uganda, *Phys. Chem. Earth*, 2011, **36**, 918–928.
27. N. Petit-Marty, M. Liu, I. Z. Tan, A. Chung, B. Terrasa, B. Guijarro, F. Ordines, S. Ramírez-Amaro, E. Massutí and C. Schunter, Declining population sizes and loss of genetic diversity in commercial fishes: A simple method for a first diagnostic, *Front. Mar. Sci.*, 2022, **9**, 872537.
28. K. L. García-Castro and E. J. Márquez, Temporal-scale assessment of population genetics of the freshwater fish *Prochilodus magdalenae* in an area impacted by construction of a dam, *Hydrobiologia*, 2023, **851**, 1513–1531.
29. D. D. Zhai, W. J. Li, H. Z. Liu, W. X. Cao and X. Gao, Genetic diversity and temporal changes of an endemic cyprinid fish species, *Ancherythroculter nigrocauda*, from the upper reaches of Yangtze River, *Zool. Res.*, 2019, **40**, 427–438.
30. M. P. Stumpf, Haplotype diversity and SNP frequency dependence in the description of genetic variation, *Eur. J. Hum. Genet.*, 2004, **12**, 469–477.
31. G. E. Rautenberg, A. F. Bonifacio, M. B. Chiappero, M. V. Amé and A. C. Hued, Genetic structure of a native Neotropical fish species: New

insights about a South American bioindicator, *Arch. Environ. Contam. Toxicol.*, 2022, **83**, 168–179.

32. D. E. Sadler, P. C. Watts and S. Uusi-Heikkilä, The riddle of how fisheries influence genetic diversity, *Fishes*, 2023, **8**, 510.
33. S. Gu, R. Wang, C. Li, J. Li and Y. Shen, Genetic diversity and population structure of the Chinese lake gudgeon (*Sarcocheilichthys sinensis*) using microsatellite markers, *Aquacult. Fish.*, 2020, **5**, 80–85.
34. S. Mahboob, K. A. Al-Ghanim, F. Al-Misned, H. F. A. Al-Balawi, A. Ashraf and N. M. A. Al-Mulhim, Genetic diversity in tilapia populations in a freshwater reservoir assayed by randomly amplified polymorphic DNA markers, *Saudi J. Biol. Sci.*, 2019, **26**, 363–367.
35. B. Sahoo, G. Das, P. Nandanpawar, N. Priyadarshini, L. Sahoo, P. K. Meher, U. K. Udit, J. K. Sundaray and P. Das, Genetic diversity and genome-scale population structure of wild Indian major carp, *Labeo catla* (Hamilton, 1822), revealed by genotyping-by-sequencing, *Front. Genet.*, 2023, **14**, 1166385.
36. T. L. Schmidt, M.-E. Jasper, A. R. Weeks and A. A. Hoffmann, Unbiased population heterozygosity estimates from genome-wide sequence data, *Methods Ecol. Evol.*, 2021, **12**, 1888–1898.
37. R. Cimmaruta, F. Scialanca, F. Luccioli and G. Nascetti, Genetic diversity and environmental stress in Italian populations of the cyprinodont fish *Aphanius fasciatus*, *Oceanol. Acta*, 2003, **26**, 101–110.
38. Y. Liu, G. Li, F. Lu, Z. Guo, S. Cai and T. Huo, Excess iron intake induced liver injury: The role of gut-liver axis and therapeutic potential, *Biomed. Pharmacother.*, 2023, **168**, 115728.
39. D. H. Evans, The fish gill: Site of action and model for toxic effects of environmental pollutants, *Environ. Health Perspect.*, 1987, **71**, 47–58.
40. S. Tao, C. Liu, R. Dawson, A. Long and F. Xu, Uptake of cadmium adsorbed on particulates by gills of goldfish (*Carassius auratus*), *Ecotoxicol. Environ. Saf.*, 2000, **47**, 306–313.
41. S. Tao, H. Li, C. Liu and K. C. Lam, Fish uptake of inorganic and mucus complexes of lead, *Ecotoxicol. Environ. Saf.*, 2000, **46**, 174–180.
42. M. Kalay, Ö. Ay and M. Canli, Heavy metal concentrations in fish tissues from the Northeast Mediterranean Sea, *Bull. Environ. Contam. Toxicol.*, 1999, **63**, 673–681.
43. G. M. M. Anwarul Hasan, M. A. Satter, A. K. Das and M. Asif, Detection of heavy metals in fish muscles of selected local fish varieties of the Shitalakshya River and probabilistic health risk assessment, *Measurement: Food*, 2022, **8**, 100065.
44. E. M. Younis, A. A. Abdel-Warith, N. A. Al-Asgah, S. A. Elthebite and Md. Mostafizur Rahman, Nutritional value and bioaccumulation of heavy metals in muscle tissues of five commercially important marine fish species from the Red Sea, *Saudi J. Biol. Sci.*, 2021, **28**, 1860–1866.
45. A. Badeenezhad, H. Soleimani, S. Shahsavani, I. Parseh, A. Mohammadpour, O. Azadbakht, P. Javanmardi, H. Faraji and K. Babakrpur Nalosi, Comprehensive health risk analysis of heavy metal pollution using water

quality indices and Monte Carlo simulation in R software, *Sci. Rep.*, 2023, **13**, 15817.

46. L. J. Phillips and J. Moya, Exposure factors resources: contrasting EPA's exposure factors handbook with international sources, *J. Exposure Sci. Environ. Epidemiol.*, 2014, **24**, 233–243.
47. P. K. Maurya, D. S. Malik, K. K. Yadav, A. Kumar, S. Kumar and H. Kamyab, Bioaccumulation and potential sources of heavy metal contamination in fish species in River Ganga basin: Possible human health risks evaluation, *Toxicol. Rep.*, 2019, **6**, 472–481.
48. J. Saria, Assessment of human health hazard due to metal uptake via fish consumption from coastal area of Tanzania, *Ethiop. J. Environ. Stud. Manage.*, 2017, **10**, 555.
49. P. Acharya, D. P. Muduli and M. Das, Assessment of heavy metal accumulation in *Penaeus monodon* and its human health implications, *Mar. Pollut. Bull.*, 2023, **188**, 114632.
50. E. Goodarzi, R. Beiranvand, H. Naemi, I. Darvishi and Z. Khazaei, Prevalence of iron deficiency anemia in Asian female population and human development index (HDI): An ecological study, *Obstet. Gynecol. Sci.*, 2020, **63**, 497–505.
51. S. Sundararajan and H. Rabe, Prevention of iron deficiency anemia in infants and toddlers, *Pediatr. Res.*, 2021, **89**, 63–73.
52. A. V. Harrison, F. R. Lorenzo and D. A. McClain, Iron and the pathophysiology of diabetes, *Annu. Rev. Physiol.*, 2023, **85**, 339–362.
53. B. Mwambene, A. J. Mwakalesi and A. N. Nehemia, Contamination and health risk assessment of heavy metals in selected highly edible fish species from Lake Nyasa, Tanzania, *Chem. Afr.*, 2023, **6**, 2695–2703.

CHAPTER 11

Epilogue and Final Remarks

MARCELO L. LARRAMENDY[a,b*] AND
GUILLERMO ELI LIWSZYC[c]

[a]Cátedra de Citología, Facultad de Ciencias Naturales y Museo, Universidad Nacional de La Plata, Calle 64 No. 3, B1904AMA La Plata, Argentina; [b]Consejo Nacional de Investigaciones Científicas y Técnicas (CONICET), Buenos Aires, Argentina; [c]Department of Pathology, University of Helsinki, Helsinki, Finland
*E-mail: marcelo.larramendy@gmail.com

In recent times, the growing impact of anthropogenic activities generates huge quantities of toxic residues that can have direct or indirect detrimental effects upon the quality of our environment. The increase in their jeopardizing effects can exert short-, medium- and long-term consequences, affecting not only human but also environmental health in general. Furthermore, these disturbances are usually capable, in most of the cases, of reducing the biodiversity of native fauna and flora, which will, in turn, encourage the resistance and emergence of new pests and diseases.

This book, *Fish Species in Environmental Risk Assessment Strategies*, intends to provide an outline, relevant examples and stimulate practical discussions on the use of this group of vertebrates, as fish, within the scientific challenges faced by the Ecotoxicology Risk Assessment (ERA) academic world. This book provides relevant tools which may be of use in the analysis prior to the implementation of decisions leading to actions which will hopefully reduce environmental health risk against environmental factors that may adversely impact upon human and other living species' health or cause ecological imbalances between them and our environment.

Issues in Toxicology No. 50
Fish Species in Environmental Risk Assessment Strategies
Edited by Guillermo Eli Liwszyc and Marcelo L. Larramendy

Published by the Royal Society of Chemistry, www.rsc.org

Our aim was to compile information from a diversity of sources into a single volume. The rationale is to give some real-life examples in order to widen the concept that the use of non-conventional animal models, far from being a scientific conundrum, may instead provide useful avenues to be explored and further developed as a way to provide help and real answers to some of the actual problems the whole planet is dealing with. Concomitantly, these concrete examples extend concepts of hazardous factors to living species worldwide that may stimulate new research ideas and trends in their respective fields.

This volume comprises three major sections, namely (1) *Fish Species as Experimental Models in Environmental Risk Assessment Studies. General aspects*; (2) *Fish Species as Experimental Models in Environmental Risk Assessment Studies*; and (3) *Fish Species as Experimental Models in Environmental Risk Assessment Studies. Epilogue and Final Remarks.*

In the first and general frame section *Fish Species as Experimental Models in Environmental Risk Assessment Studies. General aspects,* readers will find a first broad-spectrum chapter (Chapter 1). It provides background introductory information about the concept and implications of this hot topic in our current world, environmental risk assessment, while highlighting the nature of some environmental stressors and examples of the most versatile and validated methods of analyses. In general terms, the narrative focuses on the worldwide accepted concept that ERA is a scientific practice that identifies and evaluates the likelihood of an activity or substance to threaten the environment in general and/or to living organisms, natural habitats and ecosystems in particular. In other words, the term employed describes the overall process/es and/or method/s allowing the identification of hazards and risk factors that have the potentiality to cause harm (hazard identification) as well as to analyse and evaluate the risk associated with that particular hazard (risk analysis and risk evaluation). It also suggests several, if not all, appropriate ways to eliminate or minimize the hazard, or control the risk when this cannot be eliminated (risk control).

The second frame section, *i.e.*, *Fish Species as Experimental Models in Environmental Risk Assessment Studies*, includes nine chapters (Chapters 2–10) in which these vertebrates are used as biotic matrices.

Actually, fish are aquatic, craniates, gill-bearing animals throughout life whose lack of limbs with digits, if any, are in the shape of fins. In general terms, fish extract oxygen from water using gills or use an accessory breathing organ to breathe the atmospheric oxygen.

Besides, fish have two sets of paired fins, usually one or two (rarely three) dorsal fins, an anal fin and a tail fin, possess jaws, have skin that is usually covered with scales and lay eggs. Included in this definition are the living hagfish, lampreys and cartilaginous and bony fish as well as various extinct related groups. Approximately 96% of living fish species are ray-finned fish, belonging to the class Actinopterygii, with around 99% of those being Teleosts, representing over 50% of all vertebrate species.

Fish are abundant in most bodies of water. They can be found in nearly all aquatic environments, from high mountain streams (*e.g.*, char and gudgeon) to the abyssal and even hadopelagic depths of the deepest oceans (*e.g.*, cusk-eels and snailfish), although no species has yet been documented in the deepest 25% of the oceans.[1,2] With more than 34 800 species described so far, fish exhibit greater species diversity than any other group of vertebrates.[3]

Fish are an important resource for humans worldwide, especially as food. Commercial and subsistence fishermen obtain fish from wild fisheries or farm them in ponds or in cages in the ocean (in aquaculture). Fish play an important role in fighting hunger and malnutrition.[4] Fish are not only a source of proteins and healthy fats, but also a unique source of essential nutrients. The multiple benefits of fatty fish high in long-chain omega-3s, iodine, Vit D and Ca and small fish eaten whole, containing nutrients in the skin and bones, clearly illustrate seafood's irreplaceable nutritional value.[5] In many developing countries, fish are the main or only source of animal protein and are essential for providing micronutrients to vulnerable populations. Fish can also sometimes serve as a solution or palliative to existing health problems. For instance, goiter is found in areas where iodized salt is unavailable, but the consumption of fish and the natural iodine it contains could help reduce these cases.[6] Dietary patterns are also shifting in developed and middle-income countries, and an increasing emphasis on coronary and overall health has led to an increased demand for fish.[7]

Sport fishing, recreational or game fishing in the case of non-protected species, usually ends up with the catch in the plates of these sportsmen. Fish are kept as pets and also provide educational and leisure opportunities as in the case of fish exhibited in public aquaria. It is also worth mentioning that fish have had a role in the culture of mankind through the ages, serving as deities, religious symbols as well as being included in films, books, art and other forms of artistic expression.

Most fish are ectothermic, *i.e.*, cold-blooded organisms, allowing their body temperatures to vary as ambient temperatures change. Although most fish do not regulate body temperature, there some are exceptions.[8–10] The only known bony fishes (infraclass Teleostei) that exhibit endothermy belong to the suborder Scombroidei, *i.e.*, the Istiophoriformes billfishes [*i.e.*, the sailfish *Istiophorus* spp. (Istiophoridae), the marlins (Istiophoridae) and the swordfish *Xiphias gladius* (Xiphiidae)], the tunas (Scombridae) and the butterfly kingfish *Gasterochisma melampus* (Scombridae), a basal species of mackerel as well as the opah, also called moonfish, *Lampris guttatus* (Lampriformes, Lampridae).[10] The latter species was demonstrated to use "whole-body endothermy", generating heat with its swimming muscles to warm its body while countercurrent exchange (as in respiration) minimizes heat loss.[11] It is able to actively hunt prey and swim for long distances due to the ability to warm its entire body, including its heart, which is a trait typically found in only mammals and birds (clearly homeothermic vertebrate groups).[12] Finally, it should also be pinpointed that some Chondrichthyes species of Lamniformes like the porbeagle shark *Lamna nasus* (Lamnidae) and two species of thresher sharks *Alopias* spp. (Alopiidae) have been identified as having a modified circulatory system that acts as a countercurrent heat exchanger, which allows them to retain metabolic heat, thus possessing the capability to heat their bodies significantly above ambient water temperature.[13]

Fish can acoustically communicate with each other. Communication in fish involves the transmission of acoustic signals from one individual of a species to another. The production of sounds as a means of communication among fish is most often used in the context of feeding, aggression or courtship behaviour.[14,15] The sounds emitted can vary depending on the species and stimulus involved. Fish can produce either stridulatory sounds by moving components of the skeletal system, *e.g.*, bone-on-bone interactions, or can produce non-stridulatory sounds by manipulating specialized organs such as the swimbladder.[16] As some examples of the former sounds, it has been reported that the French grunt *Haemulon flavolineatum* (Perciformes, Haemulidae) produces a grunting noise, mostly pronounced when the specimens are in distress situations, by grinding its teeth together.[16] However, in the longsnout seahorse *Hippocampus reidi* (Syngnathiformes, Sygnathidae), two different categories of sounds were recorded, *i.e.*, clicks and growls. The sounds emitted by the *H. reidi* are accomplished by rubbing their coronet bone across the grooved section of their neurocranium.[17] Clicking sounds were found to be primarily produced during courtship and feeding.[18] The frequencies were noted to be on the higher end of the range during spawning periods, when the female and male fishes were less than 15 cm apart. Growl sounds were produced when the *H. reidi* encountered stressful situations, such as handling by researchers, consisting of a series of sound pulses and are emitted simultaneously with body vibrations.[18] On the other hand, as examples of non-stridulatory sounds, some fish species are able to create noise by engaging specialized muscles that contract and cause swimbladder vibrations. Among them, is the Oyster toadfish *Opsanus tau* (Batrachoidiformes, Batrachoididae), which produces loud grunting sounds by contracting muscles located along the sides of its swim bladder, known as sonic muscles.[19] Female and male toadfishes emit short-duration grunts, often as a fright response.[20] In addition to these, male toadfishes produce "boat whistle calls".[18] These calls are longer in duration, lower in frequency and are primarily used to attract mates.[21] The frequencies of the calls depend on the rate at which the sonic muscles contract.[22] Similarly, the red drum *Sciaenops ocellatus* (Acanthuriformes, Sciaenidae) produces drumming sounds by vibrating its swimbladder by the rapid contraction of sonic muscles that surround its dorsal part with a sound-repetition pattern with frequencies ranging from 100 up to >200 Hz.[23] *Sciaenops ocellatus* can also produce different calls depending on the stimuli involved. The sounds created in courtship situations are clearly different from those made during distressing events such as predatorial attacks. Unlike the males of the *S. ocellatus* specimens, the females of this species do not produce sounds and lack sound-producing (sonic) muscles.[23]

Fish reproductive biology has shown that these vertebrates can be oviparous, ovoviparous or viviparous. It is well known that over 97% of all known fish species are oviparous, in other words, meaning that the eggs develop outside the mother's body. Examples of oviparous fish include salmon (Salmoniformes), goldfish (Cypriniformes), cichlids (Cichliformes), tuna (Scombriformes) and eels (Anguilliformes). In the majority of these species, fertilization takes place outside the female's body, with the male and female fish shedding their gametes into the surrounding water. It worth mentioning that a few oviparous fish species practice internal fertilization, with the male using some sort of intromittent organ to deliver sperm into the genital opening of the female, most notably the oviparous sharks, *e.g.*, the horn shark *Heterodontus francisci* (Heterodontiformes, Heterodontidae) and oviparous rays, *e.g.*, the skates (Rajiformes). In these cases, the male possesses a pair of modified pelvic fins known as claspers. The newly hatched young of oviparous fish are called larvae. They are usually poorly formed, carry a large yolk sac for nourishment and are very different in appearance from juvenile and adult specimens. This larval period in oviparous fish is relatively short, usually not longer than several weeks, and larvae rapidly grow and change appearance and structure to become juveniles (metamorphosis). During these transitions, larvae must switch from their yolk sac to feeding on zooplankton prey, a process which typically depends on zooplankton density, which in some cases starve many larvae.

In ovoviviparous fish, the eggs develop inside the mother's body after internal fertilization but receive little or no nourishment directly from the female, depending basically on the yolk. Each embryo develops in its own egg. Examples of ovoviviparous fish include, among others, the guppies or rainbow fish (Cyprinodontiformes, Poeciliidae), the angel sharks (Squatiniformes, Squatinidae) and the lobe-finned fish coelacanths (Coelacanthiformes, Latimeriidae). It should be mentioned that aquarists commonly refer to ovoviviparous and viviparous fish as livebearers.

Finally, some species of fish are viviparous. In such species the female retains the eggs and nourishes the embryos. Typically, viviparous fish have a structure analogous to the placenta seen in mammals connecting the mother's blood supply with that of the embryos. Among this viviparous type of fish can be included, *e.g.*, the surf-perches (Actinopterygii, Embiotocidae), the splitfins (Cyprinodontiformes, Goodeidae) and the lemon shark *Negaprion brevirostris* (Carcharhiniformes, Carcharhinidae). Some viviparous fish exhibit oophagy, in which the developing embryos eat other eggs produced by the mother. This has been observed specially among sharks, such as the shortfin mako, also called bonito shark (Lamniformes, Lamnidae) and the porbeagle or the mackerel shark (Lamniformes, Lamnidae), but it has been documented for a few bony fish as well, *e.g.*, the halfbeak *Nomorhamphus ebrardtii* (Beloniformes, Zenarchopteridae).[24] Intrauterine cannibalism is an even more unusual mode of viviparity, in which the largest embryos eat weaker and smaller siblings. This behaviour is also most commonly found among sharks, such as the grey nurse shark *Carcharias taurus* (Lamniformes, Odontaspididae), but has also been reported for *N. ebrardtii*.[24]

As previously mentioned, many groups of freshwater fish extract oxygen from the air as well as from the water using a variety of different structures. Among these can be included the lungfish, a group of vertebrates belonging to the class Dipnoi, which represent the closest living relatives of the Tetrapods (which encompasses living amphibians, mammals, birds and reptiles). They have paired lungs similar to those of Tetrapods. The gouramis, also known as gouramies, group of freshwater anabantiform fish that comprise the family Osphoronemidae are also included in this group. They have a structure called the labyrinth organ that allows them to gulp air and use atmospheric oxygen.[25] This organ is a vital adaptation for fish that often inhabit warm, shallow, oxygen-poor water.[25] Finally, many catfish, such as the armoured *Corydoras* spp. (Siluriformes, Callichthyidae) extract oxygen *via* the intestine or stomach.[26]

In fish, the body shape and the arrangement of the fins are highly variable, covering such seemingly un-fishlike forms as seahorse and seadragon (Syngnathiformes, Syngnathidae), pufferfish, also known as globefish (Tetraodontiformes, Tetraodontidae), humpback anglerfish (Lophiiformes, Melanocetidae) as well as gulper eel or pelican eel (Anguilliformes, Eurypharyngidae), to name a few.

Among other peculiar characteristics, the surface of the skin of fish may be naked [*e.g.*, in moray eels (Anguilliformes, Muraenidae)] or covered with scales, a small rigid plate that grows out of the skin of a fish, produced from the mesoderm layer of the dermis, which distinguishes them from reptile scales.[27,28] These structures can also provide effective camouflage through the use of reflection and colouration as well as possible hydrodynamic advantages. Among scales, a variety of different types are usually found and defined as placoid (typical of sharks and rays), cosmoid (fossil lungfish and coelacanths), ganoid (various fossil fish) but also living including garfish, also known as needlefish (Lepisosteiformes, Lepisosteidae) and the archaic ray-finned fish bichirs (Polypteriformes, Polypteridae). Besides, other scale types, such as cycloid and ctenoid scales, are also commonly found on most of the bony fish (Osteichthyes).[26]

It is worth pinpointing the fact that there are even fish that live mostly on land or lay their eggs on land near water.[29] Among them we can include the mudskippers, which feed and interact with one another on mudflats and go underwater to hide in their burrows.[3] Mudskippers comprise 23 extant species of amphibious fish belonging to the subfamily Oxudercinae (Actinopterygii, Gobiliformes, Gobilidae). They are known for their unusual body shapes, preferences for semiaquatic habitats, limited terrestrial locomotion and jumping, but especially for the ability to survive prolonged periods of time both in and out of water. During mating seasons, the males also develop brightly coloured spots in order to attract females, which can be red, green or blue. Unlike other fish, the mudskipper's eyes protrude from the top of its flat head. Their most noticeable feature is, however, their side pectoral fins that are located more forward and under their elongated body. These fins are jointed and function similarly to limbs which allow the mudskipper to crawl from place to place. Although having the typical body form of any other gobiid fish, these front fins allow the mudskipper to actively "skip" across muddy surfaces (consequently the vernacular name) and even climb low-hanging tree branches and scrubs. It is also worth mentioning that there are three described species of the genus *Phreatobius* (Siluriformes, Heptapteridae), which has been called a true "land fish" as this worm-like catfish strictly lives among waterlogged leaf litter.[30] Last, but not least, many fish species live in underground lakes, underground rivers or aquifers and are popularly known as cavefish or cave fish.[31]

Streamlining and swimming performance varies among different species, *e.g.,* tunas (Scombriformes, Scombridae), salmon (Salmoniformes, Salmonidae) and jacks (Carangiformes, Carangidae) can cover up to 10–20 body-lengths per second to species of ray-finned fish such as Anguilliformes, *i.e.*, eels, and species of Batoidea (Elasmobranchii), a superorder of cartilaginous fish commonly known as rays that swim no more than 0.5 body-lengths per second.[26]

Fish species' diversity is nearly divided uniformly between marine (oceanic) and freshwater ecosystems. Coral reefs in the Indo-Pacific Region constitute the centre of diversity for marine fishes, whereas continental freshwater fishes are most diverse in large river basins of tropical rainforests, especially the Amazon, Congo and Mekong basins. More than 5600 fish species inhabit freshwaters from the Neotropical Region alone; this value representing about 10% of all vertebrate species on the Earth. In agreement, exceptionally rich sites in the Amazon basin, such as Cantão State Park in the state of Tocantins (Brazil), have been reported to contain more freshwater fish species than occur in all of Europe.[32]

To the best of our knowledge, the deepest living fish found in the ocean is the snailfish *Pseudoliparis belyaevi* (Scorpaeniformes, Liparidae), filmed in the Izu-Ogasawara Trench off the coast of Japan at 8336 metres in August 2022.[33] Finally, it should be mentioned that fish range in size, varying from the huge 18.8 m long whale shark *Rhincodon typus* (Orectolobiformes, Rhincodontidae) to the tiny 8 mm long stout infantfish *Schindleria brevipinguis* (Perciformes, Schindleriidae).

In brief, unlike groupings such as Birds and Mammals, Fish are not a single clade but they constitute a paraphyletic collection of taxa, including hagfishes, lampreys, sharks and rays, ray-finned fish, coelacanths and lungfish.[26] Indeed, lungfish and coelacanths are closer relatives of Tetrapods (such as Mammals, Reptiles, Amphibians and Birds) than of other fish such as ray-finned fish or sharks, so the last common ancestor of all fish is also an ancestor to Tetrapods. As paraphyletic groups are no longer recognised in modern systematic biology, the use of the term "fish" as a biological group must be employed with caution or even avoided.

The evolution of fish can be dated about 530 million years ago during the Cambrian explosion. During this time, the early Chordates developed some internal structures, *i.e.*, the skull and the vertebral column, leading to the first Craniates and Vertebrates. The first Fish lineages are represented by the

Agnatha or jawless fish. Early examples include the extinct *Haikouichthys* sp., a Craniate that lived 518 million years ago, popularly characterized as one of the earliest fishes. Cladistic analysis indicates that the animal is probably either a basal Chordate or a basal Craniate. Surprisingly, it was not formally described until 1999.[34] During the late Cambrian, eel-like jawless fish called the Conodonts and small, mostly armoured fish, known as Ostracoderms appeared. Most jawless fish are now extinct. Nevertheless, extant lampreys (Cyclostomi) may approximate ancient pre-jawed fish, which also include the extant hagfish, suggesting that this group may have split early on from other Agnatha. It is commonly accepted that the earliest jawed vertebrates probably developed during the late Ordovician period. They are first represented in the fossil record from the Silurian by two groups of fish: the armoured fish known as Placoderms, which evolved from the Ostracoderms, and the Acanthodii (or spiny sharks). The jawed fish that are still extant in modern days also appeared during the late Silurian, *i.e.*, the Chondrichthyes (or cartilaginous fish) and the Osteichthyes (or bony fish). The latter group of fish evolved into two separate groups: the Actinopterygii (or ray-finned fish) and the Sarcopterygii (which includes the lobe-finned fish).

During the Devonian period a great rise in fish variety occurred, especially among the Ostracoderms and Placoderms and also among the lobe-finned fish and early sharks. This fact has led to the Devonian being called the *Age of Fish*. It was from the lobe-finned fish that the Tetrapods evolved, the four-limbed vertebrates, represented nowadays by Amphibians, Reptiles, Mammals and Birds. Transitional tetrapods (Tetrapodomorpha) are a clade of vertebrates consisting of tetrapods and their closest Sarcopterygian relatives, which are more closely related to living Tetrapods than to living lungfish, first appeared during the early Devonian. The diversity of jawed vertebrates may indicate an evolutionary advantage; but it is unclear if the advantage of a hinged jaw is greater biting force, improved respiration, or a combination of factors. Fish do not represent a monophyletic group, but a paraphyletic one, as they exclude the Tetrapods.[35]

Fish, like many other organisms, have been greatly affected by extinction events throughout natural history. The earliest ones, the Ordovician–Silurian extinction events, *i.e.*, the first of the "big five" major mass extinction events in Earth's history, occurring roughly 445 million years ago, led to the loss of many species. The Late Devonian extinction led to the extinction of the Ostracoderms and Placoderms by the end of the Devonian, as well as other fish. This extinction event involved several extinction dealings in the Late Devonian in the history of life on Earth. The term primarily refers to a major extinction, the Kellwasser event, also known as the Frasnian–Famennian extinction,[36] which occurred around 372 million years ago at the boundary between the Frasnian stage and the Famennian stage, the last stage in the Devonian Period.[37] A second mass extinction called the Hangenberg event, also known as the end-Devonian extinction, occurred 359 million years ago,[38] bringing an end to the Famennian and Devonian period, as the world transitioned into the Carboniferous Period.[39] It should be mentioned that the spiny sharks became extinct at the Permian–Triassic extinction event. Approximately 251.9 million years ago, the Permian–Triassic extinction event, colloquially known as the Great Dying period, represents the boundary between the Permian and Triassic geologic periods and with them the Palaeozoic and Mesozoic eras, respectively.[40] So far, it constitutes the Earth's most severe known extinction event,[41,42] involving the extinction of 57% of biological families, 83% of genera, 81% of marine species[43] and 70% of terrestrial vertebrate species.[44] Furthermore, palaeontological evidences agree in demonstrating that also during this period took place the largest known mass extinction of insects.[45]

It is known that the Conodonts became extinct at the Triassic–Jurassic extinction event. It is often so-called the end-Triassic extinction, or the Mesozoic extinction event, which marks the boundary between the Triassic and Jurassic periods about 201.4 million years ago. This event represents one of the top five major extinction pinnacles of the Phanerozoic eon,[46] profoundly affecting life in the oceans as well as on land. In the seas, the entire Class of conodonts and 23–34% of marine genera disappeared.[47,48] On land, all archosauromorphs other than crocodylomorphs, pterosaurs and dinosaurs became extinct as well. Some remaining non-mammalian therapsids and many of the large temnospondyl amphibians had become extinct prior to the Jurassic as well. Nevertheless, there is still much uncertainty regarding a connection between the Triassic–Jurassic boundary and terrestrial vertebrates, due to a lack of terrestrial fossils from the Rhaetian (latest) stage of the Triassic.[49] What was left fairly untouched were plants, crocodylomorphs, dinosaurs, pterosaurs and mammals,[50,51] allowing the dinosaurs, pterosaurs and crocodylomorphs to become the dominant land animals for the next 135 million years.[52] Analyses of marine losses during this period suggests that the decline in diversity was caused more by a decrease in speciation rather than by an increase in extinctions.[53] However, a pronounced turnover in plant spores and a collapse of coral reef communities indicates that an ecological catastrophe did occur at the Triassic–Jurassic boundary.[54,55] Previous hypotheses on extinction have proposed that gradual climate or sea level change may be the cause.[56] However, the possibility of one or more asteroid strikes is not totally discarded.[57–59] One of the most well-supported and widely-held theories for the cause of the Triassic–Jurassic extinction places the blame on the start of volcanic eruptions in the Central Atlantic Magmatic Province,[60] which in theory was responsible for outputting a high amount of CO_2 into the Earth's atmosphere,[61] inducing profound global warming together with ocean acidification.[62,63]

Fish did not suffer a mass extinction at the end of the Triassic. The late Triassic, in general, did experience a gradual drop in Actinopterygiian diversity after an evolutionary explosion in the middle Triassic. Though this may have been due to falling sea levels or the Carnian pluvial event, it may instead be a result of sampling bias considering that middle Triassic fish have been more extensively studied than late Triassic fish.[64] Despite the apparent drop in diversity, Neopterygiians (which include most modern bony fish) suffered less than more "primitive" Actinopterygiians, indicating a biological turnover where modern groups of fish started to supplant earlier groups.[49] Conodonts, which were prominent index fossils throughout the Palaeozoic and Triassic, finally became extinct at the Triassic–Jurassic boundary following declining diversity.[49]

Finally, the present-day Holocene extinction, also known as the Anthropocene extinction, is the ongoing extinction event caused by humans during the Holocene epoch.[65,66] The International Union of Geological Sciences (IUGS), during its 2024 meeting, has rejected the formal use of Anthropocene as a geologic timescale. While this rejection is based on solid academic grounds, the term will certainly continue to be used, as it clearly refers to the effects of mankind upon its planet. It is a fact that language and sciences as well are shaped by its users and not by international organizations. So, academically speaking, we are still in the Holocene, which started approx. 11 700 years ago. Returning to these previously mentioned extinctions, they cover numerous different taxa of plants and animals.[67–70] Among them are included mammals,[71,72] birds,[69,73] reptiles,[74,75] amphibians,[76,77] fish[78,79] and invertebrates,[80,81] affecting not just terrestrial, but also a large number of aquatic species.[78,79,82] With widespread degradation of biodiversity hotspots, such as coral reefs and rainforests, as well as other areas, the vast majority of these extinctions are thought to be undocumented, as many species are probably unrecorded as they are undiscovered at the time of their extinction. The current extinction rate of species is estimated at 100 to 1000 times higher than natural background extinction rates and is still increasing.[74,83–85] During the past 100–200 years, the pace of biodiversity loss and species extinction has accelerated,[83,84] to the point that most conservation biologists now believe that human activity has either produced a period of mass extinction, or is on the cusp of doing so.[86,87] For this reason, after the "Big Five" mass extinctions, the Holocene extinction event has also been referred to as the Sixth Mass extinction.[66,87,88] Furthermore, considering the recent recognition of the Capitanian mass extinction event, the term Seventh Mass extinction has also been proposed for the Holocene extinction event.[89,90]

Taking into account the explosive increase in the list of pollutants detected, the scientific community is facing challenges *vis à vis* the protection of humans and wildlife from the potentially negative consequences of their effects. Laboratory research and epidemiological studies have revealed that xenobiotics exert a broad-spectrum of side-effects on fish by interfering with several parameters, including their endocrine, immune and neural systems, reproduction, development and growth. Nevertheless, inconsistent results may seem to suggest that the potential effects of pollutant exposure on the physiological parameters in fish are clearly multifactorial, involving simultaneously several and diverse biological processes, species-specific differences, sex and age, notwithstanding the types of xenobiotics involved. Another factor deserving attention is the type of interaction of more than one pollutant, while acting simultaneously with another one/s. When the conventional concentration additivity is analysed, several studies have suggested that different classes of pollutants possibly act synergistically or antagonistically based upon their concentration. However, risk assessment information is still recorded in a scattered and inadequate partial manner, including a lack of enough fish species under study, limited number of xenobiotics investigated and insufficient geographical representation. For these reasons, our knowledge of the effects of xenobiotic-induced deleterious effects on fish still remains in its first steps, although on a more positive note, the mechanistic understanding of their mode of action is progressing at a faster pace due to the technological advances. On the contrary, research on wild fish populations and their ecosystems under pollutant-induced stress is still unavailable, as far as we know. For all the above-mentioned reasons, the aim of this book is to provide some practical evidences predicting and tracing the effects of several xenobiotics at different biological organization levels, especially starting from the molecular, cellular and individual, to the population, community and ecosystem levels, due to the limited and scattered information available at time of press.

Chapter 2 provides a background overview highlighting the toxic manifestations of exposure to heavy metals, especially in fish, which have attracted a lot of attention as a good model to monitor their toxicologic, mutagenic and carcinogenic effects. Fish are considered as sensitive indicators of water pollution in our quest for early warning signals of environmental deterioration induced by other xenobiotics. Investigations directed towards the identification and development of suitable biomarkers/biosensors does make fish species an attractive, demanding, valuable and alternative tool to address problems related to ecosystem management and planning. Research outcomes from fish are now increasingly applied to solve both human as well as environmental health issues. The current book chapter focuses on recent studies conducted to identify biological markers of exposure, susceptibility and effect upon exposure to toxic elements in diverse environmental conditions. It describes the significance of morphological markers like scales and chromatophores in monitoring the aquatic pollution and risk assessment. Furthermore, morphological changes in gills have also been considered as suitable indicators of heavy metal pollution. Application of several biochemical parameters of toxicity in fish, *e.g.*, metallothionein, free radicals, glutathione and antioxidant enzymes, as biomarkers of effect have also been examined. Results on molecular and genetic toxicology induced by pollutants have been confirmed as reliable indicators of water pollution caused by heavy metals. Collectively, these biomarkers in fish are suitable sentinels of heavy metal contamination in a water body. However, specific biomarkers for each element still need to be established. In a nutshell, fish models offer tremendously exciting opportunities to explore and inspire additional openings towards the deployment of biomarkers in environmental health risk assessment.

In Chapter 3, the authors reveal the fact that seafood consumption in the Durban basin, as a geographical example, faces a hidden threat: heavy metal contamination in fish. This chapter unveils the intricate interplay of dietary preferences and specific physiological traits that influence the distribution of heavy metal in edible fish within the Durban basin. It elucidates the divergent capacities of fish gills, liver and muscles in accumulating different metals, with certain species exceeding both South African and international regulatory limits. This revelation underscores the urgent need for comprehensive monitoring and effective management strategies. The maasbanker or Atlantic horse mackerel *Trachurus trachurus* (Carangiformes, Carangidae) and the slinger seabream *Chrysoblephus puniceus* (Spariformes, Sparidae) are pinpointed as promising bioindicators for moderate heavy metal pollution, underlining the importance of species-specific considerations for accurate monitoring. Finally, avenues for future research are identified, including longitudinal studies on the accumulation of the contaminants in question, comparative analyses of bioindicators and broader ecosystem health assessments.

Following this, Chapter 4 highlights that the occurrence of ecotoxicological agents in aquatic ecosystems has increased due to the accelerated production of anthropogenic contaminants. Fresh and marine water ecosystems have been widely affected by these compounds, with several implications for fish health. In this sense, 18% of ecotoxicological effects in fish reported genetics issues, 16% of cases presented mortality and another 16% of fish cases affected by toxicants presented accumulation as the principal effects. Additionally, an interesting issue to be aware of and also to be factored in, is the modification of the gut microbiota by ecotoxic compounds, which in turn causes dysbiosis. It is known that microbial fish biomarkers have been proposed to establish and calculate a dysbiosis index generated by these ecotoxic contaminants. The most abundant ecotoxicological compounds present in the aquatic environment are pharmaceuticals and personal care products (PPCPs), principally hormone disruptors, such as 17-β-estradiol (9.7%) and its metabolites, originating from human therapy and husbandry use allowed in some countries, affecting not only wildlife, but also humans as well. The main studied fish for ecotoxicological effects are the well-known zebrafish *Danio rerio* (Cypriniformes, Cyprinidae), followed by the fathead minnow *Pimephales promelas* (Cypriniformes, Cyprinidae), the rainbow trout *Oncorhynchus mykiss* (Salmoniformes, Salmonidae), the medaka *Oryzias latipes* (Beloniformes, Adrianichthyidae) and the common carp *Cyprinus carpio* (Cypriniformes, Cyprinidae). Ecotoxicological effects in fish are of high concern due to the bioaccumulation and biomagnification of these compounds, which eventually are destined for human consumption since fish occupy a great variety of tropical niches, acting as prey and hosts for a wide range of species.

Chapter 5 focuses on the fact that fish species are threatened by marine plastic pollution, which affects their health and nutritional value, as fish provide an affordable animal protein food source. While this is a given fact *hic et nunc*, the available data so far are not sufficient to carry out a consistent assessment of the plastic potential risks to human health and draw, at least, some preliminary conclusions. The present chapter analyses the state of knowledge about plastic pollution and the use of fish species as bioindicators in the Mediterranean Sea. The analysis shows that an increase in microplastic contamination in the Mediterranean Sea would worsen the state of its fisheries' stocks, which already suffer from a lack of correct and adequate management, as well as a lack of preventive measures that has resulted in marine resources' deterioration. In addition, there is a recommendation to prioritize the assessment of plastic content and potential effects in the most consumed fish species in the Mediterranean, such as sardine, anchovy and bluefin tuna – reported to be highly contaminated by microplastics – in order to avoid possible effects on consumers.

Chapter 6 describes the usefulness of the studies of *Colossoma macropomum*, the "tambaqui" (Characiformes, Serasalmidae), which is the most consumed and produced native fish in the North of Brazil. Coupled with an intense aquaculture productivity, high rates of parasites are observed in the breeding tanks, among these, protozoa being the most common. To minimize this problem, fish farmers use a wide range of antiparasitic agents, from this armamentarium, trichlorfon stands out, an organophosphate agent with a high capacity for contamination in fish and humans. This chemical was mainly developed and used for plant production and as an external parasiticide in animal husbandry, now being applied to fish. Its metabolites being more toxic than the primary compound. In this work, we describe the main consequences of the use of this chemical agent in the tambaqui population from aquaculture. Genomic studies have shown that homeostasis of this species has been affected.

The following chapter, Chapter 7, illustrates the fact that the extensive use of herbicides worldwide is, at present, a potentially harmful aspect to be included in any environmental health risk analyses. Formulations of auxinic herbicides containing 2,4-D, dicamba and MCPA, among others, are widely employed alone or in combination with other agrochemicals. In this sense, significant concentrations of these herbicides can be found in water bodies inhabited by aquatic species, *e.g.*, fish that are employed for toxicological evaluation. In the Pampas Region, different studies were performed to evaluate lethal and sublethal effects induced by the auxinic herbicides 2,4-D and dicamba, in several aquatic species. These included *Cnesterodon decemmaculatus* (Cyprinodontiformes, Poeciliidae), *Rhamdia quelen* (Siluriformes, Heptapteridae), *Poecilia vivipara* (Cyprinodontiformes, Poeciliidae), *Jenynsialineata* (Cyprinodontiformes, Anablepidae) and *Leporinus obtusidens* (Characiformes, Anostomidae). The results obtained demonstrated that these compounds induce lethal and sublethal effects such as genotoxicity as well as behavioural, enzymatic, biochemical and histological alterations, to name a few. These observations highlight the feasibility of auxinic herbicides of inducing detrimental effects on these species that inhabit this highly cultivated area of South America.

Continuing with practical examples of non-conventional fish species as experimental models, Chapter 8 highlights that fish species are important aquatic models utilized in ecotoxicology studies; however, it is worthwhile to underpin that most of these species are found in temperate countries. In this chapter, native fish species in Nigeria that are utilized for ecotoxicology studies were reviewed. A search of the literature was conducted using Google Scholar search engine from anytime until November 7, 2023, resulting in a total of 45 articles that were included in this review. The commonly used native Nigerian fish species in laboratory and field-based (biomonitoring) ecotoxicological studies are the African sharptooth catfish *Clarias gariepinus*, the Nile tilapia *Oreochromis niloticus* and the Guinean tilapia *Coptodon guineensis*. These fish species have been demonstrated to be valid models to study the biomarkers of exposure and effect of toxicants. The non-exhaustive list includes pesticides, effluents, heavy metals, hydrocarbons, pharmaceuticals, plant and sediment extracts. These compounds serve as bioindicators of stressors or pollutants in biomonitoring programmes with indices ranging from cellular (nuclear abnormalities, haematological effects) to tissue and organ histological alterations, oxidative stress indices as well as individual/population level effects (behavioural changes). We recommend targeted studies on the culture exploitation of other native Nigerian freshwater, brackish and marine fish species as well as ratification of *C. gariepinus* and *O. niloticus* as globally recognised model fish species for use in ecotoxicological studies.

Chapter 9 remarks on the fact that over the last decade a range of studies have demonstrated that various fitness-related types of behaviours are vulnerable to disruption induced by various contaminants of emerging concern (CECs) and legacy pollutants. However, the behavioural responses of animals when exposed to pollutants/contaminants are rarely included in discussions about the regulation of chemicals. We believe this to be an important fact that deserves wider attention by the scientific community. Thus, to improve our understanding of the opportunities and challenges for behavioural ecotoxicology, this chapter presents a synthesis of the endpoints and experimental procedures used to evaluate the behavioural response of Nile tilapia *Oreochromis niloticus*, a non-conventional species in ecotoxicology, to different CECs and legacy pollutants. By focusing on the methodological possibilities that can be implemented in evaluating the behaviour of this Neotropical species, we shed light on the feasibility and reasonableness of using this animal model as a complement or alternative *vis à vis* the use of model species traditionally used in toxicology. This approach would certainly enrich our views on the matter.

In the last chapter of this frame section, *i.e.,* Chapter 10, the authors present an overview of the status of Lake Nyasa (Lake Malawi), which is the world's ninth largest and third deepest freshwater lake, hosting about 800 to 1000 freshwater fish species. As we might guess, increased pollution is one of the emerging challenges affecting fish populations of this water body. Activities such as coal mining, agriculture, forest burning and deforestation contribute to increased chemical inputs into the lake. Water pollution produces adverse effects such as mortality, inhibited reproduction and growth in aquatic organisms. Consequently, most fish species in the lake are declared by the International Union for Conservation of Nature (IUCN) as threatened, vulnerable, critically endangered or extinct. The effects of contamination can also be extended to human health through the food chain and cause various detrimental effects. Therefore, it is essential to monitor potential toxicants in the aquatic environment to protect both animal and human health. The use of bioindicators to monitor the status of aquatic environments is one of the effective and easy-to-use options. However, available information on the use of local fish species as a potential bioindicator for risk assessment in Lake Nyasa is limited, so far. Due to the usefulness of genetic indices and heavy metal accumulation levels in fish tissues collected along Lake Nyasa, they have been shortlisted as potential tools to be used for health risk assessment with a view to a more ambitious general management strategy.

The third frame section, *i.e., Fish Species as Experimental Models in Environmental Risk Assessment Studies. Epilogue and Final Remarks* is integrated in Chapter 11, where the Editors summarize how the growing impact of anthropogenic activities generate huge quantities of toxic residues that can have direct or indirect detrimental effects upon the quality of our environment. The increase in their jeopardizing effects can exert short-, medium- and long-term consequences, affecting not only human health but also environmental health in general. Furthermore, these disturbances are usually capable, in most cases, of reducing the biodiversity of native flora and fauna, which will, in turn, encourage the resistance and emergence of new pests and diseases. This book intends to provide an outline, relevant examples and stimulate practical discussions on the use of this non-conventional group of vertebrates, the fish. It includes relevant tools which may be of use in the analyses prior to the implementation of decisions leading to actions which will hopefully reduce environmental health risks against environmental factors. Our aim was to compile information from a diversity of sources into a single volume. The rationale is to give some real-life examples in order to widen the concept that the use of non-conventional animal models, far from being a scientific conundrum, may instead provide useful avenues to be explored and further developed as a way to provide help and real answers to some of the actual problems the whole planet is dealing with currently. Concomitantly, these concrete examples extend concepts of hazardous factors to living species worldwide that may stimulate new research ideas and trends in their respective important fields.

In spite of the fact of having to deal with many diverse topics, we have tried to compile this "wealth of information" to take into consideration whether different amphibian species, of one or more populations, can be selected as experimental matrices.

This book is part of a series of four volumes along similar lines published in Issues in Toxicology by the Royal Society of Chemistry (see Figure 11.1) utilizing different species (Mammals, Birds, Reptiles and Amphibians). The present volume devoted to Fish is the fourth one. First in the list is *Marsupial and Placental Mammal Species in Environmental Risk Assessment Strategies* (see Figure 11.1A),[91] followed by *Bird and Reptiles Species in Environmental Risk Assessment Strategies* (see Figure 11.1B),[92] continuing the list with *Amphibians*

Figure 11.1 Front book covers of the books published in Issues in Toxicology, The Royal Society of Chemistry *Marsupial and Placental Mammal Species in Environmental Risk Assessment Strategies*, 2022 (A), followed by *Fish and Reptiles Species in Environmental Risk Assessment Strategies*, 2023 (B), continuing with *Amphibian Species in Environmental Risk Assessment Strategies*, 2023 (C) and the current volume *Fish Species in Environmental Risk Assessment Strategies*, 2024 (D). Images kindly provided by the Royal Society of Chemistry.

Species in Environmental Risk Assessment Strategies (see Figure 11.1C),[93] concluding with *Fish Species in Environmental Risk Assessment Strategies* (see Figure 11.1D).[94]

Without running the risk of being repetitive, we would like to recap on some of the important concepts as mentioned erstwhile in the two volumes previously published in Issues in Toxicology by the Royal Society of Chemistry in 2017 (see Figure 11.2), *i.e.*, *Ecotoxicology and Genotoxicology – Non-traditional Terrestrial Models* (see Figure 11.2A)[95] and *Ecotoxicology and Genotoxicology – Non-traditional Aquatic Models* (see Figure 11.2B),[96] by the same editor group. We strongly recommend the perusal of both volumes, which are not overlapping subjects, in order to gain the full benefit of this series and have a holistic and stimulating approach to the matter.

New emerging studies in several reputable publications of common worldwide human diseases such as arterial hypertension, among others, seem to suggest that not only classical and lifestyle factors should be addressed in the search for adequate treatment.[97] From a medical perspective, it is not only a disease of affluence. One in eight deaths worldwide is due to this condition and its corresponding factors (heart, kidney disease and stroke). These studies seem to suggest that a closer look at other, up till now unrelated factors,

Figure 11.2 Front book covers of the books *Ecotoxicology and Genotoxicology – Non-traditional Terrestrial Models* (A) and *Ecotoxicology and Genotoxicology – Non-traditional Aquatic Models* (B), previously published in Issues in Toxicology, the Royal Society of Chemistry, by the same editor group in 2017. Images kindly provided by the Royal Society of Chemistry.

is required. Early-life nutrition and exposure to air pollution, heavy metals and even noise are factors implicated that may push blood pressure up later in life. Thinking out of silos and cross-linking medical with applied science will lead to a faster pace of understanding of all the factors related to the emergence of a disease and thus finding a cure.

Many scientists feel that time has come for a shift in emphasis away from Basic towards Applied Research. While Basic Research is needed in order to shed more light on the fundamentals, Applied Research is required in order to find solutions to the many problems the world faces, *e.g.*, overpopulation, excessive use of the Earth's natural resources and pollution.

The chapters included in this book are just a mere enumeration of some practical examples. There are many more species that can be used as experimental models and the list should be expanded by different research groups and academics from all over the world. Quite recently, among the list of species to be considered, some scientists have included animals that can be easily accessed and also share the same household with human beings, namely dogs, cats and other pets, including some fish spp. While on one side, they may seem to be a faithful mirror on the situation, there are certain pitfalls to be considered. One of them is nutrition, as they mostly now sadly consume only industrial feeds and not fresh produce. We shall certainly hear more on the matter as this approach is being explored.

We hope that many more scientists realise that it is more important to tackle subjects related to the status of the environment using autochthonous, non-target species, which are truly exposed to locally used xenobiotic agents. On the other hand, this research plan would also increase the chances of independent scientists and research institutions in getting access to grants and attracting the attention of local authorities, who would be more interested in financing projects that really are within their sphere of political interest and pride.

Globally, freshwater fish, as an example, are under threat, with as many as a third of global populations in danger of extinction. Populations of migratory freshwater fish have plummeted by 76% since 1970, and large fish, *i.e.*, those weighing more than 30 kg, have been all but wiped out in most rivers. The global population of megafish is down by 94% and 16 freshwater fish species were declared extinct during the last two years. The report by 16 global conservation organizations, called The World's Forgotten Fishes, stated that global populations of freshwater fish were in freefall. The problems are diverse and include pollution, overfishing and destructive fishing practices, the introduction of invasive non-native species, climate change and the disruption of river ecologies. Most of the world's rivers are now dammed in parts, have water extracted for irrigation or have their natural flows disrupted, making life difficult for freshwater fish. This is the result of the adding up of their natural habitat destruction for land use, disease, pollution and climate change. In many species, these factors have added so much pressure that they have been completely wiped out. It is our hope that with the emergence of the new legal figure of Ecocide, governments, politicians, business corporations and producers will have to parse more carefully the consequence of their actions and activities. It is certainly a first important step towards their accountability, and public pressure is mandatory in order for it to be successful, as actions by agents and politicians will be hard to fathom. Another step forward in this direction is the recently launched "Banana Index" by The Economist, a financial weekly publication. It tries to give a hint at the environmental impact of foodstuffs produced, giving consumers, economists and the public in general a rough idea of the matter. As in statistics in general, this indicator reveals a lot, but hides the essential. Nevertheless, it is a trailblazing intent to raise awareness at the time we pick our supplies from groceries and markets. There are certainly some certification schemes already in place (Rainforest Alliance, FSC, *etc.*) with their labels inserted in the packages. Most of them will have to be dramatically improved in order to lighten or reduce as much as possible their overload on the environment. Clearly the problem here is twofold. On one hand the production of foodstuffs and, on the other hand, its packaging and disposal of waste. All this should be factored in at the till. Food safety, as the present wars in Ukraine and Israel have clearly shown, is a phantom still hanging around. Ultra-processed foodstuffs labelled as "organic", can, in many cases mislead customers into making what they think are healthy choices. These products may contain "organically grown" commodities as basic raw materials, but due to the heavy industrial processes employed in their preparation, the final product can hardly support the "natural" or "organic" claim. All these situations would deserve another volume and clearly escape the purpose of this series.

The UN Biodiversity Conference Intactness Index (BII) shows a world average of 75%; this is a value certainly lower than the 90% required to be considered as the safe limit for not falling into "ecological recession".

Stone Age, Copper Age, Iron Age, Medieval Ages, Modern Age, these designate a few of the developmental stages of mankind. The Anthropocene will be probably followed by the Extinction Age, if we remain quiet bystanders. While the clock is ticking, we still have some valuable time left to be used diligently.

Many researchers have contributed to the publication of this book. We hope that it serves as a herald in order to bolster enthusiasm for the use of native or easily available local species in order to widen our knowledge on the subject. Last, but not least, we would like to especially thank the authors who have responded to our call, for their time, contribution and feedback making the compilation of this book possible. Let us hope that the Anthropocene will not be the Last Age of Humanity.

References

1. P. H. Yancey, M. E. Gerringer, J. C. Drazen, A. A. Rowden and A. Jamieson, Marine fish may be biochemically constrained from inhabiting the deepest ocean depths, *Proc. Natl. Acad. Sci. U. S. A.*, 2014, **111**, 4461–4465.
2. A. J. Jamieson, T. D. Linley, S. Eigler and T. Macdonald, A global assessment of fishes at lower abyssal and upper hadal depths (5000 to 8000 m), *Deep Sea Res., Part I*, 2021, **178**, 103642.

3. FishBase Search, 2020, https://fishbase.net.br/search.php, accessed on March 15, 2024.
4. J. Chen, M. Jayachandran, W. Bai and B. Xu, A critical review on the health benefits of fish consumption and its bioactive constituents, *Food Chem.*, 2022, **369**, 130874.
5. S. Maulu, K. Nawanzi, M. Abdel-Tawwab and H. S. Khalil, Fish nutritional value as an approach to children's nutrition, *Front. Nutr.*, 2021, **8**, 780844.
6. S. Benvenga, F. Famà, L. G. Perdichizzi, A. Antonelli, G. Brenta, F. Vermiglio and M. Moleti, Fish and the Thyroid: A Janus Bifrons relationship caused by pollutants and the omega-3 polyunsaturated fatty acids, *Front. Endocrinol.*, 2022, **13**, 891233.
7. D. Caldeira, B. Nogueira-Garcia, A. Abreu and F. J. Pinto, Fish intake and risk of cardiovascular events: An analysis of the VITAL cohort, *Eur. J. Clin. Nutr.*, 2023, **77**, 400–404.
8. B. A. Block and J. R. Finnerty, Endothermy in fishes: A phylogenetic analysis of constraints, predispositions, and selection pressures, *Environ. Biol. Fishes*, 1993, **40**, 283–302.
9. M. Haesemeyer, Thermoregulation in fish, *Mol. Cell. Endocrinol.*, 2020, **518**, 110986.
10. F. G. Carey and K. D. Lawson, Temperature regulation in free-swimming bluefin tuna, *Comp. Biochem. Physiol., Part A: Mol. Integr. Physiol.*, 1973, **44**, 375–392.
11. N. C. Wegner, O. E. Snodgrass, H. Dewar and J. R. Hyde, Whole-body endothermy in a mesopelagic fish, the opah, *Lampris guttatus*, *Science*, 2015, **348**, 786–789.
12. National Oceanic Atmospheric Administration (NOA), *Warm Blood Makes Opah an Agile Predator*, Southwest Fisheries Center, 2015, https://www.fisheries.noaa.gov/news-and-announcements/news?field_region_vocab_target_id%5B1000001126%5D=1000001126, accessed on March 15, 2024.
13. F. G. Carel, J. G. Casey, H. L. Pratt, D. Urquhart and J. E. McCosker, Temperature, heat production, and heat exchange in lamnid sharks, *Mem. South. Calif. Acad. Sci.*, 1985, **9**, 92–108.
14. S. R. Weinmann, A. N. Black, M. L. Richter, M. Itzkowitz and R. M. Burger, Territorial vocalization in sympatric damselfish: acoustic characteristics and intruder discrimination, *Bioacoustics*, 2017, **27**, 87–102.
15. L. Roberts and A. N. Rice, Vibrational and acoustic communication in fishes: The overlooked overlap between the underwater vibroscape and soundscape, *J. Acoust. Soc. Am.*, 2023, **154**, 2708–2720.
16. F. Bertucci, L. Ruppé, S. V. Wassenbergh, P. Compère and E. Parmentier, New insights into the role of the pharyngeal jaw apparatus in the sound-producing mechanism of *Haemulon flavolineatum* (Haemulidae), *J. Exp. Biol.*, 2014, **217**, 3862–3869.
17. D. J. Colson, S. N. Patek, E. L. Brainerd and S. M. Lewis, Sound production during feeding in *Hippocampus* seahorses (Syngnathidae), *Environ. Biol. Fishes*, 1998, **51**, 221–229.

18. T. P. R. Oliveira, F. Ladich, D. Abed-Navandi, A. S. Souto and I. L. Rosa, Sounds produced by the longsnout seahorse: a study of their structure and functions, *J. Zool.*, 2014, **294**, 114–121.
19. L. F. Fine, C. B. King and T. M. Cameron, Acoustical properties of the swimbladder in the oyster toadfish Opsanus tau, *J. Exp. Biol.*, 2009, **212**, 3542–3552.
20. M. L. Fine and T. D. Waybright, Grunt variation in the oyster toadfish *Opsanus tau*: effect of size and sex, *PeerJ*, 2015, **3**, e1330.
21. S. W. Ricci, D. R. Bohnenstiehl, D. B. Eggleston, M. L. Kellogg and R. P. Lyon, Oyster toadfish (Opsanus tau) boatwhistle call detection and patterns within a large-scale oyster restoration site, *PLoS One*, 2017, **12**, e0182757.
22. C. R. Skoglund, Functional analysis of swimbladder muscles engaged in sound productivity of the toadfish, *J. Cell Biol.*, 1961, **10**, 187–200.
23. E. Parmentier, J. Tock, J. C. Falguière and M. Beauchaud, Sound production in *Sciaenops ocellatus*: Preliminary study for the development of acoustic cues in aquaculture, *Aquaculture*, 2014, **432**, 204–211.
24. A. Meisner and J. Burns, Viviparity in the halfbeak genera *Dermogenys* and *Nomorhamphus* (Teleostei: Hemiramphidae), *J. Morphol.*, 1997, **234**, 295–317.
25. M. Hellweg, Gorgeous little gouramis, *Trop. Fish Hobbyist*, 2023, vol. 7.
26. D. E. Facey, B. W. Bowen, B. B. Collette and G. S. Helfman, *The Diversity of Fishes: Biology, Evolution and Ecology*, John Wiley and Sons, Inc., New York, NY, 3rd edn, 2022.
27. A. Mongera and C. Nüsslein-Volhard, Scales of fish arise from mesoderm, *Curr. Biol.*, 2013, **23**, R338–R339.
28. P. T. Sharpe, Fish scale development: Hair today, teeth and scales yesterday?, *Curr. Biol.*, 2001, **11**, R751–R752.
29. K. L. M. Martin, *Beach-spawning Fishes: Reproduction in an Endangered Ecosystem*, CRC Press, Boca Raton, FL, 2014.
30. P. A. Henderson and I. Walker, Spatial organization and population density of the fish community of the litter banks within a central Amazonian blackwater stream, *J. Fish Biol.*, 1990, **37**, 401–411.
31. A. Romero, *The Biology of Hypogean Fishes in Developments in Environmental Biology of Fishes*, Springer Dordrecht, Germany, 2001, vol. 21.
32. E. Ferreira, J. Zuanon, G. dos Santos and S. Amadio, A ictiofauna do Parque Estadual do Cantão, Estado do Tocantins, Brasil, *Biota Neotrop.*, 2011, **11**, 1–8.
33. A. J. Jamieson, P. J. Maroni, T. Bond, Y. Niyazi, J. Kolbusz, P. Arasu and H. Kitazato, New maximum depth record for bony fish: Teleostei, Scorpaeniformes, Liparidae (8336 m, Izu-Ogasawara Trench), *Deep Sea Res., Part I*, 2023, **199**, 104132.
34. D.-G. Shu, H.-L. Luo, S. Conway Morris, X.-L. Zhang, S.-X. Hu, L. Chen, J. Han, M. Zhu, Y. Li and L.-Z. Chen, Lower Cambrian vertebrates from south China, *Nature*, 1999, **402**, 42–46.
35. G. Lecointre and H. Le Guyader, *The Tree of Life: A Phylogenetic Classification*, Kelkbap Press of Harvard University Press, 2006.

36. J. A. Clack, Devonian climate change, breathing, and the origin of the Tetrapod stem group, *Integr. Comp. Biol.*, 2007, **47**, 510–523.
37. R. T. Becker and M. R. House, Kellwasser events and goniatite successions in the Devonian of the Montagne Noire with comments on possible causations, *CFS, Cour. Forschungsinst. Senckenberg*, 1986, **169**, 45–77.
38. L. Sallan and A. K. Galimberti, Body-size reduction in vertebrates following the end-Devonian mass extinction, *Science*, 2015, **350**, 812–815.
39. M. L. Caplan and R. M. Bustin, Devonian–Carboniferous Hangenberg mass extinction event, widespread organic-rich mudrock and anoxia: causes and consequences, *Palaeogeogr., Palaeoclimatol., Palaeoecol.*, 1999, **148**, 187–207.
40. H. Jurikova, M. Gutjahr, K. Wallmann, S. Flögel, V. Liebetrau, R. Posenato, L. Angiolini, C. Garbelli, U. Brand, M. Wiedenbeck and A. Eisenhauer, Permian–Triassic mass extinction pulses driven by major marine carbon cycle perturbations, *Nat. Geosci.*, 2020, **13**, 745–750.
41. D. H. Erwin, The End-Permian mass extinction, *Annu. Rev. Ecol. Evol. Syst.*, 1990, **21**, 69–91.
42. Y. Chen, S. Richoz, L. Krystyn and Z. Zhang, Quantitative stratigraphic correlation of Tethyan conodonts across the Smithian-Spathian (Early Triassic) extinction event, *Earth-Sci. Rev.*, 2019, **195**, 37–51.
43. S. M. Stanley, Estimates of the magnitudes of major marine mass extinctions in earth history, *Proc. Natl. Acad. Sci. U. S. A.*, 2016, **113**, E6325–E6334.
44. S. Sahney and M. J. Benton, Recovery from the most profound mass extinction of all time, *Proc. R. Soc. B*, 2008, **275**, 759–765.
45. C. Labandeira, The fossil record of insect extinction: New approaches and future directions, *Am. Entomol.*, 2005, **51**, 14–29.
46. P. D. Ward, J. W. Haggart, E. S. Carter, D. Wilbur, H. W. Tipper and T. Evans, Sudden productivity collapse associated with the Triassic-Jurassic boundary mass extinction, *Science*, 2001, **292**, 1148–1151.
47. G. Ryder, D. E. Fastovsky and S. Gartner, *The Cretaceous-Tertiary Event and Other Catastrophes in Earth History*, Geological Society of America, Boulder, CO, 1996, vol. 307.
48. J. J. Sepkoski, A kinetic model of Phanerozoic taxonomic diversity. III. Post-Paleozoic families and mass extinctions, *Paleobiology*, 1984, **10**, 246–267.
49. L. H. Tanner, S. G. Lucas and M. G. Chapman, Assessing the record and causes of Late Triassic extinctions, *Earth-Sci. Rev.*, 2004, **65**, 103–139.
50. M. Barbacka, G. Pacyna, Á. Kocsis, A. Jarzynka, J. Ziaja and E. Bodor, Changes in terrestrial floras at the Triassic-Jurassic boundary in Europe, *Palaeogeogr., Palaeoclimatol., Palaeoecol.*, 2017, **480**, 80–93.
51. O. Toljagić and R. J. Butler, Triassic–Jurassic mass extinction as trigger for the Mesozoic radiation of crocodylomorphs, *Biol. Lett.*, 2013, **9**, 1–4.
52. M. J. Benton, What really happened in the late Triassic?, *Hist. Biol.*, 1991, **5**, 263–278.

53. R. K. Bambach, A. H. Knoll and S. C. Wang, Origination, extinction, and mass depletions of marine diversity, *Paleobiology*, 2004, **30**, 522–542.
54. N. R. Bonis, M. Ruhl and W. R. Kürschner, Milankovitch-scale palynological turnover across the Triassic–Jurassic transition at St. Audrie's Bay, SW UK, *J. Geol. Soc.*, 2010, **167**, 877–888.
55. J. M. Pandolfi, M. John and W. Kiessling, Gaining insights from past reefs to inform understanding of coral reef response to global climate change, *Curr. Res. Environ. Sustainability*, 2014, **7**, 52–58.
56. E. H. Colbert, Tetrapod extinctions at the end of the Triassic Period, *Proc. Natl. Acad. Sci. U. S. A.*, 1958, **44**, 973–977.
57. J. P. Hodych and G. R. Dunning, Did the Manicouagan impact trigger end-of-Triassic mass extinction?, *Geology*, 1992, **20**, 51–54.
58. P. E. Olsen, N. H. Shubin and M. H. Anders, New early Jurassic tetrapod assemblages constrain Triassic–Jurassic tetrapod extinction event, *Science*, 1987, **237**, 1025–1029.
59. J. G. Spray, S. P. Kelley and D. B. Rowley, Evidence for a late Triassic multiple impact event on Earth, *Nature*, 1998, **392**, 171–173.
60. J. H. Whiteside, P. E. Olsen, T. Eglington, M. E. Brookfield and R. N. Raymond, Compound-specific carbon isotopes from Earth's largest flood basalt eruptions directly linked to the end-Triassic mass extinction, *Proc. Natl. Acad. Sci. U. S. A.*, 2010, **107**, 6721–6725.
61. Q. Jiang, F. Jourdan, H. K. H. Olierook, R. E. Merle, J. Bourdet, D. Fougerouse, B. Godel and A. T. Walker, Volume and rate of volcanic CO_2 emissions governed the severity of past environmental crises, *Proc. Natl. Acad. Sci. U. S. A.*, 2022, **119**, e2202039119.
62. J. C. McElwain, D. J. Beerling and F. I. Woodward, Fossil plants and global warming at the Triassic-Jurassic boundary, *Science*, 1999, **285**, 1386–1390.
63. M. Hautmann, Effect of end-Triassic CO_2 maximum on carbonate sedimentation and marine mass extinction, *Facies*, 2004, **50**, 257–261.
64. C. Romano, M. B. Koot, I. Kogan, A. Brayard, A. V. Minikh, W. Brinkmann, H. Bucher and J. Kriwet, Permian-Triassic Osteichthyes (bony fishes): diversity dynamics and body size evolution, *Biol. Rev. Cambridge Philos. Soc.*, 2014, **91**, 106–147.
65. R. Wagler, The Anthropocene mass extinction: An emerging curriculum theme for science educators, *Am. Biol. Teach.*, 2011, **73**, 78–83.
66. R. H. Cowie, P. Bouchet and B. Fontaine, The sixth mass extinction: fact, fiction or speculation?, *Biol. Rev. Cambridge Philos. Soc.*, 2022, **97**, 640–663.
67. A. M. Humphreys, R. Govaerts, S. Z. Ficinski, E. N. Lughadha and M. S. Vorontsova, Global dataset shows geography and life form predict modern plant extinction and rediscovery, *Nat. Ecol. Evol.*, 2019, **3**, 1043–1047.
68. J.-G. Gao, H. Liu, N. Wang, J. Yang and X.-L. Zhang, Plant extinction excels plant speciation in the Anthropocene, *BMC Plant Biol.*, 2020, **20**, 430.
69. C. J. W. McClure, I. Berkunsky, E. R. Buechley, L. Dunn, J. Johnson, J. McCabe, S. Oppel, B. W. Rolek, L. J. Sutton and R. Gumbs, Conserving the evolutionary history of birds, *Conserv. Biol.*, 2023, **37**, e14141.

70. M. Galetti, M. Moleón, P. Jordano, M. M. Pires, P. R. Guimarães Jr, T. Pape, E. Nichols, D. Hansen, J. M. Olesen, M. Munk, J. S. de Mattos, A. H. Schweiger, N. Owen-Smith, C. N. Johnson, R. J. Marquis and J.-C. Svenning, Ecological and evolutionary legacy of megafauna extinctions, *Biol. Rev. Cambridge Philos. Soc.*, 2018, **93**, 845–862.
71. J. M. Ament, C. Carbone, J. J. Crees, R. Freeman and S. T. Turvey, Anthropogenic predictors of varying Holocene occurrence for Europe's large mammal fauna, *Biol. Lett.*, 2023, **19**, 20220578.
72. Y. Malhi, C. E. Doughty, M. Galetti, F. A. Smith, J.-C. Svenning and J. W. Terborgh, Megafauna and ecosystem function from the Pleistocene to the Anthropocene, *Proc. Natl. Acad. Sci. U. S. A.*, 2016, **113**, 838–846.
73. C. Emer, M. Galetti, M. A. Pizo, P. Jordano and M. Verdú, Defaunation precipitates the extinction of evolutionarily distinct interactions in the Anthropocene, *Sci. Adv.*, 2019, **5**, eaav6699.
74. R. C. Rodríguez-Caro, E. Graciá, S. P. Blomberg, H. Cayuela, M. Grace, C. P. Carmona, H. A. Pérez-Mendoza, A. Giménez and R. Salguero-Gómez, Anthropogenic impacts on threatened species erode functional diversity in chelonians and crocodilians, *Nat. Commun.*, 2023, **14**, 1542.
75. M. R. Helmus, D. L. Mahler and J. B. Losos, Island biogeography of the Anthropocene, *Nature*, 2014, **513**, 543–546.
76. N. Nolan, M. W. Hayward, K. Klop-Toker, M. Mahony, F. Lemckert and A. Callen, Complex organisms must deal with complex threats: How does amphibian conservation deal with biphasic life cycles?, *Animals*, 2023, **13**, 1634.
77. M. Tietje and M.-O. Rödel, Contradicting habitat type-extinction risk relationships between living and fossil amphibians, *R. Soc. Open Sci.*, 2017, **4**, 170051.
78. L. Ferreira Colares, L. Fogaça de Assis Montag and B. Dunck, Habitat loss predicts the functional extinction of fish from Amazonian streams during the Anthropocene, *Sci. Total Environ.*, 2022, **838**, 156210.
79. J. S. S. Denton, Diversification patterns of lanternfishes reveal multiple rate shifts in a critical mesopelagic clade targeted for human exploitation, *Curr. Biol.*, 2018, **28**, 933–940.
80. D. Goulson, The insect apocalypse, and why it matters, *Curr. Biol.*, 2019, **29**, R967–R971.
81. D. L. Wagner, E. M. Grames, M. L. Forister, M. R. Berenbaum and D. Stopak, Insect decline in the Anthropocene: Death by a thousand cuts, *Proc. Natl. Acad. Sci. U. S. A.*, 2021, **118**, e2023989118.
82. A. J. Reid, A. K. Carlson, I. F. Creed, E. J. Eliason, P. A. Gell, P. T. J. Johnson, K. A. Kidd, T. J. MacCormack, J. D. Olden, S. J. Ormerod, J. P. Smol, W. W. Taylor, K. Tockner, J. C. Vermaire, D. Dudgeon and S. J. Cooke, Emerging threats and persistent conservation challenges for freshwater biodiversity, *Biol. Rev. Cambridge Philos. Soc.*, 2019, **94**, 849–873.
83. G. Ceballos and P. R. Ehrlich, The misunderstood sixth mass extinction, *Science*, 2018, **360**, 1080–1081.

84. S. L. Pimm, C. N. Jenkins, R. Abell, T. M. Brooks, J. L. Gittleman, L. N. Joppa, P. H. Raven, C. M. Roberts and J. O. Sexton, The biodiversity of species and their rates of extinction, distribution, and protection, *Science*, 2014, **344**, 1246752.
85. C. Finn, F. Grattarola and D. Pincheira-Donoso, More losers than winners: investigating Anthropocene defaunation through the diversity of population trends, *Biol. Rev. Cambridge Philos. Soc.*, 2023, **98**, 1732–1748.
86. K. V. Rosenberg, A. M. Dokter, P. J. Blancher, J. R. Sauer, A. C. Smith, P. A. Smith, J. C. Stanton, A. Panjabi, L. Helft, M. Parr and P. P. Marra, Decline of the North American avifauna, *Science*, 2019, **366**, 120–124.
87. A. D. Barnosky, N. Matzke, S. Tomiya, G. O. U. Wogan, B. Swartz, T. B. Quental, C. Marshall, J. L. McGuire, E. L. Lindsey, K. C. Maguire, B. Mersey and E. A. Ferrer, Has the Earth's sixth mass extinction already arrived?, *Nature*, 2011, **471**, 51–57.
88. J. C. Briggs, Emergence of a sixth mass extinction?, *Biol. J. Linn. Soc.*, 2017, **122**, 243–248.
89. L. Villier and D. Korn, Morphological disparity of ammonoids and the mark of Permian mass extinctions, *Science*, 2004, **306**, 264–266.
90. P. A. Carpenter and P. C. Bishop, The seventh mass extinction: Human-caused events contribute to a fatal consequence, *Futures*, 2009, **41**, 715–722.
91. M. L. Larramendy and G. E. Liwszyc, *Marsupial and Placental Mammal Species in Environmental Risk Assessment Strategies*, Royal Soc. Chem., Cambridge, United Kingdom, 2022.
92. G. E. Liwszyc and M. L. Larramendy, *Bird and Reptile Species in Environmental Risk Assessment Strategies*, Royal Soc. Chem., Cambridge, United Kingdom, 2023.
93. M. L. Larramendy and G. E. Liwszyc, *Amphibians Species in Environmental Risk Assessment Strategies*, Royal Soc. Chem., Cambridge, United Kingdom, 2022.
94. G. E. Liwszyc and M. L. Larramendy, *Amphibians Species in Environmental Risk Assessment Strategies*, Royal Soc. Chem., Cambridge, United Kingdom, 2023.
95. M. L. Larramendy, *Ecotoxicology and Genotoxicology: Non-traditional Terrestrial Models*, Royal Soc. Chem., Cambridge, United Kingdom, 2017.
96. M. L. Larramendy, *Ecotoxicology and Genotoxicology: Non-traditional Aquatic Models*, Royal Soc. Chem., Cambridge, United Kingdom, 2017.
97. NCD Risk Factor Collaboration, Worldwide trends in blood pressure from 1975 to 2015: A pooled analysis of 1479 population-based measurement studies with 19.1 million participants, *Lancet*, 2017, **389**, 37–55.

Subject Index

Note: Page numbers in *italics* refer to figures and those in **bold** to tables